THIRD EDITION

Virulence Mechanisms of Bacterial Pathogens

Edited by

Kim A. Brogden
National Animal Disease Center,
Agricultural Research Service,
U.S. Department of Agriculture,
Ames, Iowa

James A. Roth
Veterinary Microbiology and Preventive Medicine,
College of Veterinary Medicine,
Iowa State University, Ames, Iowa

Thaddeus B. Stanton
National Animal Disease Center,
Agricultural Research Service,
U.S. Department of Agriculture,
Ames, Iowa

Carole A. Bolin
National Animal Disease Center,
Agricultural Research Service,
U.S. Department of Agriculture,
Ames, Iowa

F. Chris Minion
Veterinary Medical Research Unit,
College of Veterinary Medicine,
Iowa State University, Ames, Iowa

Michael J. Wannemuehler
Veterinary Medical Research Unit,
College of Veterinary Medicine,
Iowa State University, Ames, Iowa

ASM
PRESS   Washington, DC

5. Relationships between Community Behavior and
Pathogenesis in *Pseudomonas aeruginosa*
Matthew R. Parsek and E. P. Greenberg
77

6. Antibiotic Resistance and Survival in the Host
Emma L. A. Macfarlane and Robert E. W. Hancock
93

II.   BACTERIAL EVASION OF HOST DEFENSE
MECHANISMS   105
Section editor: Carole A. Bolin

7. What Is the Very Model of a Modern Macrophage Pathogen?
David G. Russell
107

8. Bacterial Resistance to Antibody–Dependent Defenses
Lynette B. Corbeil
119

9. Mechanisms of Resistance to NO–Related
Antibacterial Activity
Andrés Vazquez-Torres and Ferric C. Fang
131

10. DNA Repair and Mutators: Effects on Antigenic Variation
and Virulence of Bacterial Pathogens
Thomas A. Cebula and J. Eugene LeClerc
143

III.   BACTERIAL EFFECTS ON HOST CELL
FUNCTION   161
Section editor: Michael J. Wannemuehler

11. Bacterial Toxins in Disease Production
Joseph T. Barbieri and Kristin J. Pederson
163

12. Exploitation of Mammalian Host Cell Function by
Shigella spp.
Larean D. Brandon and Marcia B. Goldberg
175

13. Bacterial Induction of Cytokine Secretion in Pathogenesis of
Airway Inflammation
Alice Prince
189

14. The Type III Secretion Pathway: Dictating the Outcome of
Bacterial-Host Interactions
Raymond Schuch and Anthony T. Maurelli
203

IV.   IDENTIFICATION, REGULATION, AND TRANSFER OF VIRULENCE GENES   225

Section editor: F. Chris Minion

15. Impact of Horizontal Gene Transfer on the Evolution of *Salmonella* Pathogenesis

Robert A. Kingsley, Renée M. Tsolis, Stacy M. Townsend, Tracy L. Norris, Thomas A. Ficht, L. Garry Adams, and Andreas J. Bäumler

227

16. Regulation of Virulence Gene Expression In Vivo

James M. Slauch

241

17. Identification of Virulence Genes In Silico: Infectious Disease Genomics

George M. Weinstock, Steven J. Norris, Erica J. Sodergren, and David Smajs

251

V.   CONCLUDING PERSPECTIVE   263

Section editor: Kim A. Brogden

18. State and Future of Studies on Bacterial Pathogenicity: Impact of New Methods of Studying Bacterial Behavior In Vivo

H. Smith

265

Index   283

ORGANIZING COMMITTEE

Kim Alan Brogden (Chair)
National Animal Disease Center, Agricultural Research Service, U.S. Department of Agriculture, Ames, IA 50010

Mark R. Ackermann
Department of Veterinary Pathology, College of Veterinary Medicine, Iowa State University, Ames, IA 50011

Carole A. Bolin
National Animal Disease Center, Agricultural Research Service, U.S. Department of Agriculture, Ames, IA 50010

Dawne Buhrow
Institute for International Cooperation in Animal Biologics, Iowa State University, Ames, IA 50011

Thomas O. Bunn
National Veterinary Services Laboratories, Animal and Plant Health Inspection Service, U.S. Department of Agriculture, Ames, IA 50010

Thomas A. Casey
National Animal Disease Center, Agricultural Research Service, U.S. Department of Agriculture, Ames, IA 50010

Jane Galyon
Institute for International Cooperation in Animal Biologics, Iowa State University, Ames, IA 50011

Louise M. Henderson
Center for Veterinary Biologics, Animal and Plant Health Inspection Service, U.S. Department of Agriculture, Ames, IA 50010

F. Chris Minion
Veterinary Medical Research Institute, College of Veterinary Medicine, Iowa State University, Ames, IA 50011

Harley W. Moon
Veterinary Medical Research Institute, Iowa State University, Ames, IA 50011

James A. Roth
Veterinary Microbiology and Preventive Medicine, College of Veterinary Medicine,
Iowa State University, Ames, IA 50011

Judy R. Stabel
National Animal Disease Center, Agricultural Research Service, U.S. Department of
Agriculture, Ames, IA 50010

Thaddeus B. Stanton
National Animal Disease Center, Agricultural Research Service, U.S. Department of
Agriculture, Ames, IA 50010

Eileen Thacker
Veterinary Medical Research Institute, Iowa State University, Ames, IA 50011

Michael J. Wannemuehler
Veterinary Medical Research Institute, College of Veterinary Medicine, Iowa State
University, Ames, IA 50011

Richard L. Zuerner
National Animal Disease Center, Agricultural Research Service, U.S. Department of
Agriculture, Ames, IA 50010

CONTRIBUTORS

L. Garry Adams
Department of Veterinary Pathobiology, College of Veterinary Medicine, Texas A&M University, College Station, TX 77843-4467

Joseph T. Barbieri
Department of Microbiology and Molecular Genetics, Medical College of Wisconsin, Milwaukee, WI 53226

Andreas J. Bäumler
Department of Medical Microbiology and Immunology, College of Medicine, Texas A&M University Health Science Center, College Station, TX 77843-1114

Larean D. Brandon
Infectious Disease Division, Massachusetts General Hospital, Boston, MA 02114

Kim Alan Brogden
Respiratory Diseases of Livestock Research Unit, National Animal Disease Center, Agricultural Research Service, U.S. Department of Agriculture, Ames, IA 50010

John H. Brumell
Biotechnology Laboratory, University of British Columbia, Vancouver, British Columbia. V6T 1Z3, Canada

Thomas A. Cebula
Division of Molecular Biological Research and Evaluation, Center for Food Safety and Applied Nutrition, U.S. Food and Drug Administration, Washington, DC 20204

Lynette B. Corbeil
Department of Pathology, University of California—San Diego School of Medicine, San Diego, CA 92103-8416

Ferric C. Fang
Departments of Medicine, Pathology, and Microbiology, University of Colorado Health Sciences Center, Denver, CO 80262

PREFACE

The field of microbial pathogenesis has expanded rapidly over the last decade. Literally thousands of research papers have been published, and new specialty journals on microbial pathogenesis have appeared. Researchers now working within the field have become increasingly specialized within specific areas. Occasionally, reviews are published that overview recent work on the major virulence determinants (e.g., extracellular products such as enzymes and toxins or cell-associated products such as capsular polysaccharides, lipopolysaccharides, and outer membrane proteins) and mechanisms of attachment, invasion, and colonization. However, these reviews also are specialized and generally are limited in scope. Periodically, the need for a text that overviews the entire field of bacterial virulence mechanisms and summarizes major advances arises. This is the third monograph to originate from a series of international symposia on virulence mechanisms of bacterial pathogens which tries to meet that need. This edition resulted from the International Symposium of Virulence Mechanisms of Bacterial Pathogens held in Ames, Iowa, on 13 to 15 September 1999. Previous monographs were published by ASM Press in 1988 and 1995 as a result of similar symposia held in Ames in 1987 and 1994. The purpose of this monograph, similar to that of previous editions, is to provide an overview of current knowledge of the wide variety of mechanisms used by bacterial pathogens to establish infection, produce disease, and persist in the host. The overall emphasis is to understand the mechanisms of host–pathogen interactions rather than to focus on specific research approaches. Within each section, internationally recognized authorities in the field reviewed ongoing concepts in bacterial invasion, colonization, and survival (section I); bacterial evasion of host defense mechanisms (section II); bacterial effects on host cell function (section III); and identification, regulation, and transfer of virulence genes (section IV). This monograph is specifically intended to be a source of information for molecular biologists wanting an understanding of how molecular mechanisms relate to the disease process; infectious disease specialists wanting to add to their understanding of the cellular and molecular basis of pathogenesis; researchers attempting to elucidate pathogenic mechanisms in bacterial diseases that are not yet well characterized; industry scientists wishing to identify promising approaches to disease prevention and therapy; and faculty and graduate students wishing to gain an overview of the subject.

Harry Smith was again asked to provide a summation to fulfill a rich tradition we have established in the previous two editions (section V). However, in his chapter, he goes beyond the task of summing the field and presents an interesting view on the rising interest in the activities of bacterial pathogens in vivo and the design of new methods for studying virulence mechanisms in the complex in vivo environment, perhaps the major development in studies of bacterial pathogenicity since this series of symposia began.

KIM A. BROGDEN
December 1999

ACKNOWLEDGMENTS

This volume resulted from the International Symposium on Virulence Mechanisms of Bacterial Pathogens held in Ames, Iowa, 12 to 15 September 1999. The symposium was sponsored by the following organizations:

Institute for International Cooperation in Animal Biologics
National Animal Disease Center, Agricultural Research Service, U.S. Department of Agriculture
National Veterinary Services Laboratories, Animal and Plant Health Inspection Service, U.S. Department of Agriculture
Center for Veterinary Biologics, Animal and Plant Health Inspection Service, U.S. Department of Agriculture
Iowa State University College of Veterinary Medicine
National Research Initiative Competitive Grants Program, U.S. Department of Agriculture
Iowa State University Biotechnology Program

The Organizing Committee thanks the following for their generous financial support of the symposium:

Becton Dickinson Microbiology Systems, Sparks, Md.
Pfizer Animal Health Biological Discovery, Groton, Conn.
Schering-Plough Animal Health, Union, N.J.

BACTERIAL INVASION, COLONIZATION, AND SURVIVAL

I

BACTERIAL ADHERENCE, COLONIZATION, AND INVASION OF MUCOSAL SURFACES

John H. Brumell and B. Brett Finlay

I

Mucosal surfaces provide an interface between the body and the external environment. Through specialization of the mucosal epithelium, specific functions can be performed, such as the absorption of nutrients by the intestine. At the same time, mucosal surfaces must provide a barrier to microorganisms that thrive in the nutrient-rich environment of the body. Despite their defenses, mucosal surfaces are susceptible to damage by pathogenic bacteria. These bacteria, though often closely genetically related, utilize their own pathogenic strategies to cause disease. In this chapter, we provide a brief overview of the pathogenic strategies of some bacteria that infect the mucosal surface of the intestinal tract. We will focus on two model systems for the study of bacterial pathogens that cause disease by colonizing (enteropathogenic *Escherichia coli* [EPEC]) or penetrating (*Salmonella* species) the intestinal epithelium.

THE MUCOSAL SURFACE AND ITS DEFENSES

Mucosal surfaces have many physiological defenses against pathogenic bacteria (61). These include entrapment in a thick blanket of mucus and clearance by peristalsis in the gut or ciliary movement in the airways. Fluid flow across the mucosal epithelium also helps to flush away pathogens and prevent their attachment to the epithelial cell surface. In addition, resident microbial flora provide competition for adhesion sites and nutrients, thereby serving as an added defense against disease-causing bacteria.

The intestinal surface is a mixed-cell population consisting of goblet cells, Paneth cells, columnar absorptive cells, and membranous (M) cells. Goblet cells secrete mucus, composed of many carbohydrates, that coats the epithelial surface. Paneth cells contribute to the mucous layer by releasing lysozyme and other antimicrobial components into the intestinal lumen. Other secreted agents that interfere with the actions of pathogenic bacteria in the gut include secreted antibodies, cationic peptides, and complement factors.

The vast majority of epithelial cells are columnar absorptive cells present on the surfaces of the absorptive villi, which serve to transport nutrients across the epithelium. These cells

John H. Brumell and B. Brett Finlay Biotechnology Laboratory, University of British Columbia, Room 237–Wesbrook Building, 6174 University Boulevard, Vancouver, British Columbia, Canada V6T 1Z3.

Virulence Mechanisms of Bacterial Pathogens, 3rd ed., Edited by K. A. Brogden et al.
©2000 ASM Press, Washington, D.C.

form a polarized barrier, with tight junctions making an impermeable seal between the apical (lumenal) surface of adjacent cells. The apical surface of absorptive cells contains fingerlike projections called microvilli, which aid in the absorption of nutrients by increasing the cell's surface area. A dense network of cell surface-expressed glycoproteins, the glycocalyx, limits access of pathogens to the epithelial surface (51).

An immunological barrier at the intestinal mucosa is provided by the gut-associated lymphoid tissues, which include Peyer's patches, isolated lymphoid follicles, the appendix, and mesenteric lymph nodes (31). Of special interest are the Peyer's patches, aggregates of subepithelial lymphoid follicles present throughout the small intestine. These dome-shaped follicles are covered by the follicle-associated epithelium, whose cell composition is different from that of absorptive villi. The predominant cell type is also the columnar absorptive cell, but the follicle-associated epithelium has very few goblet cells and therefore only minimal amounts of mucus overlay these surfaces. Interspersed between the columnar absorptive cells are M cells, which are unique to the follicle-associated epithelium. In contrast to the absorptive cells, M cells are specialized epithelial cells with few and irregular microvilli and little of the glycocalyx glycoproteins expressed at their surface. The basolateral surface of M cells is deeply invaginated, forming pockets in which lymphocytes and macrophages are found associated with the M cell. M cells have a high capacity to internalize both large and small particles from the gut lumen and transport them to their basolateral surface. As such, M cells are able to "sample" the intestinal environment, acting as a conduit for the presentation of lumenal antigens to underlying lymphoid tissues (51). This well-designed antigen delivery system provides for adaptive immunity at the mucosal surface. Ironically, the many adaptations of M cells that make them suitable for antigen presentation also make them preferred targets for pathogenic bacteria, viruses, and protozoan parasites (64).

DIFFERENT STRATEGIES OF PATHOGENIC BACTERIA AT MUCOSAL SURFACES

Adhesion to the intestinal epithelium is an important determinant of virulence for many pathogenic bacteria, allowing them to resist the fluid flow of lumenal contents. Adhesins on the bacterial surface also provide specificity for interaction with target host cells. For example, M cells of the intestinal epithelium have cell surface glycosylation patterns that vary between species and tissue location (51). Specific interactions between bacterial adhesins and their host cell receptors determine the specific location in which pathogenic bacteria can initiate disease. While adhesion is a common first step for interaction with the host, pathogens have evolved a variety of strategies to suit their needs after initial adherence with the intestinal epithelium. Some pathogens remain attached to the epithelial surface and establish a microcolony. One such colonizing pathogen is *Vibrio cholerae*, the causative agent of cholera infections, which establishes microcolonies in the small intestine at both villi and Peyer's patches. These bacteria produce cholera toxin, an enterotoxin that causes electrolyte and fluid secretion from the intestinal epithelium. The resulting diarrhea causes rapid dissemination of *V. cholerae* to the outside environment, where it can potentially infect other hosts (38).

A family of closely related pathogens that colonize the epithelial surface form attaching and effacing (A/E) lesions, including EPEC. A/E lesions are characterized by intimate adherence to epithelial cells, degeneration of microvilli, and the formation of actin-rich pedestals beneath the adherent bacteria. These lesions allow colonization of the intestinal surface and are associated with diarrhea that follows infection with these pathogens (56).

Adhesion is often a prerequisite for penetration of the mucosal surface, though different pathogens penetrate this barrier by different means and with different ends. *Yersinia enterocolitica* breaches the intestinal mucosa by M cell transcytosis at the terminal ileum of the small intestine. Once at the basolateral surface, the bacteria spread to the underlying lymphoid tissues. Characteristic of *Y. enterocolitica* is the ability to block their uptake by phagocytic cells. This allows for their extracellular replication and prevents exposure to the antimicrobial arsenal that phagocytic cells use to kill internalized bacteria (11).

Shigella flexneri, the causative agent of bacillary dysentery, penetrates the mucosal surface of the colon following uptake by M cells. To avoid contact with underlying lymphoid cells, *S. flexneri* escapes from its membrane vacuole where it is free to replicate within the nutrient-rich environment of the cytosol. A striking feature of intracellular *S. flexneri* is its ability to move rapidly within the host cell by harnessing its actin cytoskeleton. This actin-based propulsion allows *S. flexneri* to spread laterally from cell to cell. Additionally, extracellular *S. flexneri* possesses the ability to invade adjacent epithelial cells from the basolateral side (62). *Salmonella enterica* serovar Typhimurium can penetrate both M cells and columnar epithelial cells of the small intestine of mice, though M cells appear to be the preferred target at early times in the infection (37). Following entry, the M cells are destroyed and the bacteria are then able to infect neighboring absorptive cells via their basolateral domains and also underlying lymphocytes. Unlike *Y. enterocolitica* and *S. flexneri*, *Salmonella* serovar Typhimurium does not avoid interaction with phagocytic cells. By contrast, this serovar of *S. enterica* has adapted mechanisms to survive and replicate in a vacuolar compartment within these cells by subverting their killing mechanisms. Infection of macrophages in particular allows *Salmonella* serovar Typhimurium to spread beyond the mucosal surface and initiate a fatal systemic infection in susceptible mice.

TYPE III SECRETION SYSTEMS IN PATHOGENIC BACTERIA

An important feature of many gram-negative pathogenic bacteria is their ability to directly manipulate host cell signals to their own benefit. A common mechanism for manipulating host cell signals is through the action of type III secretion systems (36). These systems are needle-like structures (44) that span both bacterial membranes and can deliver their protein cargo directly into the host cell (Fig. 1). The delivered proteins, called effectors, interact with host cell signaling systems in different ways, depending on the pathogenic strategy of the infecting bacterium. In addition to delivery of proteins into the host cell (translocation), type III secretion systems also release effector proteins into the extracellular environment (secretion) under certain conditions in vitro, a feature that has enhanced the study of these systems.

Type III secretion systems appear to be the "weapon of choice" for gram-negative pathogens directly interacting with host signaling systems and are used by many plant and animal bacterial pathogens (36). Genes that encode the secretion apparatus, many (if not all) of their effectors, their chaperones, and regulators of the system are encoded within pathogenicity islands. Pathogenicity islands are defined as clusters of virulence genes present in pathogenic strains of a species and not in related nonpathogenic strains of that species (60). These virulence gene clusters have G+C contents that are distinct from those of the rest of the bacterial chromosome, suggesting that these islands were acquired by horizontal transfer from an unknown source.

Many functions have been attributed to the actions of type III secretion systems and their effectors in pathogenic bacteria. In the case of EPEC, intimate adherence with the epithelial surface is accomplished by delivery of a receptor for one of its own adhesins into the host cell membrane. Invasion of epithelial cells by *S. flexneri* and *Salmonella* serovar Typhimurium is dependent on the delivery of effectors that manipulate the host's actin cytoskeleton, causing ruffling of the cell surface and their uptake in membrane vacuoles. *Salmonella* serovar Typhimurium also possesses a second type III secretion system that allows its survival and replication within a vacuolar compartment in the host cell. Delivery of a tyrosine phosphatase into phagocytic cells by *Y. enterocolitica* blocks the actions of tyrosine kinases,

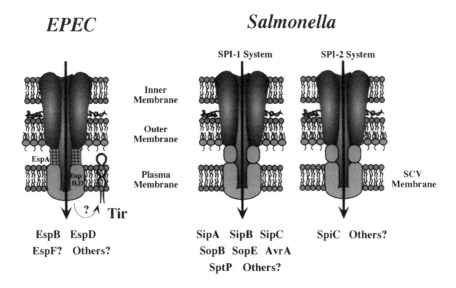

Host Cell

FIGURE 1 Type III secretion systems of EPEC and *Salmonella* serovar Typhimurium. Depicted are the needle-like complexes (44) that span both bacterial membranes of the pathogen and deliver translocated effector proteins into the host cell. These effector proteins can then manipulate host signaling systems. Note that *Salmonella* serovar Typhimurium express two type III secretion systems: one that delivers effectors across the plasma membrane (encoded within SPI-1) to mediate invasion (see text), and a separate system that mediates survival and replication within host cells (SPI-2) by delivering effector proteins across the membrane of the SCV. Artwork provided by A. Gauthier.

which normally initiate phagocytosis of bacteria. Furthermore, *S. flexneri*, *Salmonella* serovar Typhimurium, and *Y. enterocolitica* all induce apoptosis (programmed cell death) of macrophages in a manner dependent on their respective type III secretion systems (50, 59, 71).

EPEC AND REPEC: MODEL SYSTEMS FOR THE STUDY OF A/E PATHOGENS

EPEC, a major cause of infantile diarrhea in the developing world, belongs to a family of pathogens that produce A/E lesions. Also included in this family are enterohemorrhagic *E. coli* (EHEC), the causative agent of "hamburger" disease; *Citrobacter rodentium*; and *Hafnia alvei*. In addition, rabbit pathogens such as REPEC O103 and RDEC-1 have provided

valuable animal models for the study of A/E-lesion-causing bacteria.

A/E lesions are described as pedestal-like structures on the apical surface of the intestinal epithelium. Depending on the pathogen, these structures may be present on columnar absorptive cells or M cells, in the small or large intestine. A/E lesion formation results in the loss of microvilli from the epithelial surface and the formation of an actin-rich pedestal beneath the adherent bacteria. Though A/E lesions are associated with diarrhea, the actual cause of fluid release by the intestinal surface in infected animals is not known.

EPEC provides a suitable model for understanding A/E pathogens and has largely been studied in vitro by infection of epithelial tissue cell cultures. Many of the effects of EPEC seen in vivo can be mimicked in these situations and allow for genetic and cell biological analyses of these events. For example, infection of polarized MDCK cells with EPEC causes disruption of microvilli, formation of actin-rich A/E lesion structures, and eventual disruption of the monolayer (8). A typical pedestal structure induced by EPEC in vitro is shown in Fig. 2A. These pedestals resemble those seen in vivo with REPEC 0103 in rabbits (arrows in Fig. 2B).

A/E Lesion Formation

The study of EPEC in vitro has led to a three-stage model for interaction with epithelial cells and formation of A/E lesions (Fig. 2C). The first stage involves initial binding of the pathogen to the host epithelium. For EPEC, initial binding to epithelial cells of the small intestine in humans is mediated by the bundle-forming pilus (BFP), encoded on a 55- to 70-MDa plasmid common to EPEC strains (4). BFP expression also mediates self-association of EPEC and microcolony formation on the surface of epithelial cells. Rabbit A/E pathogens express various plasmid encoded fimbriae, which mediate their binding to M cells of Peyer's patches in infected animals.

After initial binding, EPEC uses a type III secretion system to deliver effector proteins, known as Esp's (*E. coli* secreted proteins), to the host cell (Fig. 1). The genes encoding this type III secretion system are found within a pathogenicity island called LEE (locus of enterocyte effacement) on the EPEC chromosome (21). Indeed, all the genes needed for A/E lesion formation by EPEC and EHEC are encoded within the LEE region (17).

The translocation process remains poorly understood. Evidence suggests that EspA helps to form the filamentous structure of the type III secretion system that interacts with the host cell membrane (43). In this way, EspA may aid in the translocation of other Esp's and possibly also act as an adhesin (16). EspB is translocated directly into the host membrane and cytosol and may also perform dual roles, in this case forming a translocation pore in the host membrane and directly initiating host signals upon delivery to the host cell cytosol (68). EspD has been identified in host cell membrane and cytosolic fractions, though little is known of its function (67). The role of EspF, a recently identified protein encoded within the LEE region of EPEC, also remains unknown (47). Knowing that other pathogenic bacteria can translocate effector proteins not encoded within the pathogenicity island of the secretion system (e.g., *Salmonella* serovar Typhimurium, see below), we anticipate that EPEC and other A/E pathogens express other translocated effectors that contribute to their pathogenesis.

Upon translocation, the Esp's activate host cell signals involved in pedestal formation. The number of host cell signals initiated by EPEC infection in vitro is impressive (14). These include activation of phospholipase C (41), a consequence of which is inositol phosphate fluxes (20) that cause the release of Ca^{2+} from intracellular stores (2). EPEC has also been demonstrated to activate protein kinases such as protein kinase C (12), myosin light chain kinase (46), and unknown tyrosine kinases (58). The exact role played by EPEC's translocated effectors in activating these host cell signals and their role in A/E lesion formation remain undetermined.

Surprisingly, the type III secretion system of EPEC also mediates delivery of a protein into the bilayer of the plasma membrane in the host cell (40). This protein, the translocated intimin receptor (Tir), then serves as a receptor for intimin, an outer membrane protein expressed by EPEC. Tir sequences are conserved among A/E pathogens and presumably perform a similar function during pathogenesis. The mechanism by which Tir inserts into the host cell membrane is unknown and presents an interesting question.

Binding of intimin to Tir on the host membrane allows intimate interaction between EPEC and the epithelial cell. In this final stage of EPEC interaction, clustering of Tir elicits dramatic cytoskeletal rearrangements within the host cell, leading to the formation of actin-rich pedestals beneath the

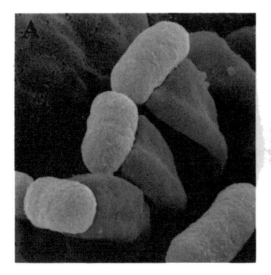

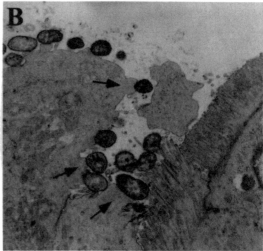

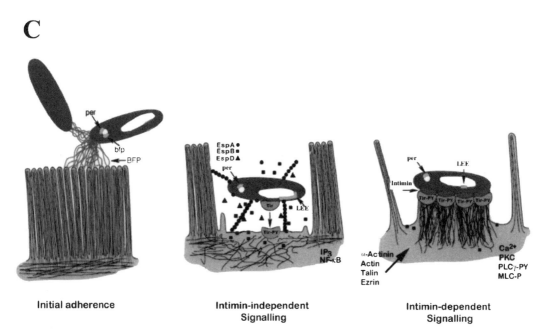

| Initial adherence | Intimin-independent Signalling | Intimin-dependent Signalling |

FIGURE 2 Models of EPEC pathogenesis. (A) Formation of pedestals by EPEC on the surface of HeLa epithelial cells in vitro. Scanning electron micrograph provided by Dr. I. Rosenshine. (B) Formation of A/E lesions in vivo by REPEC 0103. Pedestal formation on the intestinal surface is indicated with arrows. Note that on these cells, effacement of microvilli has occurred (compare to upper right hand side). TEM provided by Dr. U. Heczko and reproduced from *The Journal of Experimental Medicine*, 1998, volume 188, pages 1907–1916 (1) by copyright permission of the Rockefeller University Press. (C) Three-stage model of EPEC pedestal formation. In the first step, attachment of EPEC is mediated by bundle-forming pilus (BFP) binding to the epithelial cell surface. Next, the type III secretion system mediates delivery of both Esp's and the Tir. Intimin-independent host signals are activated and Tir is phosphorylated by an unknown tyrosine kinase. Finally, intimin binding mediates clustering of phosphorylated Tir, intimin-dependent host signals are activated, and rearrangements of the host actin cytoskeleton cause pedestal formation. Artwork provided by Dr. R. DeVinney and reprinted from *Current Opinion in Microbiology*, volume 2, pages 83–88, copyright 1999 (14), with permission from Elsevier Science.

adherent EPEC. Many regulators of the actin cytoskeleton can be found at EPEC pedestals, including α-actinin, ezrin, talin, villin, and myosin light chain (27). Interestingly, EPEC's Tir requires tyrosine phosphorylation by an unknown tyrosine kinase for pedestal formation to occur (39), whereas Tir from EHEC does not (15). These findings suggest that each A/E pathogen may utilize its own Tir molecule to influence the host cell cytoskeleton in a distinct manner.

A major question is why EPEC forms a pedestal on the mucosal surface. One possibility is that pedestal formation prevents its uptake by M cells at Peyer's patches during infection. Recent findings have suggested that EPEC can also prevent its uptake by other cells, including professional phagocytes, which have numerous receptors for the ingestion of bacteria (26). EPEC was found to inhibit its phagocytosis by a macrophage cell line in vitro, in a manner dependent on both Esp delivery and intimin-Tir interactions. Interestingly, dephosphorylation of host proteins on tyrosine residues was concomitant with the antiphagocytic effect displayed by EPEC. While this suggests that EPEC might deliver a tyrosine phosphatase to the host in a manner analogous to that of *Y. enterocolitica*, such a phosphatase remains to be found. The mechanism by which EPEC inhibits phagocytosis by macrophages and its relevance to in vivo infections is still undefined. However, the ability to avoid uptake by mucosal M cells may allow EPEC to escape the underlying lymphoid cells.

REPEC as an In Vivo Model of A/E Pathogens

There are many obvious limitations to studying EPEC infection in cell culture models. As an alternative approach, some investigators have used animal models of infection, in particular the study of REPEC in rabbits. These experiments have demonstrated the essential roles of both the type III secretion system of REPEC (13) and the secreted effector proteins EspA and EspB (1) in A/E lesion formation and production of diarrhea in infected animals.

We anticipate that future studies of REPEC will further elucidate the mechanisms by which A/E lesion-forming pathogens can cause disease. Particular questions that need to be addressed include the role of Tir, the targets of translocated effector proteins (including possibly the identity of novel Esp's and their targets), and the ultimate cause(s) of diarrhea.

Salmonella Serovar Typhimurium: a Model System for Study of Invasive Pathogens That Survive within a Vacuolar Compartment

In contrast to EPEC and other A/E pathogens that remain on the epithelial surface, the pathogenic strategy of *Salmonella* species includes penetration of the mucosal barrier and interaction with immune system cells. As such, the virulence mechanisms used by these pathogens are more complex. Indeed, it is estimated that 4% of the *Salmonella* serovar Typhimurium chromosome (about 200 genes) are virulence factors (5). These factors include, to date, five pathogenicity islands, numerous smaller pathogenicity "islets," other virulence factors on the chromosome, and at least one virulence plasmid (28, 60).

Salmonella species infect a broad range of animals and can cause different diseases in different hosts. For example *Salmonella enterica* serovar Typhi causes typhoid fever in humans, which can be fatal. By contrast, *Salmonella* serovar Typhimurium usually causes a self-limiting gastroenteritis in humans but induces a systemic disease in mice that is similar to typhoid fever. As such, the study of *Salmonella* serovar Typhimurium in mice has provided a valuable animal model for the study of these clinically relevant invasive pathogens that survive within a vacuolar compartment in host cells. Additionally, in vitro models of *Salmonella* serovar Typhimurium have allowed genetic, cell biological, and biochemical analyses of the infection process.

INVASION OF EPITHELIAL CELLS

Studies of *Salmonella* serovar Typhimurium infection in mice suggest that M cells are the preferred target of this pathogen, with inva-

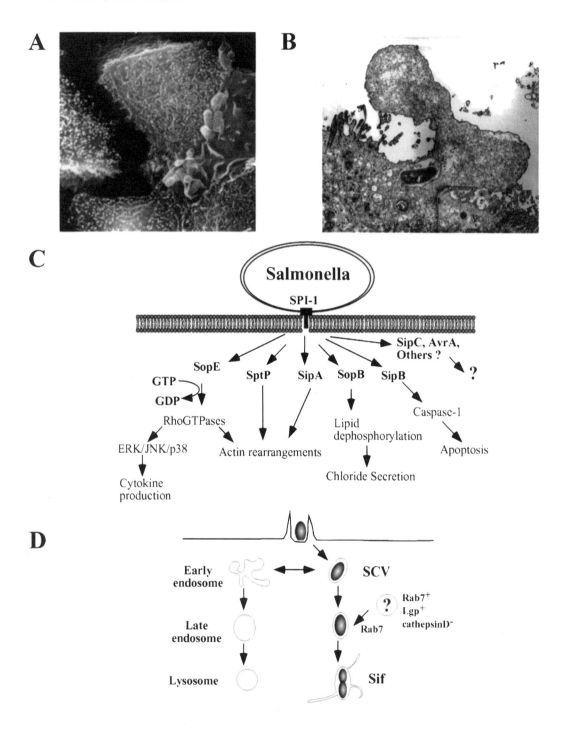

FIGURE 3 Models of *Salmonella* serovar Typhimurium pathogenesis. (A) Cell-surface ruffling of Caco-2 epithelial cells induced by *Salmonella* serovar Typhimurium. Note the loss of microvilli near bacteria and the formation of large ruffles. (B) Uptake of *Salmonella* serovar Typhimurium in large vacuoles that resemble macropinosomes in vitro. (C) Invasion of epithelial cells by delivery of translocated effectors into the host cell. Depicted are the translocated effectors of the SPI-1-encoded type III secretion system and the host

sion of these cells occurring at early times (15 min) following inoculation (37). Binding to M cells is thought to be mediated by the *lpf* fimbrial operon, though the host receptor for this adhesin is unknown (3). Interestingly, recent evidence suggests that the host receptor for *Salmonella* serovar Typhi is the cystic fibrosis transmembrane conductance regulator (CFTR), a chloride channel that is nonfunctional in patients with cystic fibrosis (54). Invasion of M cells rapidly leads to their destruction and the dissemination of *Salmonella* serovar Typhimurium to deeper tissues. M cell destruction by serovar Typhimurium may be similar to perforation of the intestine witnessed in typhoid patients infected with *Salmonella* serovar Typhi (37).

Salmonella serovar Typhimurium and other *Salmonella* serovars can also invade columnar absorptive cells at their apical surface. Since M cell models remain difficult to study in vitro, studies of *Salmonella* invasion have largely been performed with cell lines that resemble absorptive enterocytes. As depicted in Fig. 3A, invasion involves denuding of the microvilli and ruffling of the cell surface. These cell surface rearrangements lead to uptake of the bacterium in large vacuoles that resemble macropinosomes (Fig. 3B).

In contrast to other pathogenic bacteria that utilize adhesin-receptor interactions for their uptake, *Salmonella* species utilize a type III secretion system to cause host epithelial cell ruffling and subsequently their internalization (10). This secretion system, encoded within *Salmonella* pathogenicity island 1 (SPI-1), de-

livers a variety of effectors into the host cell that mediate uptake of the pathogen (Fig. 3C). These effectors directly interact with host cell signaling systems, independently of receptor-ligand interactions on the host cell surface. As such, invasion by *Salmonella* species could be considered distinct from receptor-mediated phagocytosis (cell eating) and more likened to "force feeding" of the host cell (6). *Salmonella* serovar Typhimurium mutants lacking SPI-1 activity are unable to invade and destroy M cells and are attenuated for systemic virulence when delivered orally to mice. Interestingly, injection of these mutants into the peritoneum of these animals (thereby bypassing the need for invasion of the intestine) reveals that these pathogens have normal virulence for causing systemic disease (23). Thus, SPI-1 appears to be specialized for gastrointestinal events during the infection process.

Recent progress has revealed the mechanisms by which the translocated effectors of SPI-1 mediate invasion by *Salmonella* serovar Typhimurium. Crucial to invasion is rearrangement of the actin cytoskeleton, which is accomplished by the actions of at least three effectors. SopE acts as a guanine nucleotide exchange factor for Cdc42 and Rac, Rho GTPases that regulate the actin cytoskeleton (29). Interestingly, SopE is not encoded within the SPI-1 pathogenicity island, yet it is translocated by this secretion system into the host cell (30). Downstream actions of SopE also include the activation of kinase cascades that lead to the production of proinflammatory cytokines. SptP has two functional do-

---

signaling systems that they directly interact with. These effectors have numerous effects on the host cell, including ruffling of the cell surface (thereby directing *Salmonella* serovar Typhimurium uptake), production of proinflammatory cytokines, and apoptosis. Artwork modified from *Current Biology* (6) with permission from Elsevier Science. (D) Intracellular trafficking of *Salmonella* serovar Typhimurium in host cells. Upon entry into host cells, serovar Typhimurium remain in a vacuolar compartment (SCV) that interacts transiently with early endosomes. However, these vacuoles do not undergo further processing within the endosomal system and do not fuse with mature lysosomes. Instead, the SCV appear to interact with an unknown compartment that mediates delivery of lysosomal glycoproteins (such as LAMP-1) but not degradative lysosomal enzymes such as cathepsin D (48). After several hours, intracellular *Salmonella* serovar Typhimurium begins to replicate and long, contiguous tubules (Sifs) are formed. Artwork provided by O. Steele-Mortimer and modified from *Cellular Microbiology* (65) with permission from Blackwell Science Ltd.

mains, a tyrosine phosphatase domain and a region with homology to both exotoxin S from *Pseudomonas aeruginosa* and YopE from *Yersinia* spp. Both domains of SptP initiate cytoskeletal rearrangements in the host cell by unknown mechanisms (22). SipA is an actin-binding protein that inhibits depolymerization of actin filaments, thereby stabilizing actin polymers at the site of *Salmonella* serovar Typhimurium interaction with the host cell (70). SopB from *S. enterica* serovar Dublin is a phosphatase, capable of dephosphorylating polyphosphoinositides and inositol phosphates (52). Its phosphatase activity in host cells is thought to activate Ca^{2+}-dependent chloride channels and thereby contribute to fluid secretion induced by *Salmonella* serovar Dublin in bovine calf ileal loop models of infection. Interestingly, SopB is encoded within a separate pathogenicity island, SPI-5, which appears to be required for enteropathogenicity (69). The SopB homolog in *Salmonella* serovar Typhimurium, SigD, was identified as an invasion gene in SPI-1 mutants (35). This suggests that SopB/SigD may be secreted by other mechanisms into host cells and may also play a role in invasion.

Invasion of macrophages by *Salmonella* serovar Typhimurium using the SPI-1 secretion system (as opposed to phagocytosis) has been shown to induce rapid apoptosis of the infected cell. Recently, it was shown that SipB, a translocated effector encoded on SPI-1 that also helps to form a pore in the host plasma membrane during secretion, is a key mediator of this effect. SipB binds to and activates caspase-1, a proapoptotic protease that converts the precursor form of interleukin-1β to its mature form (33). The cytotoxic effect of invasive *Salmonella* serovar Typhimurium may be relevant to infection at Peyer's patches, where macrophages are found to underlie target M cells. Apoptosis of infected columnar absorptive cells has also been observed, but it occurs much more slowly and less frequently than that seen in macrophages (42), prompting the question of why each cell responds differently to invasion by *Salmonella* serovar Typhimurium.

Colonization of Macrophages by *Salmonella* Serovar Typhimurium in a Systemic Mouse Model of Infection

Following invasion of host cells in vitro, *Salmonella* serovar Typhimurium is localized within a membrane compartment known as the *Salmonella*-containing vacuole (SCV). These bacteria are capable of survival and replication within the SCV, eventually killing the host cell and being released into the extracellular medium to infect other cells. The ability of *Salmonella* serovar Typhimurium to survive and replicate within macrophages, which possess an arsenal of antimicrobial defenses, indicates that these are the host cells colonized during the systemic phase of disease. Indeed, mutants that are incapable of survival in macrophages in vitro are avirulent in the mouse model of infection (19). To analyze the localization of *Salmonella* serovar Typhimurium in infected mice, a low-dose model of infection was used (57). This model entails the injection of approximately 100 CFU of wild-type serovar Typhimurium into the tail vein of experimental mice and examination of liver thick sections (30 μm) by confocal microscopy at different times following injection. A major advantage of this method is that it precludes the need for an artificially high inoculum at late infection times to visualize bacteria in the organs and allows a much larger area of study than conventional electron microscopy. Shortly after appearance in the liver, *Salmonella* serovar Typhimurium cells were localized at inflammatory foci where infiltrating neutrophils could be seen. At the later stages of infection, three-dimensional confocal projections were used to reveal that the bacteria reside within macrophages. Infection was found to induce apoptosis of both neutrophils and macrophages within the liver, possibly allowing evasion of host immune cell responses in these tissues (57). Thus, *Salmonella* serovar Typhimurium remains a facultative intracellular pathogen during initial infection of epithelial cells in the gut and throughout the duration of infection of solid organs such as the liver.

INTRACELLULAR TRAFFICKING OF THE SCV

Intracellular trafficking of the SCV differs from that of other internalized particles and may represent a key mechanism for survival and replication within host cells (Fig. 3D). Shortly following invasion, interaction with early endosomes is evidenced by delivery of early endosome antigen-1 and the transferrin receptor (65). Within 30 min, however, SCVs become uncoupled from the endocytic pathway of the host. As such, the SCV do not fuse with lysosomes and avoid exposure to the harsh degradative enzymes that they contain. While vacuoles containing *Salmonella* serovar Typhimurium do not fuse with lysosomes, they do acquire lysosomal glycoproteins (such as LAMP-1).

Recent findings suggest that lysosomal glycoproteins are acquired through interaction with an unknown compartment and that Rab7 mediates this processing of the SCV (48). Uncoupling from the endocytic pathway is witnessed only with live *Salmonella* serovar Typhimurium and is similar in both epithelial (25) and macrophage cell lines (49, 55), suggesting that these pathogens actively utilize a common mechanism for altering intracellular trafficking. In both cell types, invasion is followed by a lag phase of 3 to 4 h, after which the intracellular bacteria begin replication. At this time, *Salmonella* serovar Typhimurium are localized within tubular structures known as Sifs (*Salmonella*-induced filaments). These filamentous structures can be visualized in cells by immunostaining for lysosomal glycoproteins and appear to require bacterial replication for their formation (24). The role of these structures for *Salmonella* serovar Typhimurium pathogenesis remains unknown.

MECHANISMS FOR SURVIVAL AND REPLICATION OF INTRACELLULAR *SALMONELLA* SEROVAR TYPHIMURIUM

Many factors that are required for survival and replication within host cells and causing disease have been identified in *Salmonella* serovar Typhimurium (28, 60). These include factors that mediate nutrient biosynthesis within the host cell (34), resistance to reactive intermediates of oxygen and nitrogen in the SCV (18, 45), and DNA repair (7). Of particular interest for the study of typhoid is the recent discovery of a type III secretion system within the second pathogenicity island of *Salmonella* serovar Typhimurium (SPI-2) (53, 63). Many genes encoded within SPI-2 are induced following invasion of host cells and play a critical role in allowing replication of the bacteria within the SCV: type III secretion system apparatus or putative translocated effector mutants are highly attenuated in mouse models of infection (9, 32). Recent evidence suggests that one putative effector of SPI-2, SpiC, is translocated into the cytosol of infected macrophages and may influence intracellular trafficking of the SCV (66). We anticipate that the functions of other SPI-2 effectors will be as diverse and complex as those of the SPI-1 effectors.

CONCLUSIONS

Pathogenic bacteria have evolved different strategies to initiate infection at mucosal surfaces. Following adhesion to the intestinal epithelium, a common feature is the use of type III secretion systems to deliver effector proteins that directly influence host cell signaling systems to suit the pathogenic strategy of the particular pathogen. We hope that future in vitro studies will further elucidate the mechanisms by which these effector proteins "hijack" the host cell. These findings will necessitate confirmation in vivo, where complex processes such as diarrhea production and immune responses can be studied.

ACKNOWLEDGMENTS

We thank the members of our laboratory who have contributed to this field. Artwork was kindly provided by A. Gauthier, R. DeVinney, U. Heczko, O. Steele-Mortimer, and I. Rosenshine, as acknowledged in the figure legends.

Work in the lab of B.B.F. is supported by operating grants from the Medical Research Council of

Canada, the Canadian Bacterial Disease Centre of Excellence, and a Howard Hughes International Research Scholar award. J.H.B. is supported by a Natural Science and Engineering Research Council of Canada fellowship, and is an honorary fellow of the Killam Memorial Foundation.

REFERENCES

1. **Abe, A., U. Heczko, R. G. Hegele, and B. B. Finlay.** 1998. Two enteropathogenic *Escherichia coli* type III secreted proteins, EspA and EspB, are virulence factors. *J. Exp. Med.* **188:** 1907–1916.

2. **Baldwin, T. J., W. Ward, A. Aitken, S. Knutton, and P. H. Williams.** 1991. Elevation of intracellular free calcium levels in HEp-2 cells infected with enteropathogenic *Escherichia coli. Infect. Immun.* **59:**1599–1604.

3. **Baumler, A. J., R. M. Tsolis, and F. Heffron.** 1996. The lpf fimbrial operon mediates adhesion of Salmonella typhimurium to murine Peyer's patches. *Proc. Natl. Acad. Sci. USA* **93:** 279–283.

4. **Bieber, D., S. W. Ramer, C. Y. Wu, W. J. Murray, T. Tobe, R. Fernandez, and G. K. Schoolnik.** 1998. Type IV pili, transient bacterial aggregates, and virulence of enteropathogenic Escherichia coli [see comments]. *Science* **280:** 2114–2118.

5. **Bowe, F., C. J. Lipps, R. M. Tsolis, E. Groisman, F. Heffron, and J. G. Kusters.** 1998. At least four percent of the *Salmonella typhimurium* genome is required for fatal infection of mice. *Infect. Immun.* **66:**3372–3377.

6. **Brumell, J. H., O. Steele-Mortimer, and B. B. Finlay.** 1999. Bacterial invasion: force feeding by Salmonella. *Curr. Biol.* **9:**R277–R280.

7. **Buchmeier, N. A., C. J. Lipps, M. Y. So, and F. Heffron.** 1993. Recombination-deficient mutants of Salmonella typhimurium are avirulent and sensitive to the oxidative burst of macrophages. *Mol. Microbiol.* **7:**933–936.

8. **Canil, C., I. Rosenshine, S. Ruschkowski, M. S. Donnenberg, J. B. Kaper, and B. B. Finlay.** 1993. Enteropathogenic *Escherichia coli* decreases the transepithelial electrical resistance of polarized epithelial monolayers. *Infect. Immun.* **61:** 2755–2762.

9. **Cirillo, D. M., R. H. Valdivia, D. M. Monack, and S. Falkow.** 1998. Macrophage-dependent induction of the Salmonella pathogenicity island 2 type III secretion system and its role in intracellular survival. *Mol. Microbiol.* **30:**175–188.

10. **Collazo, C. M., and J. E. Galan.** 1997. The invasion-associated type-III protein secretion system in Salmonella—a review. *Gene* **192:**51–59.

11. **Cornelis, G. R.** 1998. The Yersinia Yop virulon, a bacterial system to subvert cells of the primary host defense. *Folia Microbiol.* **43:**253–261.

12. **Crane, J. K., and J. S. Oh.** 1997. Activation of host cell protein kinase C by enteropathogenic *Escherichia coli. Infect. Immun.* **65:**3277–3285.

13. **De Rycke, J., E. Comtet, C. Chalareng, M. Boury, C. Tasca, and A. Milon.** 1997. Enteropathogenic *Escherichia coli* O103 from rabbit elicits actin stress fibers and focal adhesions in HeLa epithelial cells, cytopathic effects that are linked to an analog of the locus of enterocyte effacement. *Infect. Immun.* **65:**2555–2563.

14. **DeVinney, R., D. G. Knoechel, and B. B. Finlay.** 1999. Enteropathogenic Escherichia coli: cellular harassment. *Curr. Opin. Microbiol.* **2:**83–88.

15. **DeVinney, R., M. Stein, D. Reinscheid, A. Abe, S. Ruschkowski, and B. B. Finlay.** 1999. Enterohemorrhagic *Escherichia coli* O157: H7 produces Tir, which is translocated to the host cell membrane but is not tyrosine phosphorylated. *Infect. Immun.* **67:**2389–2398.

16. **Ebel, F., T. Podzadel, M. Rohde, A. U. Kresse, S. Kramer, C. Deibel, C. A. Guzman, and T. Chakraborty.** 1998. Initial binding of Shiga toxin-producing Escherichia coli to host cells and subsequent induction of actin rearrangements depend on filamentous EspA-containing surface appendages. *Mol. Microbiol.* **30:** 147–161.

17. **Elliott, S. J., L. A. Wainwright, T. K. McDaniel, K. G. Jarvis, Y. K. Deng, L. C. Lai, B. P. McNamara, M. S. Donnenberg, and J. B. Kaper.** 1998. The complete sequence of the locus of enterocyte effacement (LEE) from enteropathogenic Escherichia coli E2348/69. *Mol. Microbiol.* **28:**1–4.

18. **Fang, F. C., M. A. DeGroote, J. W. Foster, A. J. Baumler, U. Ochsner, T. Testerman, S. Bearson, J. C. Giard, Y. Xu, G. Campbell, and T. Laessig.** 1999. Virulent Salmonella typhimurium has two periplasmic Cu, Zn-superoxide dismutases. *Proc. Natl. Acad. Sci. USA* **96:**7502–7507.

19. **Fields, P. I., R. V. Swanson, C. G. Haidaris, and F. Heffron.** 1986. Mutants of Salmonella typhimurium that cannot survive within the macrophage are avirulent. *Proc. Natl. Acad. Sci. USA* **83:**5189–5193.

20. **Foubister, V., I. Rosenshine, and B. B. Finlay.** 1994. A diarrheal pathogen, enteropathogenic Escherichia coli (EPEC), triggers a flux of inositol phosphates in infected epithelial cells. *J. Exp. Med.* **179:**993–998.

21. **Frankel, G., A. D. Phillips, I. Rosenshine, G. Dougan, J. B. Kaper, and S. Knutton.**

1998. Enteropathogenic and enterohaemorrhagic Escherichia coli: more subversive elements. *Mol. Microbiol.* **30:**911–921.

22. **Fu, Y., and J. E. Galan.** 1998. The Salmonella typhimurium tyrosine phosphatase SptP is translocated into host cells and disrupts the actin cytoskeleton. *Mol. Microbiol.* **27:**359–368.

23. **Galan, J. E., and R. D. Curtiss.** 1989. Cloning and molecular characterization of genes whose products allow Salmonella typhimurium to penetrate tissue culture cells. *Proc. Natl. Acad. Sci. USA* **86:**6383–6387.

24. **Garcia-del Portillo, F., and B. B. Finlay.** 1994. *Salmonella* invasion of nonphagocytic cells induces formation of macropinosomes in the host cell. *Infect. Immun.* **62:**4641–4645.

25. **Garcia-del Portillo, F., and B. B. Finlay.** 1995. Targeting of Salmonella typhimurium to vesicles containing lysosomal membrane glycoproteins bypasses compartments with mannose 6-phosphate receptors. *J. Cell Biol.* **129:**81–97.

26. **Goosney, D. L., J. Celli, B. Kenny, and B. B. Finlay.** 1999. Enteropathogenic *Escherichia coli* inhibits phagocytosis. *Infect. Immun.* **67:**490–495.

27. **Goosney, D. L., M. de Grado, and B. B. Finlay.** 1999. Putting E. coli on a pedestal: a unique system to study signal transduction and the actin cytoskeleton. *Trends Cell Biol.* **9:**11–14.

28. **Groisman, E. A., and H. Ochman.** 1997. How Salmonella became a pathogen. *Trends Microbiol.* **5:**343–349.

29. **Hardt, W. D., L. M. Chen, K. E. Schuebel, X. R. Bustelo, and J. E. Galan.** 1998. S. typhimurium encodes an activator of Rho GTPases that induces membrane ruffling and nuclear responses in host cells. *Cell* **93:**815–826.

30. **Hardt, W. D., H. Urlaub, and J. E. Galan.** 1998. A substrate of the centisome 63 type III protein secretion system of Salmonella typhimurium is encoded by a cryptic bacteriophage. *Proc. Natl. Acad. Sci. USA* **95:**2574–2579.

31. **Hein, W. R.** 1999. Organization of mucosal lymphoid tissue, p. 1–15. *In* J.-P. Kraehenbuhl and M. R. Neutra (ed.), *Defences of Mucosal Surfaces: Pathogenesis, Immunity and Vaccines.* Springer-Verlag, Heidelberg, Germany.

32. **Hensel, M., J. E. Shea, S. R. Waterman, R. Mundy, T. Nikolaus, G. Banks, A. Vazquez-Torres, C. Gleeson, F. C. Fang, and D. W. Holden.** 1998. Genes encoding putative effector proteins of the type III secretion system of Salmonella pathogenicity island 2 are required for bacterial virulence and proliferation in macrophages. *Mol. Microbiol.* **30:**163–174.

33. **Hersh, D., D. M. Monack, M. R. Smith, N. Ghori, S. Falkow, and A. Zychlinsky.** 1999.

The Salmonella invasin SipB induces macrophage apoptosis by binding to caspase-1. *Proc. Natl. Acad. Sci. USA* **96:**2396–2401.

34. **Hoiseth, S. K., and B. A. Stocker.** 1981. Aromatic-dependent Salmonella typhimurium are non-virulent and effective as live vaccines. *Nature* **291:**238–239.

35. **Hong, K. H., and V. L. Miller.** 1998. Identification of a novel *Salmonella* invasion locus homologous to *Shigella ipgDE. J. Bacteriol.* **180:**1793–1802.

36. **Hueck, C. J.** 1998. Type III protein secretion systems in bacterial pathogens of animals and plants. *Microbiol. Mol. Biol. Rev.* **62:**379–433.

37. **Jones, B. D., and S. Falkow.** 1996. Salmonellosis: host immune responses and bacterial virulence determinants. *Annu. Rev. Immunol.* **14:**533–561.

38. **Kaper, J. B., J. G. Morris, Jr., and M. M. Levine.** 1995. Cholera. *Clin. Microbiol. Rev.* **8:**48–86. [Erratum, *Clin. Microbiol. Rev.* **8:**316.]

39. **Kenny, B.** 1999. Phosphorylation of tyrosine 474 of the enteropathogenic Escherichia coli (EPEC) Tir receptor molecule is essential for actin nucleating activity and is preceded by additional host modifications. *Mol. Microbiol.* **31:**1229–1241.

40. **Kenny, B., R. DeVinney, M. Stein, D. J. Reinscheid, E. A. Frey, and B. B. Finlay.** 1997. Enteropathogenic E. coli (EPEC) transfers its receptor for intimate adherence into mammalian cells. *Cell* **91:**511–520.

41. **Kenny, B., and B. B. Finlay.** 1997. Intimin-dependent binding of enteropathogenic *Escherichia coli* to host cells triggers novel signaling events, including tyrosine phosphorylation of phospholipase C-gamma1. *Infect. Immun.* **65:**2528–2536.

42. **Kim, J. M., L. Eckmann, T. C. Savidge, D. C. Lowe, T. Witthoft, and M. F. Kagnoff.** 1998. Apoptosis of human intestinal epithelial cells after bacterial invasion. *J. Clin. Invest.* **102:**1815–1823.

43. **Knutton, S., I. Rosenshine, M. J. Pallen, I. Nisan, B. C. Neves, C. Bain, C. Wolff, G. Dougan, and G. Frankel.** 1998. A novel EspA-associated surface organelle of enteropathogenic Escherichia coli involved in protein translocation into epithelial cells. *EMBO J.* **17:**2166–2176.

44. **Kubori, T., Y. Matsushima, D. Nakamura, J. Uralil, M. Lara-Tejero, A. Sukhan, J. E. Galan, and S. I. Aizawa.** 1998. Supramolecular structure of the Salmonella typhimurium type III protein secretion system. *Science* **280:**602–605.

45. **Lundberg, B. E., R. E. Wolf, Jr., M. C. Dinauer, Y. Xu, and F. C. Fang.** 1999. Glucose 6-phosphate dehydrogenase is required for *Sal*-

monella typhimurium virulence and resistance to reactive oxygen and nitrogen intermediates. *Infect. Immun.* **67:**436–438.

46. **Manjarrez-Hernandez, H. A., B. Amess, L. Sellers, T. J. Baldwin, S. Knutton, P. H. Williams, and A. Aitken.** 1991. Purification of a 20 kDa phosphoprotein from epithelial cells and identification as a myosin light chain. Phosphorylation induced by enteropathogenic Escherichia coli and phorbol ester. *FEBS Lett.* **292:**121–127.

47. **McNamara, B. P., and M. S. Donnenberg.** 1998. A novel proline-rich protein, EspF, is secreted from enteropathogenic Escherichia coli via the type III export pathway. *FEMS Microbiol. Lett.* **166:**71–78.

48. **Méresse, S., O. Steele-Mortimer, B. B. Finlay, and J. P. Gorvel.** 1999. The rab7 GTPase controls the maturation of Salmonella typhimurium-containing vacuoles in HeLa cells. *EMBO J.* **18:**4394–4403.

49. **Mills, S. D., and B. B. Finlay.** 1998. Isolation and characterization of Salmonella typhimurium and Yersinia pseudotuberculosis-containing phagosomes from infected mouse macrophages: Y. pseudotuberculosis traffics to terminal lysosomes where they are degraded. *Eur. J. Cell Biol.* **77:**35–47.

50. **Monack, D. M., B. Raupach, A. E. Hromockyj, and S. Falkow.** 1996. Salmonella typhimurium invasion induces apoptosis in infected macrophages. *Proc. Natl. Acad. Sci. USA* **93:**9833–9838.

51. **Neutra, M. R.** 1999. M cells in antigen sampling in mucosal tissues, p. 17–32. *In* J.-P. Kraehenbuhl and M. R. Neutra (ed.), *Defences of Mucosal Surfaces: Pathogenesis, Immunity and Vaccines,* 1st ed. Springer-Verlag, New York, N.Y.

52. **Norris, F. A., M. P. Wilson, T. S. Wallis, E. E. Galyov, and P. W. Majerus.** 1998. SopB, a protein required for virulence of Salmonella dublin, is an inositol phosphate phosphatase [see comments]. *Proc. Natl. Acad. Sci. USA* **95:**14057–14059.

53. **Ochman, H., F. C. Soncini, F. Solomon, and E. A. Groisman.** 1996. Identification of a pathogenicity island required for Salmonella survival in host cells. *Proc. Natl. Acad. Sci. USA* **93:**7800–7804.

54. **Pier, G. B., M. Grout, T. Zaidi, G. Meluleni, S. S. Mueschenborn, G. Banting, R. Ratcliff, M. J. Evans, and W. H. Colledge.** 1998. Salmonella typhi uses CFTR to enter intestinal epithelial cells. *Nature* **393:**79–82.

55. **Rathman, M., L. P. Barker, and S. Falkow.** 1997. The unique trafficking pattern of *Salmonella typhimurium*-containing phagosomes in murine

macrophages is independent of the mechanism of bacterial entry. *Infect. Immun.* **65:**1475–1485.

56. **Raupach, B., J. Mecsas, U. Heczko, S. Falkow, and B. B. Finlay.** 1999. Bacterial epithelial cell cross talk. *Curr. Top. Microbiol. Immunol.* **236:**137–161.

57. **Richter-Dahlfors, A., A. M. J. Buchan, and B. B. Finlay.** 1997. Murine salmonellosis studied by confocal microscopy: Salmonella typhimurium resides intracellularly inside macrophages and exerts a cytotoxic effect on phagocytes in vivo. *J. Exp. Med.* **186:**569–580.

58. **Rosenshine, I., S. Ruschkowski, M. Stein, D. J. Reinscheid, S. D. Mills, and B. B. Finlay.** 1996. A pathogenic bacterium triggers epithelial signals to form a functional bacterial receptor that mediates actin pseudopod formation. *EMBO J.* **15:**2613–2624.

59. **Ruckdeschel, K., A. Roggenkamp, V. Lafont, P. Mangeat, J. Heesemann, and B. Rouot.** 1997. Interaction of *Yersinia enterocolitica* with macrophages leads to macrophage cell death through apoptosis. *Infect. Immun.* **65:**4813–4821.

60. **Salama, N. R., and S. Falkow.** 1999. Genomic clues for defining bacterial pathogenicity. *Microb. Infect.* **1:**615–619.

61. **Salyers, A. A., and D. D. Whitt.** 1994. *Bacterial Pathogenesis: A Molecular Approach.* ASM Press, Washington, D.C.

62. **Sansonetti, P. J., G. Tran Van Nhieu, and C. Egile.** 1999. Rupture of the intestinal epithelial barrier and mucosal invasion by Shigella flexneri. *Clin. Infect. Dis.* **28:**466–475.

63. **Shea, J. E., M. Hensel, C. Gleeson, and D. W. Holden.** 1996. Identification of a virulence locus encoding a second type III secretion system in Salmonella typhimurium. *Proc. Natl. Acad. Sci. USA* **93:**2593–2597.

64. **Siebers, A., and B. B. Finlay.** 1996. M cells and the pathogenesis of mucosal and systemic infections. *Trends Microbiol.* **4:**22–29.

65. **Steele-Mortimer, O., S. Méresse, J.-P. Gorvel, B.-H. Toh, and B. B. Finlay.** 1999. Biogenesis of Salmonella typhimurium-containing vacuoles in epithelial cells involves interactions with the early endocytic pathway. *Cell. Microbiol.* **1:**33–51.

66. **Uchiya, K., M. A. Barbieri, K. Funato, A. H. Shah, P. D. Stahl, and E. A. Groisman.** 1999. A salmonella virulence protein that inhibits cellular trafficking. *EMBO J.* **18:**3924–3933.

67. **Wachter, C., C. Beinke, M. Mattes, and M. A. Schmidt.** 1999. Insertion of EspD into epithelial target cell membranes by infecting en-

teropathogenic Escherichia coli. *Mol. Microbiol.* **31:**1695–1707.

68. **Wolff, C., I. Nisan, E. Hanski, G. Frankel, and I. Rosenshine.** 1998. Protein translocation into host epithelial cells by infecting enteropathogenic Escherichia coli. *Mol. Microbiol.* **28:**143–155.

69. **Wood, M. W., M. A. Jones, P. R. Watson, S. Hedges, T. S. Wallis, and E. E. Galyov.** 1998. Identification of a pathogenicity island re-

quired for Salmonella enteropathogenicity. *Mol. Microbiol.* **29:**883–891.

70. **Zhou, D., M. S. Mooseker, and J. E. Galan.** 1999. Role of the S. typhimurium actin-binding protein SipA in bacterial internalization. *Science* **283:**2092–2095.

71. **Zychlinsky, A., M. C. Prevost, and P. J. Sansonetti.** 1992. Shigella flexneri induces apoptosis in infected macrophages. *Nature* **358:**167–169.

BACTERIAL EVASION OF HOST-DERIVED ANTIMICROBIAL PEPTIDES ON MUCOSAL SURFACES

Kim Alan Brogden

2

The mammalian mucosa, composed of epithelial tissues, is in continual contact with the external environment (64). These tissues constitute an enormous surface area (66, 116), and it is here that microorganisms (including pathogens) frequently make their first contact with the host, giving them the potential to cause systemic or local disease. A variety of mechanisms have evolved at mucosal surfaces to prevent microbial invasion and damage. These include nonimmune and immune mechanisms (64). Predominant among the nonimmune mechanisms are antimicrobial proteins and peptides present in the granules of phagocytes (Table 1) and in mucosal fluids (Table 2). These include large proteins such as bactericidal/permeability-increasing protein (27), cathepsin G (70, 111), elastase (105, 111), azurocidin (18, 31), lactoferrin (53, 106), lysozyme (90), and small peptides such as the defensins (57) and protegrins (69). The small peptides form a structurally diverse family, ranging from α-helical molecules to dicyclic, β-sheet-containing peptides stabilized with 3-disulfide bonds (9, 56, 69).

Despite these antimicrobial agents, microorganisms persist and proliferate on the mucosa by using various host-bacterial interactions. One type of interaction is seen with classical pathogens such as *Salmonella enterica* serovar Typhimurium (17), *Yersinia* species (17), and *Brucella abortus* (84). These organisms use an array of determinants to infect the mucosal surface and invade the underlying tissues. Then they are phagocytized (100). In the phagosomes, they resist lysosomal antimicrobial peptides. A different type of interaction is that between the host and an extensive, diverse, endogenous microbial flora. These gram-positive and gram-negative organisms are often not pathogenic, yet colonize host surfaces effectively; numbers can be as high as 10^8 CFU/ml of oral secretion (8) or 10^{11} CFU/g of intestinal content (34). The concentration and composition of this flora depend upon the mucosal site as well as the individual. A third type of interaction is that of pathogens in latent or carrier states such as mucosal colonizers like *Pasteurella haemolytica* (97), *Pasteurella multocida* (1), *Haemophilus somnus* (118), and *Neisseria meningitidis* (21). Latency is a suspended state or period of inactivity. The pathogen persists in the host without concurrent damage but has the potential to reactivate and cause disease (76). In

Kim Alan Brogden Respiratory Diseases of Livestock Research Unit, National Animal Disease Center, Agricultural Research Service, U.S. Department of Agriculture, 2300 Dayton Road, P.O. Box 70, Ames, IA 50010.

Virulence Mechanisms of Bacterial Pathogens, 3rd ed., Edited by K. A. Brogden et al.
©2000 ASM Press, Washington, D.C.

TABLE 1 Antimicrobial proteins in neutrophil granules

Protein	Relative mass (kDa)	Reference(s)
Azurophilic granules		
Myeloperoxidase	150	30
BPI (CAP57)	60	27
BP55	55	111
CAP37	37	77, 93, 94
BP30	30	111
Elastase	29	105, 111
Cathepsin G	24–29	70, 111
Azurocidin	29	18, 31
Bactenecin 5 (Bac5)	5	119
Bactenecin 7 (Bac7)	7	99
Proteinase 3		111
Defensins	4	57
Specific granules		
Lactoferrin	80	53, 106
Lysozyme	12	90

the carrier state, organisms may or may not induce a chronic low-grade infection (76). The difference between the carrier state and latency is the ability of the organism in the former to transmit to another host and cause disease without first inducing disease in the initial host, as is necessary for transmission from the latent state. The carrier state is due to a pathophysiologic defect in the host that allows the organism to persist and cause ongoing damage to the host (76). This occurs in infections by *Pseudomonas aeruginosa*, *Staphylococcus aureus*, and *Haemophilus influenzae* in cystic fibrosis (19). In all these situations, the respective bacteria evade killing by antimicrobial peptides. The bacterial mechanisms involved may be physical subversion and "hiding" from the peptides in the mucosal fluid, alteration of bacterial surfaces and mem-

TABLE 2 Antimicrobial proteins in mucosal fluids

System and protein	Relative mass (kDa)	Reference
Respiratory tract		
Anionic peptides	<1	14
β-Defensins	4	63
α-Defensins	4	90
Surfactant proteins	8	51
Lysozyme	12	3
Secretory leukoprotease inhibitor	12	43
Lactoferrin	80	106
Salivary mucin glycoprotein MG2		2
Salivary histatin 5		82
Cathelicidins		6
Intestinal tract		
Secretory leukoprotease inhibitor	12	28
Lysozyme	12	28
α-Defensins	4	80
β-Defensins	4	46
Cryptdin	4	25
Urogenital tract		
Secretory leukoprotease inhibitor	12	28
Lysozyme	12	28
α-Defensins	4	81
β-Defensins	4	47
Cryptdin	4	36
hCAP18/LL-37		29
RK-1 defensin-like peptide		117

branes to prevent peptide interaction, or enzymatic degradation of peptides. Chronic infection due to a physiologic host dysfunction also occurs; the antimicrobial peptides may be inactivated (102) or in low concentrations (13).

The small antimicrobial peptides found in phagocytic cells, epithelial cells, and on mucosal surfaces and their mechanisms of antimicrobial killing have been reviewed recently (9, 41, 56, 67, 69). Here, I summarize them as background to discussing the numerous strategies that microorganisms use to avoid being killed by them. This is an exciting area of research. Identification of mechanisms for resistance to antimicrobial peptides can provide insight as to how microorganisms interact with the innate immune system to either produce progressive infection or enter into commensal, latency, or carrier states. The latter are important but infrequently addressed aspects of pathogenicity (101).

TYPES OF ANTIMICROBIAL PEPTIDES AND SITES OF ANTIMICROBIAL EXPRESSION

Originally, the source of antimicrobial peptides was used for classification (9). Peptide groups were based on the animal species, tissue, or cell type from which they were obtained. Then, similarities among peptide action and structures became apparent. These peptides can be direct translational products, posttranslationally modified molecules, or even protein degradation products (9, 69). Some are charge-neutralizing propieces of larger zymogens (12). They are now grouped according to similarities in sequence homology, functional similarity, and three-dimensional structure (9, 41, 69). Major categories are summarized in Table 3.

Anionic Antimicrobial Peptides

In 1968, Rudolph Galask and Irvin Snyder (32) found that amniotic fluid inhibited the growth of several bacterial species. Subsequently, a low molecular mass fraction with broad-spectrum antimicrobial activity was isolated (87, 89). It was an anionic peptide requiring zinc for antimicrobial activity (88). A similar peptide also requiring zinc was isolated from human lung lavage fluid (26, 54). In my laboratory, sheep pulmonary surfactant killed bacteria when incubated in normal serum (11). The activity is due to small (721.6 to 823.8 Da) anionic peptides: H-GDDDDDD-OH, H-DDDDDDD-OH, and H-GADDDDD-OH (14). These peptides are very similar to the charge-neutralizing propeptides of sheep trypsinogen (20), which when synthesized were shown to be antimicrobial (12). Recently, anionic peptidelike molecules were also detected in human respiratory samples (13).

Anionic antimicrobial peptide fragments have been isolated from protein digests. For example, aprotinin (75) and α-lactalbumin (74), digested with pepsin, trypsin, or chymotrypsin, yielded three anionic peptides, which were antimicrobial for gram-positive bacteria but less so for gram-negative bacteria.

Cationic Antimicrobial Peptides

In the early 1980s, Hans Boman and Robert Lehrer independently isolated and purified the first families of insect cecropins and mammalian defensins, respectively (9, 33). Now, over 300 different antimicrobial peptides have been identified or predicted from nucleic acid sequences. Most of them are cationic molecules that can be purified by use of cationic exchange columns, but they show considerable variation in structure (9, 41, 56, 67, 69). Within each group below there are direct translational peptides and those that occur as posttranslationally modified molecules. The latter include cathelicidins, a diverse group differing greatly in sequence, structure, and size. They have a common N-terminal pre-proregion of about 100 residues that is homologous to that of the cysteine protease inhibitor cathelin (120) and a highly variable C terminus containing the cationic antimicrobial domain. After synthesis, the C terminus is cleaved off, forming the mature antimicrobial peptide. The peptides are anti-

TABLE 3 Structural classes of cationic peptides

Anionic peptides
- Small peptides from human amniotic fluid and both human and sheep bronchoalveolar lavage fluid rich in glutamic and aspartic acids

Cationic peptides

Amphipathic helical molecules
- Peptides from insects (cecropin, andropin, moricin, ceratotoxin, and melittin), amphibians (magainin, dermaseptin, bombinin, brevinin-2, esculentins-1 and 2, and PGLa), and mammals (seminalplasmin)
- Cathelicidins include C18, PMAP, SMAP, BMAP, SC5, and LL37

Proline-rich molecules
- Peptides from insects including the proline- and arginine-rich apidaecin, abaecin, and hymenoptaecin and the glycine- and proline-rich coleoptericin and holotricin 2
- Cathelicidins include Bac5 and Bac7 from ruminants (45 to 49% proline and 24 to 29% arginine) and PR-39 from pigs
- Peptides rich in other amino acids include indolicidin (rich in tryptophan) and prophenin (rich in proline and phenylalanine)

Cysteine-stabilized sheet molecules
- Peptides include the α- and β-defensins, cryptdin, gallanicin, tachyplesin, thionin, polyphemusin, and drosomycin
- Cathelicidins include protegrins

Fragments of biomolecules
- Peptides from antimicrobial proteins include CAP37 (residues 20 through 44), lysosome (residues 98 through 112), human lactoferrin (residues 20 through 37), bovine lactoferrin (residues 19 through 36)
- Peptides from nonantimicrobial proteins include gastric inhibitory polypeptide (residues 7 through 42), diazepam-binding inhibitor (residues 32 through 86), casocidin I (from casein)

microbial in micromolar concentrations against a variety of bacteria and may be potent modulators of inflammation (9).

LINEAR AND AMPHIPATHIC PEPTIDES THAT FORM AN α-HELIX

Peptides in this group are generally 20 to 40 residues in size; are linear, helical peptides without cysteine residues; and may or may not have a hinge or "kink" in the middle. If they assume an α-helical structure, they become amphiphilic, which is necessary to permeabilize the bacterial cell membrane. There are a number of peptides in this group (Table 3) and some form the family of cecropins (A, B, D, and P1). Cecropins have a strong basic N terminus and a hydrophobic stretch at the C terminus, joined by a hinge region. Cecropin A, examined by circular dichroism (103) and nuclear magnetic resonance spectroscopy (44), was shown to have two helical regions extending from residues 5 to 21 and 24 to 37 interspaced by an Ala-Gly hinge. The N terminus is a perfect amphipathic helix with a clear and approximately equal separation of the hydrophilic and hydrophobic faces. Cecropin P1, isolated from porcine small intestine, forms a helical molecule with a hingelike sequence of Ser-Glu-Gly. Cecropins are highly active against gram-negative bacteria and some gram-positive bacteria.

Also in this group are cathelicidins that have a high content of the basic amino acids arginine and lysine. C18 is a 21-residue peptide derived from the C terminus region of CAP18, a cathelicidin expressed in rabbit neutrophils. Also included are myeloid antimicrobial peptides from pigs (PMAP); sheep (SMAP), and cattle (BMAP); SC5 from sheep bone-marrow cells; and LL37 from human neutrophils (107).

LINEAR PEPTIDES, RICH IN PROLINE

Peptides in this group are generally 40 to 80 residues in size, are linear without cysteine residues, and may form extended coils. Usually they have a high proportion of certain residues, particularly proline, arginine, and glycine, and some contain large amounts of tryptophan. Many of the peptides in this group (Table 3) are formed by insects. They have high concentrations of proline and arginine (apidaecin, abaecin, and hymenoptaecin) or glycine and proline (coleoptericin and holotricin 2). Also in this group are cathelicidins Bac5, Bac7, and PR-39 with high concentrations of proline (45 to 49%) and arginine (24 to 29%). Bac5 and Bac7 are stored as proproteins of 20 and 16 kDa, respectively, in large cytoplasmic granules present in the neutrophils of cattle, goats, and sheep (96). They become activated by proteases in the azurophils when these granules fuse with the probactenecin-containing granules. Also included are prophenin, a peptide from pigs that is high in proline and phenylalanine (42), and indolicidin, a peptide from cattle that is high in tryptophan.

CYSTEINE-STABILIZED PEPTIDES WITH A β-SHEET STRUCTURE

Peptides in this group (Table 3) are generally 16 to 40 residues in size, contain cysteine, have two or more disulfide bonds, and form a stabilized β-sheet structure. Good examples are the defensins, which constitute a large portion of protein in mammalian granulocytes (up to 17%, depending upon species). They have conserved amino acids at residues 6 (arginine), 15 (arginine), and 24 (glycine). In mammals, the defensins can be split into two families, the α-defensins and the β-defensins, differing in the number of residues, the location of the cysteine residues, and the order of disulfide bonding.

The α-defensins are the smaller of the two. They have 29 to 35 residues and are in the azurophil granules of neutrophils, macrophages, and intestinal Paneth cells in humans and rodents.

The β-defensins are larger and range from 38 to 42 residues. They are arginine rich, with broad antimicrobial, antiviral, and cytotoxic activity (33). Some are chemotactic and/or opsonic, and they may modulate hormonal responses (33). Recently, β-defensins have been extensively examined as important molecules deterring microbial infections of mucosal surfaces. In cattle, two β-defensins, tracheal antimicrobial peptide (TAP) and lingual antimicrobial peptide (LAP), are in columnar cells of the pseudostratified epithelium throughout the conducting airway and tongue (23, 24, 91). TAP genes have also been identified in sheep (48). β-Defensin mRNA is found to be widely expressed in numerous exposed epithelia and is higher in tissues that are constantly exposed to and colonized by *P. haemolytica* (104). This suggests that β-defensins are an integral component of the rapidly mobilized local defense inflammatory response.

Also in this group are cathelicidins. These include the protegrins from porcine leukocytes, which have 16 residues and contain four cysteines stabilized by two intramolecular disulfide bonds. Two antiparallel β sheets are linked by a β turn, and there is sequence similarity to the defensins, particularly in 8 to 10 residues containing three cysteines. Protegrins have a broad spectrum of antimicrobial activity.

PEPTIDES AS FRAGMENTS OF LARGER PROTEINS

Larger antimicrobial proteins have been fragmented experimentally into smaller peptides to search for the smallest sequence representing the antimicrobial domain (Table 3). CAP37 (residues 20 through 44) and lysosome (residues 98 through 112) are good examples. Also, antimicrobial peptides may be isolated as cationic fragments from enzymatic digests of larger nonantimicrobial proteins. Gastric inhibitory polypeptide (residues 7 through 42), diazepam-binding inhibitor (residues 32 through 86), and casocidin I (from casein) are good examples (9), although not known to occur naturally.

MECHANISM OF LYTIC ACTIVITY

Antimicrobial peptides are attracted electrostatically to the charged surfaces of both gram-negative and gram-positive bacteria. It is possible that both cationic and anionic forms exist in the host to cope with phase variation in surface charge that some microorganisms use as a primary strategy to overcome antimicrobial peptide activity.

Cationic Antimicrobial Peptides

To induce lysis, a peptide must undergo a series of events (Fig. 1). First, the peptide must be attracted to the membrane surface, usually by electrostatic interaction between it and a net anionic charge on the surface of the bacterium. Second, it must disrupt and penetrate the outer membrane, which often results in the formation of blebs in the outer envelope (Fig. 2). Third, it must enter the periplasm and build up to an antimicrobial concentration. Fourth, the peptide must bind to the cell membrane. Finally, it must enter the membrane and form lytic pores. The exact mechanism of antimicrobial action and the time required for killing differ significantly among antimicrobial peptides. Killing can occur almost instantaneously or within 30 to 60 min. The nature of membrane pores also varies with the peptide. Human neutrophil defensin HNP-2, a noncovalent dimer, forms large, 20-Å multimeric pores (57), whereas rabbit NP-1 creates large, transient defects in the phospholipid bilayer (45, 57). Then there are special mechanisms. Insect defensins can induce leakage of essential intracellular contents, including potassium from bacteria; magainins disrupt the free energy metabolism in *E. coli*; and PR-39 kills bacteria by a mechanism that does not result in membrane lysis (37). The metabolic state of the bacteria can influence the outcome. For example, permeabilization of the outer and inner bacterial membranes by human defensins depends upon active bacterial metabolism and bacterial growth (55). Also, the phospholipid composition of target cell membranes is important (45). This may explain why *Serratia marcescens*, an organism with unique membrane phospholipids, is more resistant than other organisms.

Anionic Antimicrobial Peptides

The mechanism of killing for these peptides is not known but is thought to be different from that of the cationic peptides. Initially, the peptide must be attracted to the membrane surface. Since these peptides require zinc (14, 26, 54, 88) and complex with it (10), a cationic salt bridge may be formed to overcome the

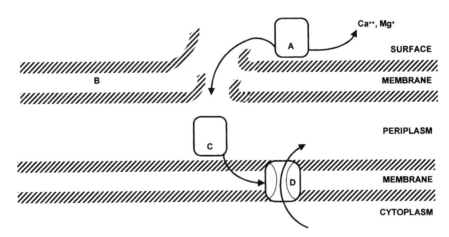

FIGURE 1 The lysis of bacterial membranes by antimicrobial peptides. (A) Peptides may competitively displace divalent cations (e.g., magnesium and calcium) from their binding sites on the LPS in the outer leaflet of the outer membrane. (B) This distorts the outer membrane, often resulting in the formation of blebs. (C) The peptide moves into the periplasm and makes contact with the cytoplasmic membrane. (D) The peptide then penetrates the membrane, creating lethal, lytic pores.

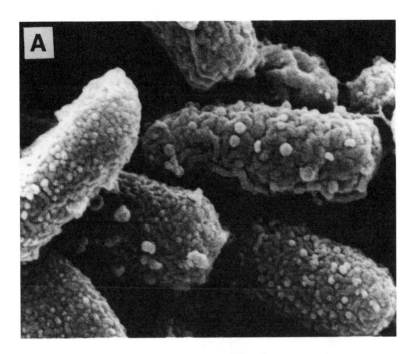

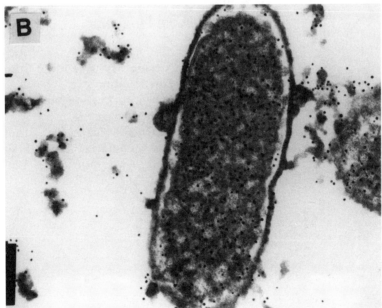

FIGURE 2 *P. aeruginosa* PAO1 incubated with a synthetic cathelicidin. (A) Scanning electron micrograph showing the presence of blebs induced by the interaction of the peptide with the outer envelope. (B) Transmission electron micrograph of a thin section of PAO1. Specific antibody and protein A colloidal gold labeling can be used to demonstrate the rapid penetration of the peptide into the bacterial cell. Note the extensive outer membrane material that has been "scrubbed" from the bacterial cell surface. Panel A courtesy of Hong Peng Jia, Paul McCray, Jr., and Brian Tack, Departments of Pediatrics, and Microbiology, University of Iowa College of Medicine, Iowa City, Iowa.

net negative charge on the microbial surface. Second, the peptide penetrates the outer membrane without inducing any morphological changes (Fig. 3) (14). Third, it must bind and penetrate the inner cell membrane. Leenhouts et al. present an interesting mechanism that may apply (54a). The peptide attaches electrostatically to the bacterial membrane by the positive charge found at the N terminus (or by the attached zinc). Once there, the locally lower interfacial pH of the anionic phospholipids protonates the available carboxyl groups, resulting in a more hydrophobic molecule that is pulled into the lipid bilayer. The pore-forming domains of colicins require a low pH and negatively charged lipids for membrane insertion (72, 108). Once intracellular, the peptide may then attach to ribosomes and inhibit ribonuclease activity similar to that seen with polymers of aspartic acid (58, 92, 109). Ultimately, the cytoplasmic protein precipitates and settles out, suggesting an internal mechanism of protein inactivation (Fig. 3). Killing occurs within 30 min (14).

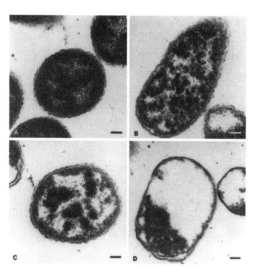

FIGURE 3 Transmission electron micrograph of *P. haemolytica* serotype A1 strain 82-25 incubated in zinc saline solution (A) and zinc saline solution containing 0.5 mM anionic peptide (B–D). Note the distended outer envelope (B) and flocculated intracellular constituents (C and D) in cells incubated with anionic peptide. Bars, 0.5 μM.

MICROBIAL SUBVERSION OF INNATE ANTIMICROBIAL PEPTIDES

Many pathogenic organisms are susceptible in vitro to antimicrobial peptides, but in vivo can exist in environments containing the same concentrations of antimicrobial peptides. Recent work has identified a number of resistance strategies these organisms can use to circumvent antimicrobial peptide killing (Table 4). These are described below. Some organisms may use more than one strategy.

Penetration of Susceptible Microorganisms into Epithelia

Rapid penetration of epithelial cells reduces the time of contact between microorganisms and antimicrobial peptides in mucosal secretions. This occurs with *N. meningitidis*, *Salmonella* serovar Typhimurium, *H. influenzae* (52), and perhaps *S. enterica* serovar Arizonae (16, 65). Although the concept is simple, the mechanisms these organisms use to rapidly attach to and penetrate epithelial cells are not (52).

Coating of Susceptible Organisms by Host Products

Some organisms on mucosa contain molecules on their outer surfaces that are similar to host constituents. For example, choline can be a surface constituent of mucosal pathogens such as *Streptococcus pneumoniae*, *H. influenzae*, *N.*

TABLE 4 Mechanisms of bacterial evasion of host-derived antimicrobial peptides on mucosal surfaces

Mechanism of bacterial evasion
"Hiding" within epithelial cells
Coating themselves with host material
Releasing endotoxin
Altering outer membrane proteins
Altering LPS structure
Altering teichoic acids
Transporting the peptide in or out of the cell
Degrading the peptide
Utilizing the peptide to adhere to host cells
Taking advantage of the host condition
Inactivating peptides during viral infection

meningitidis, Mycoplasma species, and *P. aeruginosa* (113–115). It is incorporated into the lipopolysaccharide (LPS) of *Haemophilus* species (115). Choline is a cationic molecule. In the bacterial surface, it might induce resistance to cationic antimicrobial peptides. This coating would not prevent penetration of antimicrobial peptides by steric hindrance, but it would reduce the net anionic surface charge, thus repelling them (112).

Release of Endotoxin

Many gram-negative organisms release endotoxin during infection (22, 86). It is possible that these extracellular products combine with and inactivate antimicrobial peptides targeted for the organism itself. For example, *S. enterica* serovar Montevideo smooth LPS, preincubated with cationic peptides, causes 70 to 80% reductions in the activities of magainins 1 and 2, melittin, CAP18, cecropin A and P1, and polymyxin B and a 47% reduction in the antimicrobial activity of lactoferricin B (62). *B. abortus* smooth LPS, preincubated with cationic peptides, also causes 50 to 66% reductions in the antimicrobial activity of magainins 1 and 2, melittin, cecropin P1, and polymyxin B and a 15% reduction in the antimicrobial activity of CAP18, cecropin A, and lactoferricin B (62). In our laboratory, preincubation of ovine *P. haemolytica* 82-25 serotype A1 LPS with a sheep myeloid antimicrobial peptide (SMAP29) decreases the MIC of the SMAP29 from 2.5 to 10.0 µg/ml for *P. haemolytica* 82-25 (Brogden and Tack, unpublished observation).

Bacterial Adaptation

Direct adaptation by gram-negative and gram-positive microorganisms to become resistant in an environment containing antimicrobial peptides is the most characterized strategy. This adaptation involves modifications in LPS structure, outer membrane proteins, and teichoic acid structure.

OUTER MEMBRANE PROTEINS

In gram-negative bacteria, alterations in outer membrane proteins increase their resistance to antimicrobial peptides. For example, Visser et al. (110) demonstrated the resistance of virulent *Yersinia enterocolitica* to antimicrobial peptides was correlated with the presence of a 70-kb plasmid, designated pYVe. pYVe$^+$ *Y. enterocolitica* was less susceptible to antimicrobial peptides in extracts of human neutrophil granules than pYVe$^-$ *Y. enterocolitica*. The pYVe plasmid codes for calcium dependence for growth, the production of at least two outer membrane proteins called *Yersinia* adhesin A (YadA) and *Yersinia* lipoprotein A (YlpA), and the secretion of 11 proteins, referred to as Yops (110). Of these, YadA confers resistance to the killing of *Y. enterocolitica* by antimicrobial peptides in extracts of human neutrophil granules. YadA is maximally expressed at 37°C. It forms a polymeric fibrillar matrix covering the outer membrane of the organism and has distinct hydrophobic domains, which play a role in the attachment of the organism to host cells and to the extracellular matrix proteins fibronectin, laminin, and collagen. YadA also contributes to the resistance of the bactericidal action of human complement. Other plasmid-encoded factors may also be involved in resistance to antimicrobial peptides. YadA does not protect *Y. enterocolitica* entirely against the antimicrobial effect of granule extract, and YadA does not reduce the antimicrobial activity of lysozyme or defensins against *Y. enterocolitica*.

Deficiencies in the outer membrane proteins can lead to an increased susceptibility to antimicrobial peptides, but these mechanisms are less defined. For example, the absence of the porin OmpC from the outer membrane of *E. coli* increases its susceptibility to cecropin D (98).

LPS O-SPECIFIC SIDE CHAINS AND CORE

LPS form a large, unique class of molecules characteristic of gram-negative bacteria (59). Associated with protein, they are in the outer leaflet of the outer membrane of the bacterial cell (59) and are among surface components to interact with cationic antimicrobial peptides. LPS is a long, covalently linked hetero-polysaccharide subdivided into the core and O-specific chain regions linked to a lipid region, lipid A (59) (Fig. 4). It is amphoteric because of the presence of carboxyl, phosphoryl, and ethanolamine residues.

A certain amount of innate resistance to antimicrobial peptides is already built into the structure and composition of the LPS O-specific side chains and core. This could explain the variation in MICs of antimicrobial peptides for a variety of gram-negative bacteria (79). Generally, enteric and nonenteric organisms with smooth LPS (i.e., with long O-specific side chains) are more resistant to cationic peptides than organisms with rough LPS (i.e., with short O-specific side chains). In an assessment of the susceptibility of smooth serovar Typhimurium and a series of serovar Typhimurium rough-LPS mutants (Ra to Re) to magainin 2, the latter showed an ordered increase in sensitivity to magainin 2 as the depth of the rough lesion in the LPS increased (61).

The LPS of *Bordetella pertussis* does not contain an extended O-specific side chain. Typically, two bands are seen in silver-stained acrylamide gels. Band A corresponds to a charged trisaccharide containing *N*-acetylglucosamine (GlcNAc), 2,3-dideoxy-2,3-di-*N*-acetylmannosaminuronic acid (2,3-diNAcmanA), and *N*-acetyl-*N*-methylfucosamine (FucNAcMe), all linked to the LPS core region. Band B corresponds to the core region lacking the trisaccharide. *Bordetella bronchiseptica* LPS is smooth with O-specific side chains linked to the above trisaccharide. Transposon (Tn5) insertions in *B. bronchiseptica*, which renders it highly susceptible to cationic peptides, occur in genes (*wlbA* and *wlbL*) that code for a dehydrogenase involved in the biosynthesis of 2,3-diNAcManA and a 2,6-dideoxy-galactose derivative of FucNAcMe, respectively (7). Inactivation of genes *wlbA* and *wlbL* also converts the smooth LPS to a rough phenotype (7).

To further demonstrate the role of O-specific side chain in antimicrobial resis-tance, Martinez de Tejada et al. incubated antimicrobial peptides with viable *B. abortus* 45/20 (antimicrobial resistant organism) previously coated with smooth LPS from *B. abortus* S19 (antimicrobial resistant organism) and LPS from serovar Montevideo (antimicrobial sensitive organism) (62). The strain 19 LPS increased resistance, whereas serovar Montevideo LPS decreased resistance to the peptides to levels close to those observed for serovar Montevideo (62). The lack of outer surface ionic groups for electrostatic binding was thought to be the reason for strain 45/20 resistance.

LPS LIPID A

As early as 1982, it was known that gram-negative bacteria could modify the surface charge of lipid A by adding or omitting substitutions in order to adapt to the ionic environment (59). For example, in about 50% of serovar Typhimurium lipid A molecules, the phosphate group linked to glucosamine I is substituted by a phosphorylethanolamine residue with a free amino group (Fig. 4). In about 30 to 60% of the lipid A molecules, the phosphate bound to glucosamine II is substituted by 4-amino-4-deoxy-L-arabinose, the linkage being through C1. These and other differences such as forming heptaacylated lipid A by adding palmitate and replacing myristate on lipid A with 2-OH myristate (40) could affect innate resistance to antimicrobial peptides.

The mechanism used by *Salmonella* species to alter their lipid A and resist killing by antimicrobial peptides is induced by activation of the two-component regulatory system called PhoP-PhoQ. This regulon comprises a sensor kinase PhoQ and a transcriptional activator, PhoP (39), triggered in response to environmental signals including changes in extracellular magnesium or calcium levels, pH, or other signals after infection or phagocytosis (40). This system simultaneously activates or represses more than 40 different genes, termed PhoP-activated genes (*pag/pqa*) and PhoP-repressed genes (*prg/pqr*) (5). For example, the

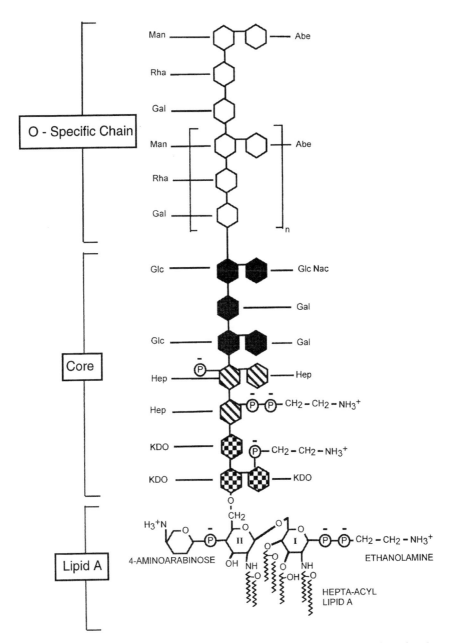

FIGURE 4 An LPS molecule containing a long, covalently linked heteropolysaccharide that is subdivided into the core and O-specific chain regions linked to a lipid region, lipid A (59). The sites altered for increased antimicrobial resistance include the O-specific side chain, phosphate groups attached to the core region, and lipid A. The changes in lipid A include (i) adding aminoarabinose to lipid A phosphate groups, (ii) forming heptaacylated lipid A by adding palmitate, and (iii) replacing myristate on lipid A with 2-OH myristate. Adapted from Luderitz et al. (59, 60).

PhoP-PhoQ regulon controls genes in macrophage survival (*pagC*), invasion of mammalian cells and protein secretion (*prgHIJK*), nonspecific acid phosphatase (*phoN*), and magnesium transporters (*mgtA* and *mgtCB*) (5). PhoP-PhoQ also regulates the PmrA-PmrB two-component regulatory system, which in turn controls genes *pmrE/ugd* and the *pmrF* operon. The *pmrF* operon is strongly regulated by the PmrA-PmrB operon. Inactivation of genes in the PhoP-PhoQ regulon and assessment of the sensitivity of mutants to antimicrobial peptides have revealed both the genes and the respective modifications of lipid A involved in antimicrobial peptide resistance (5, 39).

The first alteration in lipid A involves the addition of aminoarabinose to the negatively charged phosphate bound in ester linkage to the C-4 of glucosamine II (59). This adds a positive charge that reduces electrostatic interaction with cationic antimicrobial peptides. The *Salmonella enterica* serovar Typhi *pqaB* gene and the serovar Typhimurium *pmrF* operon, which have a high level of homology (about 98%), both are involved with 4 aminoarabinose-lipid A modifications and polymyxin B resistance. Both *pqaB* and ORF5 encode a bactoprenol-4 aminoarabinose-lipid A transferase. Inactivation of *pmrF* and *pqaB* results in mutants sensitive to polymyxin. Likewise, the serovar Typhi PhoP-constitutive phenotype (PhoP^c) has increased resistance to polymyxin B compared with that of wild-type Ty2. Therefore, the predicted function of the *pmrF* operon gene is to encode the enzymes required for the biosynthesis of the 4 aminoarabinose modification of the lipid A component of LPS (38). Interestingly, *Y. enterocolitica*, *Proteus vulgaris*, *E. coli*, *Klebsiella pneumoniae*, and *Burkholderia cepacia* add similar covalent modifications to LPS.

The second alteration in lipid A that could affect resistance to antimicrobial peptides involves the conversion of myristate to 2-hydroxymyristate (39). Serovar Typhimurium PhoP mutants that help to delineate the composition of lipid A responsible for cationic antimicrobial peptide resistance have been made.

In the PhoP^c phenotype, a replacement of amino acid 48 of PhoQ with isoleucine in the Pho-24 allele locks serovar Typhimurium in a state of *pag* activation and *prg* repression (39). These mutants are deficient in invasion of mammalian cells, form spacious phagosomes within phagocytes, are defective for macrophage survival, and are attenuated for mice (5). Deletion of PhoP or PhoQ results in a PhoP null phenotype (PhoP$^-$) (39). These mutants have activated *prg* and repressed *pag*. Mutants that are defective in macrophage survival and attenuated for virulence in mice are also sensitive to cationic antimicrobial peptides (5, 40). Comparison of the fatty acid composition of serovar Typhimurium wild type, PhoP^c, and PhoP$^-$ strains with their sensitivity to cationic antimicrobial peptides has shown some striking correlations. PhoP^c strains, resistant to cationic antimicrobial peptides, contain significantly more 2-OH C14:0 and 3-OH C14:0 than both the wild-type and PhoP$^-$ strains (39). PhoP$^-$ strains, sensitive to cationic antimicrobial peptides, do not contain 2-OH C14:0 and have considerably less 3-OH C14:0 (39).

The third alteration in lipid A that could affect resistance to antimicrobial peptides is an increase in C16:0 (39). PhoP^c strains, resistant to cationic antimicrobial peptides, contain more than twice as much C16:0 as susceptible serovar Typhimurium wild-type and PhoP$^-$ strains (39). *pagP* is required for the regulated addition of palmitate to lipid A and promotes resistance to cationic antimicrobial peptides (40) in response to a low magnesium growth environment. PagP contains putative membrane domains and may work with *cspE* and *crcB* gene products to increase lipid A acylation in serovar Typhimurium via activation of lipid A transferases in a manner similar to that described for *E. coli* (40). In *E. coli*, gene products HtrB and MsbB are thought to be inner membrane proteins that function at a relatively late stage in lipid A biosynthesis to transfer C16:0 to lipid A. The increase of 2-OH C14:0 and C16:0 in the lipid A fatty acids is thought to decrease the fluidity of the outer membrane by increasing hydrophobic interactions between an increased number of lipid A acyl tails. When antimicrobial peptides come in contact with this layer, the increased hydrophobic moment retards or abolishes peptide insertion and pore formation.

Although the above models could explain the increased resistance of *Salmonella* species to antimicrobial peptides, Guo et al. (40) felt that none of the above models was fully responsible for PhoP-PhoQ regulated antimicrobial resistance. As-yet-undefined resistance mechanisms may exist, such as a role of an outer membrane protease (below).

TEICHOIC ACIDS

Gram-positive bacteria, often on mucosa, tolerate high concentrations of cationic antimicrobial peptides via the *dlt* operon (78). This operon encodes four proteins, DltA, DltB, DltC, and DltD, involved in the transport of D-alanine from the cytoplasm to the surface teichoic acid (Fig. 5). The teichoic acid backbone is highly charged by deprotonized phosphate groups, and esterification with D-alanine reduces the net negative surface charge by introduction of amino groups. This is thought to induce resistance. Disruption of the *dlt* genes in *Staphylococcus xylosus* and *S. aureus* by transposon mutagenesis results in mutants with decreased amounts of D-alanine in lipoteichoic and teichoic acids and increased susceptibility to cationic antimicrobial peptides. In wild-type strains, 75 and 95% of alditol phosphate residues in lipoteichoic acids were esterified with D-alanine in *S. aureus* and *S. xylosus*, respectively, and 51 and 15% of teichoic acids were esterified with D-alanine (78). However, in *dlt* mutants, no alanine was detected in either lipoteichoic acids or teichoic acids for both organisms, indicating inactivation of the pathway for D-alanine incorporation into teichoic acid. The mutants were sensitive to a variety of antimicrobial peptides, but resistance was restored by complementation with a plasmid containing the *dlt* operon.

Antimicrobial Peptide Transporters

Antimicrobial resistance mechanisms have been reported for transport of antimicrobial peptides both into the cell (via the ATP-binding cassette transporter) or away from the cell (via the resistance-nodulation-division [RND] efflux pump). Both mechanisms require energy and active transport of peptide for antimicrobial resistance.

ATP-BINDING CASSETTE (ABS) TRANSPORTER

Other membrane and periplasmic proteins involved in membrane transport contribute to resistance of gram-negative bacteria to cationic antimicrobial peptides (37, 73). Resistance is related to a *sapABCDF* multicistronic operon that codes for five proteins exhibiting sequence identity with eukaryotic and prokaryotic transport proteins (ABC transporter family) and uses energy from ATP to import or export a variety of solutes (37, 73). This mechanism appears to be common among a number of gram-negative bacteria (Fig. 6A). Initially, it was identified when serovar Typhimurium mutants were screened for hypersensitivity to protamine. Hypersensitive *sap* mutants (sensitive to antimicrobial peptide) have transposon insertions within a resistance locus that codes for five proteins called SapA (61 kDa), SapB (36 kDa), SapC (about 30 kDa), SapD (38 kDa), and SapF (about 30 kDa). SapA contains a signal sequence and has a predicted periplasmic location. SapB and SapC have predicted hydrophobic regions that correspond to transmembrane domains. SapD and SapF are similar to several members of the ATP-binding cassette family (73). All five Sap proteins appear to be required for resistance, and the inactivation of any one gene increases sensitivity to antimicrobial peptides by varying degrees (inactivation of SapA results in the highest level of sensitivity and inactivation of SapD results in the lowest level). The exact mechanism mediating resistance is not known. However, the Sap system is thought to capture and transport antimicrobial peptides through the target cytoplasmic membrane. They are then thought to be either digested and inactivated by periplasmic peptidases or transported intracellularly to initiate a regulatory cascade that activates other resistance

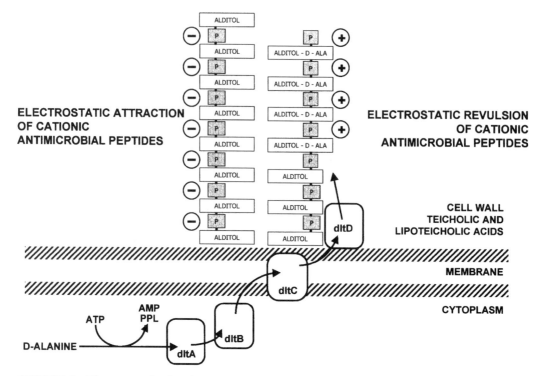

FIGURE 5 The proposed role of the *dlt* operon in reducing the net negative cell charge by adding D-alanine to the teichoic acid, thus charge repelling cationic host antimicrobial peptides (78). A D-alanine-D-alanyl carrier protein ligase (*dltA*) activates D-alanine in the bacterial cytoplasm by hydrolysis of ATP and transfers it to the phosphopantetheine cofactor of the specific D-alanine carrier protein (*dltC*). The hydrophobic protein *dltB* then picks up the D-alanine and transfers it across the cytoplasmic membrane. *dltD* and the presence of a putative *N*-terminal signal peptide then catalyzes the esterification of teichoic acid alditol groups with D-alanine, resulting in the introduction of positive charges into the negatively charged teichoic acids. Adapted from Peschel et al. (78).

mechanisms (37, 73). The process must involve energy-requiring transport of peptides, because a mutation in the ATP-binding site of SapD (that abolishes ATP hydrolysis yet retains binding of ATP) renders *Salmonella* susceptible to protamine (37).

RND EFFLUX PUMP

Another mechanism of resistance to antimicrobial peptides involves the RND efflux pumps. They are used by bacteria to remove toxic, foreign compounds from the cytoplasmic membrane (68, 95). One such pump, termed *mtr* (multiple transferable resistance), is present in *Neisseria gonorrhoeae* (95). This energy-dependent efflux pump confers enhanced resistance to a variety of membrane-

damaging compounds, including antimicrobial hydrophobic agents, bile salts, and fatty acids. It removes antimicrobial peptides nonselectively (Fig. 6B). A three-gene single transcriptional unit (*mtrCDE*) encodes the *mtr* efflux pump. The gonococcal proteins MtrC, MtrD, and MtrE are similar to efflux pump proteins of *E. coli* and *P. aeruginosa*. Isogenic transformant strains of gonococci, bearing insertionally inactivated *mtrC*, *-D*, or *-E* genes, are more susceptible to antimicrobial peptides than the parental strain. For example, deletion in the *mtrD* gene causes increased sensitivity of gonococci to protegrin-1. This suggests a role for the *mtr* efflux system in gonococcal resistance to antimicrobial peptides on urethral mucosal surfaces.

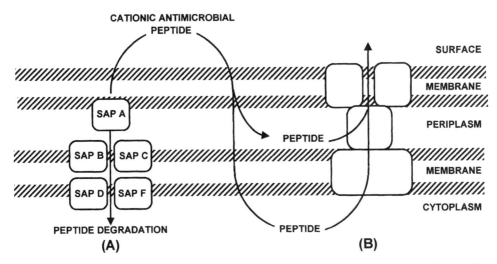

FIGURE 6 Antimicrobial resistance mechanisms involving the transport of antimicrobial peptides into the cell (via the ATP-binding cassette transporter) or away from the cell (via the RND efflux pump). Both mechanisms require energy and active transport of peptide for antimicrobial resistance. (A) SapABCDF transporter transports toxic peptides into the cytoplasm. SapA contains a signal sequence and has a predicted periplasmic location. SapB and C have predicted hydrophobic regions that correspond to transmembrane domains. SapD and F are similar to several members of the ATP-binding cassette family. The transporter is thought to capture and transport antimicrobial peptides through the target cytoplasmic membrane. The peptide is then thought to be either digested and inactivated by periplasmic peptidases or transported intracellularly to initiate a regulatory cascade that activates other resistance mechanisms. (B) The RND efflux pump used by bacteria to remove toxic, foreign compounds from the cytoplasmic membrane. Peptides cross the membrane and enter into either the periplasm or cytoplasm. Here the materials are collected and transported through an mtrC, D, and E pore spanning both membranes. Adapted from Groisman (37) and Nikaido (68).

The *mtr* efflux pump requires the proton motive force (PMF) for operation. To confirm the efflux pump as the mechanism of resistance, carbonyl cynanide-*m*-chlorophenylhydrazone was used to inactivate the PMF of gonococci. As a result, the *mtrD* mutant accumulated at least 50% more iodinated protegrin-1 than the parental strain over a 35-min period. This indicated that the *mtr* efflux pump was capable of clearing antimicrobial peptides away from the cytoplasmic membrane, thereby promoting gonococcal survival.

Enzymatic Degradation of Antimicrobial Peptides

The antimicrobial peptide resistance mechanisms described above often do not account for all the resistance seen in gram-negative bacteria (40), and increasing evidence suggests

that degradation by proteolytic enzymes may also be involved (37, 40, 85). Likewise, the mechanism of antimicrobial resistance dependent upon SapABCDF protein transport may also require terminal digestion by proteolytic enzymes (37, 73).

Early evidence suggested that bacterial resistance to magainins could correlate with the presence of proteases on the bacterial surface (50), and a 110-kDa magaininase was isolated shortly thereafter (37, 83). Immediately downstream and also divergently transcribed from the PmrF putative multicistronic operon is *pmrD* (38), a gene mapped in both *E. coli* and serovar Typhimurium. In serovar Typhimurium, PmrD confers resistance to polymyxin but not to other antimicrobial peptides isolated from human neutrophil granules (85). PmrD, an 85-amino-acid protein, has a weak sequence homology to a viral protease and is thought to contribute to resistance in gram-negative bacteria. PmrD requires PmrA for resistance to polymyxin, and *pmrD* is necessary for resistance to polymyxin in *pmrA*-deficient mutants.

Recent evidence suggests that *P. aeruginosa* PAO1 may also produce a proteinase capable of degrading cationic antimicrobial peptides (4). To find peptide-resistance genes, a bank of transposon insertion mutants was screened for susceptibility to protamine and LL-37. Linkage of the transposon insertion to the LL-37-susceptible phenotype of a *mucD* mutant was confirmed via F116-mediated transduction into wild-type PAO1 and complementation assays. MucD is a periplasmic protease. The *mucD* mutant was susceptible to LL-37, and inserting a plasmid carrying a wild-type copy of mucD restored resistance of the mutant to LL-37. The *mucD* mutant was also susceptible to antimicrobial airway surface fluid from cultured human lung epithelia (102). However, periplasmic extracts from both the wild-type and *mucD* mutant showed proteolytic activity against LL-37. Hence, proteolysis has not yet been unequivocally proven as a mechanism of peptide resistance. Aspedon and Groisman (4) suggest that MucD may exert some control over the composition of the LPS such that in the absence of MucD, the altered LPS increases the outer membrane permeability toward antimicrobial peptides.

Role of Antimicrobial Peptide in Adherence to Mucosa

Defensins isolated from neutrophils (or from sputum samples from patients with chronic obstructive pulmonary disease) do not kill *H. influenzae*. However, defensins stimulate its adherence to human bronchial epithelial cells in both a time- and dose-dependent manner, suggesting that antimicrobial peptides are determinants of colonization (35). In assays with human lung epithelial cell lines, adherence of *H. influenzae* was maximal after 3 h, resulting in about a 1.8 $\log_{10}$ increase in bound cells in the presence of defensins. Adherence also increased with defensin concentration until a

maximal effect was observed at ≥ 10 μg/ml. At the concentrations tested under the assay conditions, defensins did not kill *H. influenzae*; survival rates generally were 91%. The exact mechanism for resistance is not yet known. Enhanced adherence by defensins was not affected by adding poly-L-aspartic acid (20 μg/ml), *H. influenzae* capsular polysaccharide (20 μg/ml), or divalent cations, indicating a mechanism other than surface charge. Defensins form complexes with and are inactivated by α1-proteinase inhibitor (71), and preincubation of defensins with α1-PI resulted in a 90% inhibition of *H. influenzae* adherence.

Host Pathophysiologic Inactivation of Innate Antimicrobial Peptide Activity

Little is known about the in situ requirements necessary for peptide efficacy or the effects of metabolic, physiologic, or genetic dysfunction on peptide antimicrobial activity. Certainly genetic abnormalities in neutrophils, the site of many antimicrobial proteins and peptides (Table 1), should increase the susceptibility of individuals to infection. Individuals with Chediak-Higashi syndrome or chronic granulomatous disease are good examples; they suffer severe recurrent infections caused by both gram-positive and gram-negative bacteria.

The concentration of anionic peptidelike molecules, detected in humans, is significantly lower in patients with cystic fibrosis (CF), suggesting a physiologic deficiency in peptide concentration that may predispose them to respiratory infections (13).

Antimicrobial peptides are sensitive to ionic conditions, and activity correlates with a number of environmental conditions, including divalent cation concentration, salt concentration, and buffer composition (9). In CF, defects in the CF transmembrane conductance regulator result in abnormal ionic transport across epithelial cell surfaces (19). Recent evidence suggests that the increase in NaCl in pulmonary secretions may inactivate the antimicrobial peptides present and predispose patients with CF to colonization with *H. influ-*

enzae, S. aureus, and *P. aeruginosa* (102). The identification and characterization of the human airway antimicrobial factors as well as their mechanisms of action may reveal the mechanism underlying the CF predisposition to infection.

Possible Inactivation of Antimicrobial Peptides during Viral Infection

Respiratory viral infections increase the susceptibility of the host to secondary bacterial infection. Overall, viral infection is thought to create a microenvironment consisting of necrotic cells and proteinaceous fluid in the lung, which favors bacterial growth. It can interfere with the mucociliary clearance mechanisms of the respiratory tract and depress the capacity of resident lung macrophages to take up and kill bacteria (49). Since many respiratory viruses injure and destroy epithelial cells that produce antimicrobial peptides, it is attractive to speculate that they reduce the concentrations of antimicrobial peptides in respiratory infections, thus favoring the growth of opportunistic respiratory flora. In animal models of pneumonia, ovine viruses such as PIV-3, BRSV, and OAV-6 can all increase susceptibility of sheep to *P. haemolytica* (15), an organism very susceptible to ovine-derived antimicrobial peptides.

CONCLUSIONS

A variety of nonimmune and immune mechanisms have evolved at mucosal surfaces to prevent microbial invasion and damage. Predominant among the nonimmune mechanisms is the presence of a multiple peptide-containing constitutive and inducible antimicrobial barrier in the granules of phagocytes and mucosal fluids. Despite this barrier, microorganisms still persist and proliferate in these environments. This is due to a number of evasive strategies they have developed to avoid killing by antimicrobial peptides. Bacteria may "hide" from the peptides in the mucosal fluid by burrowing in epithelial cells, release endotoxin to absorb peptides targeted for bacterial cells, alter the electrostatic charge on their surfaces and membranes to prevent peptide interaction, utilize peptides for attachment and colonization, or enzymatically digest peptides. Identification of these resistance mechanisms gives us insight regarding how microorganisms initially interact with the innate immune system and become primary pathogens or enter into commensal, latency, or carrier states in susceptible hosts.

REFERENCES

1. **Ackermann, M. R., N. F. Cheville, and J. E. Gallagher.** 1991. Colonization of the pharyngeal tonsil and respiratory tract of the gnotobiotic pig by a toxigenic strain of *Pasteurella multocida* type D. *Vet. Pathol.* **28:**267–274.
2. **Antonyraj, K. J., T. Karunakaran, and P. A. Raj.** 1998. Bactericidal activity and poly-L-proline II conformation of the tandem repeat sequence of human salivary mucin glycoprotein (MG2). *Arch. Biochem. Biophys.* **356:**197–206.
3. **Arima, H., H. R. Ibrahim, T. Kinoshita, and A. Kato.** 1997. Bactericidal action of lysozymes attached with various sizes of hydrophobic peptides to the C-terminal using genetic modification. *FEBS Lett.* **415:**114–118.
4. **Aspedon, A., and E. A. Groisman.** 1999. Involvement of a protease in antimicrobial peptide resistance in *Pseudomonas aeruginosa,* abstr. A-67, p. 14. *In Proceedings of the 99th General Meeting of the American Society for Microbiology,* American Society for Microbiology, Washington, D.C.
5. **Baker, S. J., J. S. Gunn, and R. Morona.** 1999. The *Salmonella typhi* melittin resistance gene *pqaB* affects intracellular growth in PMA-differentiated U937 cells, polymyxin B resistance and lipopolysaccharide. *Microbiology* **145:**367–378.
6. **Bals, R., X. R. Wang, M. Zasloff, and J. M. Wilson.** 1998. The peptide antibiotic LL-37/hCAP-18 is expressed in epithelia of the human lung where it has broad antimicrobial activity at the airway surface. *Proc. Natl. Acad. Sci. USA* **95:**9541–9546.
7. **Banemann, A., H. Deppisch, and R. Gross.** 1998. The lipopolysaccharide of *Bordetella bronchiseptica* acts as a protective shield against antimicrobial peptides. *Infect. Immun.* **66:**5607–5612.
8. **Bartlett, J. G.** 1981. Bacteriological diagnosis of pulmonary infections, p. 707–745. *In* M. A. Sackner (ed.), *Diagnostic Techniques in Pulmonary Disease,* vol. 16. Marcel Dekker, Inc., New York, N.Y.
9. **Boman, H. G.** 1995. Peptide antibiotics and their role in innate immunity. *Annu. Rev. Immunol.* **13:**61–92.

10. **Bottari, E.** 1990. Zinc(II) complexes with aspartate and glutamate. *J. Coord. Chem.* **21:**215–224.

11. **Brogden, K. A.** 1992. Ovine pulmonary surfactant induces killing of *Pasteurella haemolytica, Escherichia coli,* and *Klebsiella pneumoniae* by normal serum. *Infect. Immun.* **60:**5182–5189.

12. **Brogden, K. A., M. Ackermann, and K. M. Huttner.** 1997. Small, anionic, and charge-neutralizing propeptide fragments of zymogens are antimicrobial. *Antimicrob. Agents Chemother.* **41:**1615–1617.

13. **Brogden, K. A., M. R. Ackermann, P. B. McCray, Jr., and K. M. Huttner.** 1999. Differences in the concentrations of small, anionic, antimicrobial peptides in bronchoalveolar lavage fluid and in respiratory epithelia of patients with and without cystic fibrosis. *Infect. Immun.* **67:**4256–4259.

14. **Brogden, K. A., A. J. De Lucca, J. Bland, and S. Elliott.** 1996. Isolation of an ovine pulmonary surfactant-associated anionic peptide bactericidal for *Pasteurella haemolytica. Proc. Natl. Acad. Sci. USA* **93:**412–416.

15. **Brogden, K. A., H. D. Lehmkuhl, and R. C. Cutlip.** 1998. *Pasteurella haemolytica* complicated respiratory infections in sheep and goats. *Vet. Res.* **29:**233–254.

16. **Brogden, K. A., J. T. Meehan, and H. D. Lehmkuhl.** 1994. *Salmonella arizonae* infection and colonisation of the upper respiratory tract of sheep. *Vet. Rec.* **135:**410–411.

17. **Brubaker, R. R.** 1991. Factors promoting acute and chronic diseases caused by yersiniae. *Clin. Microbiol. Rev.* **4:**309–324.

18. **Campanelli, D., P. A. Detmers, C. F. Nathan, and J. E. Gabay.** 1990. Azurocidin and a homologous serine protease from neutrophils. Differential antimicrobial and proteolytic properties. *J. Clin. Invest.* **85:**904–915.

19. **Davis, P. B., M. Drumm, and M. W. Konstan.** 1996. Cystic fibrosis. *Am. J. Respir. Crit. Care Med.* **154:**1229–1256.

20. **de Haen, C., H. Neurath, and D. C. Teller.** 1975. The phylogeny of trypsin-related serine proteases and their zymogens. New methods for the investigation of distant evolutionary relationships. *J. Mol. Biol.* **92:**225–259.

21. **DeVoe, I. W.** 1982. The meningococcus and mechanisms of pathogenicity. *Microbiol. Rev.* **46:**162–190.

22. **DeVoe, I. W., and J. E. Gilchrist.** 1973. Release of endotoxin in the form of cell wall blebs during in vitro growth of *Neisseria meningitidis. J. Exp. Med.* **138:**1156–1167.

23. **Diamond, G., M. Zasloff, H. Eck, M. Brasseur, W. Maloy, and C. Bevins.** 1992. A novel antimicrobial peptide from mammalian tracheal mucosa. *Chest* **101:**47S.

24. **Diamond, G., M. Zasloff, H. Eck, M. Brasseur, W. L. Maloy, and C. L. Bevins.** 1991. Tracheal antimicrobial peptide, a cysteine-rich peptide from mammalian tracheal mucosa: peptide isolation and cloning of a cDNA. *Proc. Natl. Acad. Sci. USA* **88:**3952–3956.

25. **Eisenhauer, P. B., S. S. L. Harwig, and R. I. Lehrer.** 1992. Cryptdins: antimicrobial defensins of the murine small intestine. *Infect. Immun.* **60:**3556–3565.

26. **Ellison, R. T., III, D. Boose, and F. M. LaForce.** 1985. Isolation of an antibacterial peptide from human lung lavage fluid. *J. Infect. Dis.* **151:**1123–1129.

27. **Elsbach, P., J. Weiss, and O. Levy.** 1994. Integration of antimicrobial host defenses: role of the bactericidal/permeability-increasing protein. *Trends Microbiol.* **2:**324–328.

28. **Franken, C., C. J. Meijer, and J. H. Dijkman.** 1989. Tissue distribution of antileukoprotease and lysozyme in humans. *J. Histochem. Cytochem.* **37:**493–498.

29. **Frohm Nilsson, M., B. Sandstedt, O. Sorensen, G. Weber, N. Borregaard, and M. Stahle-Backdahl.** 1999. The human cationic antimicrobial protein (hCAP18), a peptide antibiotic, is widely expressed in human squamous epithelia and colocalizes with interleukin-6. *Infect. Immun.* **67:**2561–2566.

30. **Gabay, C., and I. Kushner.** 1999. Acute-phase proteins and other systemic responses to inflammation. *N. Engl. J. Med.* **340:**448–454.

31. **Gabay, J. E., and R. P. Almeida.** 1993. Antibiotic peptides and serine protease homologs in human polymorphonuclear leukocytes: defensins and azurocidin. *Curr. Opin. Immunol.* **5:**97–102.

32. **Galask, R. P., and I. S. Snyder.** 1968. Bacterial inhibition by amniotic fluid. *Am. J. Obstet. Gynecol.* **102:**949–955.

33. **Ganz, T., M. E. Selsted, and R. I. Lehrer.** 1990. Defensins. *Eur. J. Haematol.* **44:**1–8.

34. **Gorbach, S. L.** 1986. Bengt E. Gustafsson memorial lecture. Function of the normal human microflora. *Scand. J. Infect. Dis. Suppl.* **49:**17–30.

35. **Gorter, A. D., P. P. Eijk, S. van Wetering, P. S. Hiemstra, J. Dankert, and L. van Alphen.** 1998. Stimulation of the adherence of *Haemophilus influenzae* to human lung epithelial cells by antimicrobial neutrophil defensins. *J. Infect. Dis.* **178:**1067–1074.

36. **Grandjean, V., S. Vincent, L. Martin, M. Rassoulzadegan, and F. Cuzin.** 1997. Antimicrobial protection of the mouse testis: synthesis of defensins of the cryptdin family. *Biol. Repro.* **57:**1115–1122.

37. **Groisman, E. A.** 1994. How bacteria resist killing by host-defense peptides. *Trends Microbiol.* **2:** 444–448.

38. **Gunn, J. S., K. B. Lim, J. Krueger, K. Kim, L. Guo, M. Hackett, and S. I. Miller.** 1998. PmrA-PmrB-regulated genes necessary for 4-aminoarabinose lipid A modification and polymyxin resistance. *Mol. Microbiol.* **27:**1171–1182.

39. **Guo, L., K. B. Lim, J. S. Gunn, B. Bainbridge, R. P. Darveau, M. Hackett, and S. I. Miller.** 1997. Regulation of lipid A modifications by *Salmonella typhimurium* virulence genes phoP-phoQ. *Science* **276:**250–253.

40. **Guo, L., K. B. Lim, C. M. Poduje, M. Daniel, J. S. Gunn, M. Hackett, and S. I. Miller.** 1998. Lipid A acylation and bacterial resistance against vertebrate antimicrobial peptides. *Cell* **95:** 189–198.

41. **Hancock, R. E. W.** 1997. Peptide antibiotics. *Lancet* **349:**418–422.

42. **Harwig, S. S., V. N. Kokryakov, K. M. Swiderek, G. M. Aleshina, C. Zhao, and R. I. Lehrer.** 1995. Prophenin-1, an exceptionally proline-rich antimicrobial peptide from porcine leukocytes. *FEBS Lett.* **362:**65–69.

43. **Hiemstra, P. S., R. J. Maassen, J. Stolk, R. Heinzel-Wieland, G. J. Steffens, and J. H. Dijkman.** 1996. Antibacterial activity of antileukoprotease. *Infect. Immun.* **64:**4520–4524.

44. **Holak, T. A., A. Engstrom, P. J. Kraulis, G. Lindeberg, H. Bennich, T. A. Jones, A. M. Gronenborn, and G. M. Clore.** 1988. The solution conformation of the antibacterial peptide cecropin A: a nuclear magnetic resonance and dynamical simulated annealing study. *Biochemistry* **27:**7620–7629.

45. **Hristova, K., M. E. Selsted, and S. H. White.** 1997. Critical role of lipid composition in membrane permeabilization by rabbit neutrophil defensins. *J. Biol. Chem.* **272:**24224–24233.

46. **Huttner, K. M., D. J. Brezinski-Caliguri, M. M. Mahoney, and G. Diamond.** 1998. Antimicrobial peptide expression is developmentally regulated in the ovine gastrointestinal tract. *J. Nutr.* **128:**297S–299S.

47. **Huttner, K. M., C. A. Kozak, and C. L. Bevins.** 1997. The mouse genome encodes a single homolog of the antimicrobial peptide human beta-defensin 1. *FEBS Lett.* **413:**45–49.

48. **Iannuzzi, L., D. S. Gallagher, G. P. Di Meo, G. Diamond, C. L. Bevins, and J. E. Womack.** 1996. High-resolution FISH mapping of β-defensin genes to river buffalo and sheep chromosomes suggests a chromosome discrepancy in cattle standard karyotypes. *Cytogenet. Cell Genet.* **75:**10–13.

49. **Jakab, G. J.** 1982. Viral-bacterial interactions in pulmonary infection. *Adv. Vet. Sci. Comp. Med.* **26:**155–171.

50. **Juretic, D., H. C. Chen, J. H. Brown, J. L. Morell, R. W. Hendler, and H. V. Westerhoff.** 1989. Magainin 2 amide and analogues. Antimicrobial activity, membrane depolarization and susceptibility to proteolysis. *FEBS Lett.* **249:** 219–223.

51. **Kaser, M. R., and G. G. Skouteris.** 1997. Inhibition of bacterial growth by synthetic SP-B1-78 peptides. *Peptides* **18:**1441–1444.

52. **Ketterer, M. R., J. Q. Shao, D. B. Hornick, B. Buscher, V. K. Bandi, and M. A. Apicella.** 1999. Infection of primary human bronchial epithelial cells by *Haemophilus influenzae*: macropinocytosis as a mechanism of airway epithelial cell entry. *Infect. Immun.* **67:**4161–4170.

53. **Kuwata, H., T. T. Yip, C. L. Yip, M. Tomita, and T. W. Hutchens.** 1998. Bactericidal domain of lactoferrin: detection, quantitation, and characterization of lactoferricin in serum by SELDI affinity mass spectrometry. *Biochem. Biophys. Res. Commun.* **245:**764–773.

54. **LaForce, F. M., and D. S. Boose.** 1984. Effect of zinc and phosphate on an antibacterial peptide isolated from lung lavage. *Infect. Immun.* **45:**692–696.

54a. **Leenhouts, J. M., P. W. J. van den Wijngaard, A. I. P. M. de Kroon, and B. de Kruijff.** 1995. Anionic phospholipids can mediate membrane insertion of the anionic part of a bound peptide. *FEBS Lett.* **370:**189–192.

55. **Lehrer, R. I., A. Barton, K. A. Daher, S. S. Harwig, T. Ganz, and M. E. Selsted.** 1989. Interaction of human defensins with *Escherichia coli*. Mechanism of bactericidal activity. *J. Clin. Invest.* **84:**553–561.

56. **Lehrer, R. I., and T. Ganz.** 1999. Antimicrobial peptides in mammalian and insect host defense. *Curr. Opin. Immunol.* **11:**23–27.

57. **Lehrer, R. I., T. Ganz, and M. E. Selsted.** 1991. Defensins: endogenous antibiotic peptides of animal cells. *Cell* **64:**229–230.

58. **Littauer, U. Z., and M. Sela.** 1962. An ultracentrifugal study of the efficiency of some macromolecular inhibitors of ribonuclease. *Biochim. Biophys. Acta* **61:**609–611.

59. **Luderitz, O., M. A. Freudenberg, C. Galanos, V. Lehmann, E. T. Rietschel, and D. H. Shaw.** 1982. *Lipopolysaccharides of Gram-Negative Bacteria*, vol. 17. Academic Press, Inc., New York, N.Y.

60. **Luderitz, O., K.-I. Tanamoto, C. Galanos, O. Westphal, U. Zahringer, E. T. Rietschel, S. Kusumoto, and T. Shiba.** 1983. Structural principles of lipopolysaccharides and biological

properties of synthetic partial structures, p. 3–17. *In* L. Anderson and F. M. Unger (ed.), *Bacterial Lipopolysaccharides*. American Chemical Society, Washington, D.C.

61. **Macias, E. A., F. Rana, J. Blazyk, and M. C. Modrzakowski.** 1990. Bactericidal activity of magainin 2: use of lipopolysaccharide mutants. *Can. J. Microbiol.* **36:**582–584.

62. **Martinez de Tejada, G., J. Pizarro-Cerda, E. Moreno, and I. Moriyon.** 1995. The outer membranes of *Brucella* spp. are resistant to bactericidal cationic peptides. *Infect. Immun.* **63:** 3054–3061.

63. **McCray, P. B., Jr., and L. Bentley.** 1997. Human airway epithelia express a β-defensin. *Am. J. Respir. Cell Mol. Biol.* **16:**343–349.

64. **McNabb, P. C., and T. B. Tomasi.** 1981. Host defense mechanisms at mucosal surfaces. *Annu. Rev. Microbiol.* **35:**477–496.

65. **Meehan, J. T., K. A. Brogden, C. Courtney, R. C. Cutlip, and H. D. Lehmkuhl.** 1992. Chronic proliferative rhinitis associated with *Salmonella arizonae* in sheep. *Vet. Pathol.* **29:**556–559.

66. **Menache, M. G., L. M. Hanna, E. A. Gross, S. R. Lou, S. J. Zinreich, D. A. Leopold, A. M. Jarabek, and F. J. Miler.** 1997. Upper respiratory tract surface areas and volumes of laboratory animals and humans: considerations for dosimetry models. *J. Toxicol. Environ. Health* **50:** 475–506.

67. **Nicolas, P., and A. Mor.** 1995. Peptides as weapons against microorganisms in the chemical defense system of vertebrates. *Annu. Rev. Microbiol.* **49:**277–304.

68. **Nikaido, H.** 1996. Multidrug efflux pumps of gram-negative bacteria. *J. Bacteriol.* **178:**5853–5859.

69. **Nissen-Meyer, J., and I. F. Nes.** 1997. Ribosomally synthesized antimicrobial peptides: their function, structure, biogenesis, and mechanism of action. *Arch. Microbiol.* **167:**67–77.

70. **Odeberg, H., and I. Olsson.** 1975. Antibacterial activity of cationic proteins from human granulocytes. *J. Clin. Invest.* **56:**1118–1124.

71. **Panyutich, A. V., P. S. Hiemstra, S. van Wetering, and T. Ganz.** 1995. Human neutrophil defensin and serpins form complexes and inactivate each other. *Am. J. Respir. Cell Mol. Biol.* **12:**351–357.

72. **Parker, M., A. D. Tucker, D. Tsernoglou, and F. Pattus.** 1990. Insights into membrane insertion based on studies of colicins. *Trends Biochem. Sci.* **15:**126–129.

73. **Parra-Lopez, C., M. T. Baer, and E. Groisman.** 1993. Molecular genetic analysis of a locus required for resistance to antimicrobial peptides in *Salmonella typhimurium*. *EMBO J.* **12:**4053–4062.

74. **Pellegrini, A., U. Thomas, N. Bramaz, P. Hunziker, and R. von Fellenberg.** 1999. Isolation and identification of three bactericidal domains in the bovine alpha-lactalbumin molecule. *Biochim. Biophys. Acta* **1426:**439–448.

75. **Pellegrini, A., U. Thomas, N. Bramaz, S. Klauser, P. Hunziker, and R. von Fellenberg.** 1996. Identification and isolation of the bactericidal domains in the proteinase inhibitor aprotinin. *Biochem. Biophys. Res. Commun.* **222:** 559–565.

76. **Penn, C. W.** 1992. Chronic infections, latency and the carrier state, p. 107–125. *In* C. E. Hormaeche, C. W. Penn, and C. J. Smyth (ed.), *Molecular Biology of Bacterial Infection*. Cambridge University Press, Dublin.

77. **Pereira, H. A., I. Erdem, J. Pohl, and J. K. Spitznagel.** 1993. Synthetic bactericidal peptide based on CAP37: a 37-kDa human neutrophil granule-associated cationic antimicrobial protein chemotactic for monocytes. *Proc. Natl. Acad. Sci. USA* **90:**4733–4737.

78. **Peschel, A., M. Otto, R. W. Jack, H. Kalbacher, G. Jung, and F. Gotz.** 1999. Inactivation of the dlt operon in *Staphylococcus aureus* confers sensitivity to defensins, protegrins, and other antimicrobial peptides. *J. Biol. Chem.* **274:** 8405–8410.

79. **Peterson, A. A., A. Haug, and E. J. McGroarty.** 1986. Physical properties of short- and long-O-antigen-containing fractions of lipopolysaccharide from *Escherichia coli* 0111:B4. *J. Bacteriol.* **165:**116–122.

80. **Porter, E. M., E. van Dam, E. V. Valore, and T. Ganz.** 1997. Broad-spectrum antimicrobial activity of human intestinal defensin 5. *Infect. Immun.* **65:**2396–2401.

81. **Quayle, A. J., E. M. Porter, A. A. Nussbaum, Y. M. Wang, C. Brabec, K. P. Yip, and S. C. Mok.** 1998. Gene expression, immunolocalization, and secretion of human defensin-5 in human female reproductive tract. *Am. J. Pathol.* **152:**1247–1258.

82. **Raj, P. A., E. Marcus, and D. K. Sukumaran.** 1998. Structure of human salivary histatin 5 in aqueous and nonaqueous solutions. *Biopolymers* **45:**51–67.

83. **Resnick, N. M., W. L. Maloy, H. R. Guy, and M. Zasloff.** 1991. A novel endopeptidase from Xenopus that recognizes alpha-helical secondary structure. *Cell* **66:**541–554.

84. **Riley, L. K., and D. C. Robertson.** 1984. Ingestion and intracellular survival of *Brucella abortus* in human and bovine polymorphonuclear leukocytes. *Infect. Immun.* **46:**224–230.

85. **Roland, K. L., C. R. Esther, and J. K. Spitz-nagel.** 1994. Isolation and characterization of a gene, *pmrD*, from *Salmonella typhimurium* that confers resistance to polymyxin when expressed in multiple copies. *J. Bacteriol.* **176:**3589–3597.

86. **Russell, R. R. B.** 1976. Free endotoxin—a review. *Microbios Lett.* **2:**125–135.

87. **Sachs, B. P.** 1979. Activity and characterization of a low molecular fraction present in human amniotic fluid with broad spectrum antibacterial activity. *Br. J. Obstet. Gynecol.* **86:**81–86.

88. **Schlievert, P., W. Johnson, and R. P. Galask.** 1976. Bacterial growth inhibition by amniotic fluid. VI. Evidence for a zinc-peptide antibacterial system. *Am. J. Obstet. Gynecol.* **125:** 906–910.

89. **Schlievert, P., W. Johnson, and R. P. Galask.** 1976. Isolation of a low-molecular weight antibacterial system from human amniotic fluid. *Infect. Immun.* **14:**1156–1166.

90. **Schnapp, D., and A. Harris.** 1998. Antibacterial peptides in bronchoalveolar lavage fluid. *Am. J. Respir. Cell Mol. Biol.* **19:**352–356.

91. **Schonwetter, B. S., E. D. Stolzenberg, and M. A. Zasloff.** 1995. Epithelial antibiotics induced at sites of inflammation. *Science* **267:**1645–1648.

92. **Sela, M.** 1962. Inhibition of ribonuclease by co-polymers of glutamic acid and aromatic amino acids. *J. Biol. Chem.* **237:**418–421.

93. **Shafer, W. M., L. E. Martin, and J. K. Spitz-nagel.** 1984. Cationic antimicrobial proteins isolated from human neutrophil granulocytes in the presence of diisopropyl fluorophosphate. *Infect. Immun.* **45:**29–35.

94. **Shafer, W. M., L. E. Martin, and J. K. Spitz-nagel.** 1986. Late intraphagosomal hydrogen ion concentration favors the in vitro antimicrobial capacity of a 37-kilodalton cationic granule protein of human neutrophil granulocytes. *Infect. Immun.* **53:**651–655.

95. **Shafer, W. M., X. Qu, A. J. Waring, and R. I. Lehrer.** 1998. Modulation of *Neisseria gonorrhoeae* susceptibility to vertebrate antibacterial peptides due to a member of the resistance/nodulation/division efflux pump family. *Proc. Natl. Acad. Sci. USA* **95:**1829–1833.

96. **Shamova, O., K. A. Brogden, C. Zhao, T. Nguyen, V. N. Kokryakov, and R. I. Lehrer.** 1999. Purification and properties of proline-rich antimicrobial peptides from sheep and goat leukocytes. *Infect. Immun.* **67:**4106–4111.

97. **Shoo, M. K., A. Wiseman, E. M. Allan, R. G. Dalgleish, H. A. Gibbs, A. B. Al Hendi, and I. E. Selman.** 1990. Distribution of *Pasteurella haemolytica* in the respiratory tracts of carrier calves and those subsequently infected experimentally with *Dictyocaulus viviparus*. *Res. Vet. Sci.* **48:**383–385.

98. **Siden, I., and H. G. Boman.** 1983. *Escherichia coli* mutants with an altered sensitivity to cecropin D. *J. Bacteriol.* **154:**170–176.

99. **Skerlavaj, B., D. Romeo, and R. Gennaro.** 1990. Rapid membrane permeabilization and inhibition of vital functions of gram-negative bacteria by bactenecins. *Infect. Immun.* **58:**3724–3730.

100. **Smith, H.** 1995. The revival of interest in mechanisms of bacterial pathogenicity. *Biol. Rev. Camb. Philos. Soc.* **70:**277–316.

101. **Smith, H.** 1995. The state and future of studies on bacterial pathogenicity, p. 335–357. *In* J. A. Roth, C. A. Bolin, K. A. Brogden, F. C. Minion, and M. J. Wannemuehler (ed.), *Virulence Mechanisms of Bacterial Pathogens, 2nd ed.* ASM Press, Washington, D.C.

102. **Smith, J. J., S. M. Travis, E. P. Greenberg, and M. J. Welsh.** 1996. Cystic fibrosis airway epithelia fail to kill bacteria because of abnormal airway surface fluid. *Cell* **85:**229–236.

103. **Steiner, H.** 1982. Secondary structure of the cecropins; antibacterial peptides from the moth *Hyalophora cecropia*. *FEBS Lett.* **137:**283–287.

104. **Stolzenberg, E. D., G. M. Anderson, M. R. Ackermann, R. H. Whitlock, and M. Zasloff.** 1997. Epithelial antibiotic induced in states of disease. *Proc. Natl. Acad. Sci. USA* **94:**8686–8690.

105. **Thorne, K. J., R. C. Oliver, and A. J. Barrett.** 1976. Lysis and killing of bacteria by lysosomal proteinases. *Infect. Immun.* **14:**555–563.

106. **Tomita, M., H. Wakabayashi, and W. Bellamy.** 1994. Antimicrobial peptides of lactoferrin. *Adv. Exp. Med. Biol.* **357:**209–218.

107. **Turner, J., Y. Cho, N. N. Dinh, A. J. Waring, and R. I. Lehrer.** 1998. Activities of LL-37, a cathelin-associated antimicrobial peptide of human neutrophils. *Antimicrob. Agents Chemother.* **42:**2206–2214.

108. **van der Goot, F. G., N. Didat, F. Pattus, W. Dowhan, and L. Letellier.** 1993. Role of acidic lipids in the translocation and channel activity of colicins A and N in *Escherichia coli* cells. *Eur. J. Biochem.* **213:**217–221.

109. **Vandendriessche, L.** 1956. Inhibitors of ribonuclease activity. *Arch. Biochem. Biophys.* **65:** 347–353.

110. **Visser, L. G., P. S. Hiemstra, M. T. Van Den Barselaar, P. A. Ballieux, and R. Van Furth.** 1996. Role of *yadA* in resistance to killing of *Yersinia enterocolitica* by antimicrobial polypeptides of human granulocytes. *Infect. Immun.* **64:**1653–1658.

111. **Wasiluk, K. R., K. M. Skubitz, and B. H. Gray.** 1991. Comparison of granule proteins from human polymorphonuclear leukocytes which are bactericidal toward *Pseudomonas aeruginosa*. *Infect. Immun.* **59:**4193–4200.

112. **Weiser, J. N.** 1999. Adaptation of respiratory tract pathogens to innate and acquired immunity. *Pediatr. Pulmon.* **19**(Suppl.):126–127.

113. **Weiser, J. N., J. B. Goldberg, N. Pan, L. Wilson, and M. Virji.** 1998. The phosphorylcholine epitope undergoes phase variation on a 43-kilodalton protein in *Pseudomonas aeruginosa* and on pili of *Neisseria meningitidis* and *Neisseria gonorrhoeae*. *Infect. Immun.* **66:**4263–4267.

114. **Weiser, J. N., N. Pan, K. L. McGowan, D. Musher, A. Martin, and J. Richards.** 1998. Phosphorylcholine on the lipopolysaccharide of *Haemophilus influenzae* contributes to persistence in the respiratory tract and sensitivity to serum killing mediated by C-reactive protein. *J. Exp. Med.* **187:**631–640.

115. **Weiser, J. N., M. Shchepetov, and S. T. Chong.** 1997. Decoration of lipopolysaccharide with phosphorylcholine: a phase-variable characteristic of *Haemophilus influenzae*. *Infect. Immun.* **65:**943–950.

116. **Wiebe, B. M., and H. Laursen.** 1995. Human lung volume, alveolar surface area, and capillary length. *Microsc. Res. Tech.* **32:**255–262.

117. **Wu, E. R., R. Daniel, and A. Bateman.** 1998. RK-2: a novel rabbit kidney defensin and its implications for renal host defense. *Peptides* **19:**793–799.

118. **Yarnall, M., and L. B. Corbeil.** 1989. Antibody response to *Haemophilus somnus* Fc receptor. *J. Clin. Microbiol.* **27:**111–117.

119. **Zanetti, M., G. Del Sal, P. Storici, C. Schneider, and D. Romeo.** 1993. The cDNA of the neutrophil antibiotic Bac5 predicts a pro-sequence homologous to a cysteine proteinase inhibitor that is common to other neutrophil antibiotics. *J. Biol. Chem.* **268:**522–526.

120. **Zanetti, M., R. Gennaro, and D. Romeo.** 1995. Cathelicidins: a novel protein family with a common proregion and a variable C-terminal antimicrobial domain. *FEBS Lett.* **374:**1–5.

CONSEQUENCES OF BACTERIAL INVASION INTO NONPROFESSIONAL PHAGOCYTIC CELLS

Jeffrey B. Lyczak and Gerald B. Pier

3

Bacterial pathogens are faced with the seemingly difficult task of persisting within their host in the face of the latter's formidable array of defense mechanisms. To achieve this goal, microbial pathogens have evolved a diverse spectrum of survival strategies. At one extreme of this spectrum is the strategy of persisting only on the outer surfaces of the host. This strategy is exemplified by oral colonization by microbes such as *Porphyromonas gingivalis* and nasopharyngeal colonization by *Streptococcus* spp. By existing solely at these sites, the microbe does minimal (if any) damage to the tissues of its host and thus avoids drawing the attention of its host's immune system.

At the other extreme of the spectrum is the rather aggressive strategy employed by various intracellular pathogens such as *Salmonella* spp., which survive their host's immune defenses by attacking the very cells which mediate host immunity. Regardless of the exact strategy employed by a bacterial pathogen, any bacterium that requires access to the deep tissues of its host must possess mechanisms to penetrate the more superficial tissues, which in

most cases consist of the host's epithelium. Bacterial invasion into host epithelium was first reported in the mid-1980s (39, 73), and the study of this phenomenon has progressed in recent years to provide a more mechanistic analysis of how invasion proceeds. From the viewpoint of the bacterium, the host's epithelium can be perceived as the first of many barriers that must be crossed if colonization is to proceed to deeper tissue invasion. Bacterial entry into host epithelium can therefore be considered an active process, involving both microbial and host physiology, anatomy, and sensing mechanisms. The outcome could be either containment of the microbe to minimize or prevent substantive damage or local and even systemic dissemination of the microbe, leading to more severe disease.

The host epithelium is specially structured to create a differentially permeable barrier that will allow the exchange of necessary components between the host and its environment. An obvious problem with this design is that to carry out its intended function, the epithelial barrier must exist in close contact with the environment and with any potential pathogens that may exist in this milieu. Therefore, epithelial barriers have evolved special defenses, which exclude potential pathogens while still allowing the exchange of essential

Jeffrey B. Lyczak and Gerald B. Pier   Channing Laboratory, Brigham and Women's Hospital, Boston, MA 02115.

Virulence Mechanisms of Bacterial Pathogens, 3rd ed., Edited by K. A. Brogden et al.
©2000 ASM Press, Washington, D.C.

compounds. In this respect, internalization of bacteria by epithelial cells is an arm of the host defense, trapping bacteria and removing them from the mucosal site and in some cases helping to initiate an acquired immune response against them (63).

An overview of bacterial uptake by non-professional antigen-presenting cells must consider the relevant processes with regard to both the entry and survival of the bacterial pathogen and the response of the defending host. This process encompasses interactions of many tissues with a wide variety of microorganisms, so it is clear that there will be few general examples of mechanisms applicable to the majority of pathogen-host interactions. Rather, examples from various models of infection will be used to illustrate how multiple and relevant processes affect this interaction. More to the point, the interaction of a host and a pathogen must be viewed as a continuum of interactions, such that at one end of the spectrum a given interaction (e.g., ingestion of a pathogen by epithelial cells) could be critical for host resistance to infection (e.g., by removal of the pathogen from the epithelial surface), while at the other end of the spectrum the interaction could be part of the process resulting in severe pathology (e.g., damage to the epithelium from desquamation of epithelial cells with ingested bacteria, giving additional microbial cells at an infected site access to deeper tissues). Understanding the biological mechanisms triggered during these processes and defining how these events are part of a more global context of host-pathogen physiology should be the underpinnings for development of better intervention strategies to tilt the balance of this interaction in favor of microbial elimination and heightened host resistance to infection.

SITES OF INFECTION AND MECHANISMS OF INNATE DEFENSE

The adaptive immune response, while able to efficiently and specifically deal with an almost infinite number of foreign antigens, is severely handicapped by the 1- to 2-week lag period required to mount a response after first exposure to a foreign substance. Additionally, the frequency with which foreign antigens are encountered and the sheer quantities of such antigens make specific responses both impractical from a bioenergetic standpoint and potentially hazardous to the host because of the large amount of collateral damage that would occur to "innocent bystander" host cells. Thus, it is apparent that the host has a serious need for a battery of nonspecific clearance mechanisms to continuously rid itself of potential pathogens. Nowhere in the body is this need as great as at the mucosal surfaces. The combined surface area of the mucosal surfaces in the human is at least 200 times as great as that of the skin (108), thus presenting a large potential target for inhaled or ingested microbial pathogens. In both scenarios, if the pathogen survives elimination by mucous and mucous elements such as antimicrobial factors and can readily compete with the indigenous microflora, then the first cells of the host to encounter the microbe are the epithelial cells that line the respiratory, genitourinary, and gastrointestinal tracts. From the standpoint of the microbe, these epithelial cells represent a potential site of colonization or a gateway into the deeper tissues of the host. From the standpoint of host defense, this exposed epithelium is among the first lines of defense, whose protective function is challenged by the need for the epithelium to also be a thin and readily permeable barrier to effectively carry out physiological functions of nutrient and waste exchange. Therefore, the host has developed an eclectic array of nonspecific clearance mechanisms that vary from one epithelial site to another.

One of the more elaborate nonspecific clearance mechanisms of the host is found in the airways, which are continuously subjected to challenge with airborne foreign agents. The principal mechanism of nonspecific clearance in the airways is the mucociliary clearance system (Fig. 1). This system consists of the ciliated apical surface of the airway epithelial cells and a biphasic mucous layer that covers it. The

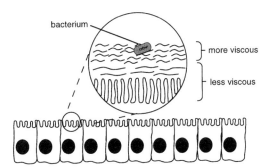

FIGURE 1 Schematic diagram of the mucociliary clearance mechanism of the airway epithelium. The apical surface of the epithelial cells comprises many hairlike cilia, which beat in a synchronized fashion. This membrane surface is also covered with a biphasic mucous layer: the lower, or periciliary, layer is more fluid while the upper layer is more viscous. Bacteria get trapped in the viscous layer and are carried upward due to the ciliary beating and are eventually expectorated and swallowed.

cilia beat in a synchronized fashion to move the mucous layer upward to the pharynx. Microbes that get caught in the mucous layer get carried along and are eventually expectorated and swallowed. It has been suggested (69) that the thickness and viscosity of the fluid "periciliary" mucous layer affect the ability of the beating cilia to move the mucous layer along the surface of the epithelium. In addition to the mucociliary clearance system, it has been postulated that antibacterial factors such as lysozyme, lactoferrin, and defensins play an important role in the defense of the lung (1, 4, 34, 95, 102). However, if microbial pathogens avoid clearance by this defense system, they then encounter the epithelial cell layer, which may play a critical role in protection due to the ability of these cells to internalize microbes, particularly bacteria.

A role for epithelial cell internalization of bacteria as part of the innate host defense system became apparent from reports showing that the epithelial cell ion channel protein, the cystic fibrosis transmembrane conductance regulator (CFTR), is essential both for conductance of chloride ions across the epithelial apical plasma membrane and for removal of

the bacterium *Pseudomonas aeruginosa* from the airway mucosal surfaces (Fig. 2) (83, 85). Since cystic fibrosis (CF) patients (who bear genetic defects leading to either a lack of any CFTR protein in their cells or a nonfunctional CFTR protein) are prone to chronic lung infection with *P. aeruginosa*, it was reasoned that binding of the bacteria by the CFTR protein and the subsequent internalization of the bacteria by epithelial cells constitute a host defense mechanism that serves to control the level of bacteria in the normal healthy lung. The actual mechanism by which this internalization event contributes to host protection is still unclear. One hypothesis (83, 85) is that the internalization of bacteria by epithelial cells serves as a means of physically removing the bacteria from the epithelial surface. According to this model, *P. aeruginosa* that binds to CFTR on the epithelial surface is internalized by the epithelial cell, after which the epithelial cell desquamates and is expectorated from the lungs. More recent evidence (C. L. Cannon and G. B. Pier, unpublished data) suggests that following internalization of the bacteria, the epithelial cell also undergoes programmed cell death (apoptosis). Epithelial cells become positive for terminal deoxynucleotidyltransferase-mediated dUTP-biotin nick-end labeling (TUNEL) staining (suggesting that double-stranded DNA breaks are being generated in the nucleus of the epithelial cell), and at least two proteins (lamin A and nucleolin) implicated (99, 112) in cell-cycle/apoptotic signaling pathways are phosphorylated (J. B. Lyczak and G. B. Pier, unpublished data). Inactivation of this clearance mechanism (through mutation of CFTR) may allow *P. aeruginosa* to persist for longer periods in the airway, perhaps giving the bacteria time to undergo other alterations (e.g., a shift to a mucoid phenotype [58, 96]) that favor the establishment of chronic infection.

The effect and significance of epithelial cell apoptosis in this defense mechanism are not clear. Apoptosis may be necessary to trigger desquamation of bacterium-laden epithelial

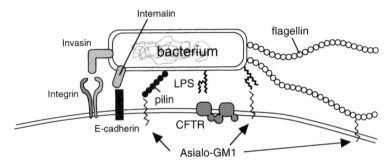

FIGURE 2 A summary of epithelial cell receptors, which mediate adhesion and invasion of bacteria into epithelial cells. The bacterium shown is a generic diagram and contains elements of both gram-positive and gram-negative bacteria. LPS, lipopolysaccharide.

cells. Alternatively, apoptosis of bacterium-laden epithelial cells may be a mechanism for later presentation of bacterial antigens to T lymphocytes, as described below.

In light of the role of the CFTR protein in epithelial-*P. aeruginosa* interactions, it is also interesting that the expression level of CFTR protein increases during the course of infection (117). The increase in CFTR protein is not accompanied by a significant increase in CFTR mRNA, but is evident at the protein level by 10 to 15 min postinfection (J. B. Lyczak, T. S. Zaidi, and G. B. Pier, unpublished data). Together, these results suggest that the mechanism that regulates CFTR protein during infection acts very late in the synthesis/maturation of CFTR, or perhaps even by mobilizing preformed CFTR protein to the epithelial cell plasma membrane. Supporting the latter option, it has been shown (114) that preformed CFTR protein is stored in vesicles residing immediately beneath the apical plasma membrane of epithelial cells. Whether the nascent CFTR protein expressed on epithelial cells during infection is sterically available to interact with bacteria and whether this regulatory event affects the ongoing course of infection should be addressed in future experiments.

Other researchers in the field of airway colonization by *P. aeruginosa* have pointed to adherence of the bacterium to epithelial cells

as a factor that predisposes the host to infection by this microorganism. The work of Prince and colleagues (90) and Panjwani and colleagues (79) has demonstrated that *P. aeruginosa* adheres to the glycosphingolipid asialo-GM1 on host epithelial cells in a specific fashion. The bacterial ligands for asialo-GM1 are reported to be pilin (59), the major structural protein of the pilus; flagellin (24), the structural protein of flagella; and possibly lipopolysaccharide (LPS) (38) (Fig. 2). More recent studies have shown that certain physiological and pathological conditions, such as wound healing and cell migration (15) and aberrations in CFTR expression (5), can alter the quantity and distribution of asialo-GM1 on epithelial cells. These findings invite speculation that such conditions favor binding of *P. aeruginosa* to the epithelium and that this binding event leads to an accumulation of the microorganism in the airways. The relative contributions of bacterial adherence and bacterial internalization by epithelial cells to the infectious process are not yet fully understood. It has recently been demonstrated (10) that *P. aeruginosa* binding to asialo-GM1 on the surface of an epithelial cell line can promote internalization of bacteria by the epithelial cells. While interesting, the relevance of these data is not entirely clear, as epithelial cells of CF patients, whose epithelial cells may express above-normal amounts of asialo-GM1 (5), do not internalize *P. aeruginosa* efficiently (84, 85). Another major problem with the hypothesis that asialo-GM1 levels contribute to the hypersusceptibility of CF patients to *P. aeruginosa* infection is that many bacterial pathogens have been reported to adhere to asialo-GM1 (55), indicating a lack of specificity for *P. aeruginosa* for this part of the pathogenic process. In addition, Prince and colleagues were unable to show increased adherence of *P. aeruginosa* to cells from CF patients with at least one mutation in the CFTR gene other than that seen in the most common allele, the ΔF508 mutation (118). Since CF patients who are not homozygous for the ΔF508 CFTR allele often have a clinical course indistinguishable from that in the ΔF508 homozygotes, it is difficult to understand how increased adherence to epithelial cells contributes to disease only in ΔF508 CFTR homozygotes while other CF patients get the same infection and have the same clinical course without this component of the bacterial-host interaction.

Additional evidence for the importance of host cell binding, internalization, and desquamation of bacterial pathogens in protecting the host from infection is seen in an experimental model of bladder infection by *Escherichia coli*. (68). Desquamation of the epithelium occurs rapidly after bacterial attachment and was found to be initiated by a caspase-dependent apoptotic mechanism. Thus, as a host defense system, internalization of *E. coli* by bladder epithelial cells appears in some respects to be similar to the internalization of *P. aeruginosa* by the epithelial cells of the human airway. Further experiments will show how extensive the similarities between these two experimental models of infection are.

In addition to the epithelial cell receptors discussed above, many other receptors for bacteria have been identified on epithelial cell surfaces. Figure 2 depicts several of these but is by no means a complete representation. Additional quantitative experiments are needed to shed further light on bacterial adherence and internalization by epithelial cells and on

the importance of these events to the overall outcome of bacterial-epithelial interactions.

Initiation of the Immune Response by Epithelial Cells

As a host defense mechanism, the mere internalization of bacteria by epithelial cells would be expected to provide incomplete protection at best; a potent immune response requires the recruitment of additional immune effector cells. The importance of such effector cells in the protection of the host from *P. aeruginosa* infection is underscored by the recent resurgence of *P. aeruginosa* infections in individuals with human immunodeficiency virus (HIV) infection and AIDS (28, 52). *P. aeruginosa* is a common environmental pathogen that is effectively countered by a number of innate immune defenses and therefore rarely causes disease in healthy individuals. The correlation of clinical *P. aeruginosa* infection with loss of CD4$^+$ T-cell function in AIDS patients strongly suggests an important role for the acquired immune response, even with innate immunity likely being fairly effective protection against this microorganism.

Epithelial cells assist in the recruitment of immune effector cells through several routes, as shown in Fig. 3. The first of these mechanisms is the production and secretion of various cytokines by infected epithelial cells. The presence of chemotactic substances in infected epithelial tissues has been documented since the mid-1980s (20, 107, 110), and more recent work has established the identity of some of these chemotactic factors as well as their role in host defense and disease (29, 74, 80). The chemotactic factors which are important in drawing inflammatory cells to the infected tissue are cytokines, the most important of which are tumor necrosis factor alpha (TNF-α) (51, 94, 101), interleukin 1-beta (IL-1β) (21, 44, 94), and interleukin 8 (IL-8) (22, 56). All of these cytokines have been demonstrated to be produced by epithelial cells during the course of infection (21, 22, 103, 115). Of these cytokines, IL-1β and TNF-α are potent chemoattractants for antigen-presenting cells

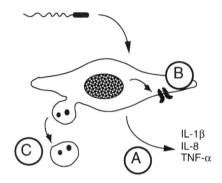

FIGURE 3 Mechanisms by which epithelial cells recruit and activate immune effector cells following epithelial cell infection. (A) Infected epithelial cells secrete proinflammatory cytokines such as IL-1β and TNF-α (which are chemotactic for a number of immune effector cells). (B) Infected epithelial cells begin to express class II MHC proteins on their surface, allowing them to function as "nonprofessional" APCs for helper T lymphocytes. (C) Infected epithelial cells undergo apoptosis. Apoptotic bodies released from dead cells are subsequently phagocytosed by dendritic cells, which then present antigens to T lymphocytes.

(APCs), whereas IL-8 is an attractant of polymorphonuclear leukocytes (PMNs, or neutrophils). The role of these cytokines in APC and PMN recruitment has been verified by the specific inhibition of cytokine activity by blocking antibodies. For example, a 1994 report by Jorens and coworkers (49) demonstrated that supernatant of *P. aeruginosa* cultures contains an activity which, when added to tracheal epithelial cells, causes the epithelial cells to induce neutrophil migration. Furthermore, the presence of blocking monoclonal antibody directed against IL-8 significantly inhibited the migration.

Nonprofessional Phagocytes and the Acquired Immune Response

The APCs of the vertebrate immune system, which typically encompass B lymphocytes, the cells of the monocyte-macrophage lineage, and dendritic cells, have specialized mechanisms to internalize foreign antigens and then present those antigens to T lymphocytes in a highly efficient manner (37, 45, 64). However, despite their specialized abilities, in many

cases these APCs lie beneath the epithelial surface and therefore are not the first cells of the host to encounter a foreign invading microbe. First contact with potential pathogens is usually made by the epithelium itself, and while epithelial cells do not normally have the ability to present antigens to T cells to initiate an immune response, increasing evidence suggests that under the influence of the innate inflammatory response, epithelial cells may have limited antigen-presenting capabilities.

The potential importance of epithelial cells as APCs has been appreciated since the late 1980s (104), and this observation is consistent with the finding that epithelial cells of the thymus present antigens to T-cell precursors as a normal part of T-cell maturation (18). While the APC function of epithelial cells was originally defined on the basis of in vitro mixed lymphocyte reactions (104), this function has since been demonstrated to be relevant to the host response to exogenous antigens (23, 76) and has been implicated in the development of autoimmune conditions (9).

In peripheral tissues, the acquisition of antigen-presenting function by epithelial cells is accompanied by the expression of both the class II major histocompatibility complex (MHC) structure and the B7 costimulatory molecule. When expressed together, these molecules allow epithelial cells to activate helper T cells (i.e., CD4$^+$ T cells) in an antigen-specific manner. It has been shown that the acquisition of APC function by epithelial cells requires a cytokine milieu, which is typical of intense inflammation (3, 109, 116). It is therefore not surprising that this function of epithelial cells is enhanced during bacterial infection (62, 116). More recent discoveries have revealed interesting features of this function of epithelial cells. It has been reported (41) (Fig. 4) that class II MHC expression by epithelial cells requires prior endocytosis of antigen via the epithelial cell's apical plasma membrane. Moreover, once antigen is processed and complexed with class II MHC proteins, the nascent MHC-peptide complexes are trafficked exclusively to the ba-

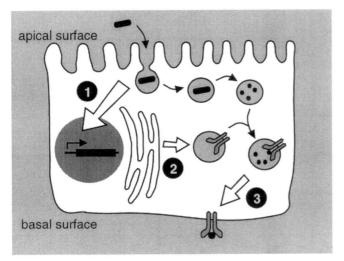

FIGURE 4 Expression of class II MHC by epithelial cells following internalization of bacteria. Step 1: synthesis of class II MHC can be induced by bacterial internalization through the apical plasma membrane or by exposure to a milieu of proinflammatory cytokines. Step 2: class II MHC is synthesized and intersects endocytic vacuoles containing endocytosed antigen. Step 3: vesicles containing complexes of bacterial antigens and nascent MHC protein are trafficked exclusively to the basolateral membrane of the epithelial cell, thus ensuring antigen presentation to T lymphocytes in the underlying submucosal tissue.

solateral surface of the epithelial cell. Thus, the antigen-presenting characteristics of epithelial cells seem to be specifically suited to sampling foreign antigens from the mucosal lumen and then presenting those antigens to T cells that reside below the epithelium, in the submucosa. Also interesting in light of these findings is the report (43) that the repertoire of antigenic peptides generated by epithelial cells differs markedly from that generated from the same original protein antigen by a classical APC. It would therefore be expected that the immune response stimulated by epithelial APCs would be qualitatively different from that initiated by a classical APC.

Another important facet of the antigen-presenting capacity of infected epithelial cells is their expression of cell adhesion molecules. It is well established that such adhesion molecules are essential to tether APCs to T lymphocytes such that the T-cell antigen receptor has the opportunity to scan the APC's MHC molecules for cognate antigenic peptides (14, 100). Thus, even if an APC presents a stimulatory antigenic peptide, antigen recognition and T-cell activation cannot occur if the T cell and APC cannot adhere to one another. The expression of intercellular adhesion molecules by epithelial cells has been studied extensively and has largely focused on expression of the

molecule ICAM-1 (intercellular adhesion molecule 1). The consensus of this work is that epithelial cells express ICAM-1 constitutively, but only at very low levels (16). Additionally, less than 20% of unstimulated epithelial cells have detectable quantities of this adhesion molecule. The cause for this cell-to-cell variability in ICAM-1 expression is not entirely understood; however, it has been reported that in the case of intestinal (50) and alveolar (8) epithelial cells, ICAM-1 expression is related to the maturational state of the cell. The fact that only a portion of cells in most unstimulated samples express this adhesion molecule may reflect the distribution of maturational states in a random sample of cells.

The expression pattern of ICAM-1 by epithelial cells changes drastically when the epithelium is inflamed (111, 113) or is exposed to bacterial products such as lipopolysaccharide (2, 61). Together, these data suggest that bacterial infection of epithelial tissues results in an enhancement of cell adhesion molecules, either directly via bacterial products or indirectly via the elaboration of inflammatory cytokines by the infected epithelium. Further work has shown that the induction of ICAM-1 expression occurs at least partially at the level of ICAM-1 mRNA (61, 70).

Lastly, infected epithelial cells may promote an acquired immune response by another pathway that leads to the activation of T lymphocytes. They may do this by undergoing apoptosis, or programmed cell death. Studies involving several experimental systems of infection suggest that infected epithelial cells can be induced to enter the apoptotic pathway (40, 48, 53). Further, there is evidence that the cell fragments resulting from apoptosis (apoptotic bodies) can be phagocytosed by dendritic cells and that the bacterial antigens can be processed and presented to T lymphocytes (87, 88). In the case of *P. aeruginosa* infection, it has been shown that epithelial cell apoptosis is triggered by a bacterial porin (since a purified preparation of this outer membrane protein was able to induce epithelial cell apoptosis [6]) and by at least two different proteins delivered into the host epithelial cell by the bacterium (40). In the case of *Helicobacter pylori* infection of the gastric mucosa, the mechanism by which apoptosis is induced was attributed to alterations in the relative expression of the proapoptotic protein Bax and the antiapoptotic protein Bcl-2 by infected epithelial cells (54).

FRIENDLY FIRE AND INNOCENT BYSTANDERS: COLLATERAL DAMAGE TO HOST CELLS

The mechanisms described above are all designed to initiate a protective immune response following invasion of epithelial cells by bacteria. However, in some cases the immune response that develops can actually favor the progression of disease or exacerbate its pathological consequences. A classic example of such immune-mediated pathology is chronic neutrophil activation. Such chronic activation is hypothesized to be responsible for the extensive tissue damage that occurs in the airways of individuals with CF (65). It has also been hypothesized that over time, such cumulative damage is the major contributor to loss of respiratory function and clinical disease (35, 36). These theories predict a vicious circle whereby infection triggers neutrophil activa-

tion, which damages host tissue. This damage then allows further microbial growth. Heightened inflammation in response to the additional bacterial burden further compromises host defenses because of additional activation of neutrophils. Another consequence may be sufficient damage to the tissue to allow a range of normally poorly pathogenic microbes to establish an infection. This model has been put forth (65) as an explanation for why individuals with chronic conditions such as CF become susceptible to infection by a multitude of weakly pathogenic microorganisms later in life.

Related to the role of neutrophils in chronic tissue damage is the suggested role of these immune cells in permitting bacterial access to deep tissues of the host. It has been demonstrated (72, 81) that the diapedesis of neutrophils between epithelial cells of the small intestine creates temporary disruptions in the tight junctions between epithelial cells and that bacterial pathogens such as *Salmonella* spp. can utilize such disruptions to gain passage across the epithelial layer (Fig. 5). Thus, under certain conditions, a protective response to infection may require that some arms of the immune response (e.g., neutrophils) remain inactive. Several biological agents that are ca-

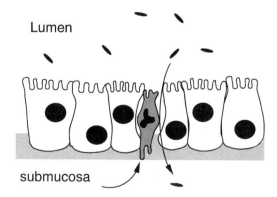

FIGURE 5 Diapedesis of neutrophils (shown in gray) disrupts tight junctions between epithelial cells (shown in white). The resulting disruption may provide an access point for bacteria (shown in black), allowing them to cross the epithelium and reach deeper tissues.

pable of inhibiting neutrophil chemotaxis have been identified (19, 71, 98). Such reagents may prove to be invaluable tools for further dissecting these biological events and for developing strategies aimed at modulating them for the benefit of the host.

BACTERIAL SURVIVAL TACTICS

A large factor in the success of any pathogenic microbe is its ability to conceal itself from the immune system of its host. The vertebrate immune system has evolved over millions of years to utilize many diverse effector mechanisms, some of which are detailed above, to exclude or inactivate foreign invaders. Given this need of the potential pathogen to defeat or circumvent such a wide range of host defenses, it is useful to consider that the invasion of host epithelial cells may create a safe haven from many of these defenses. Once inside a host epithelial cell, the potential pathogen becomes invisible to the host's humoral immune system as well as to other soluble host defense molecules (e.g., defensins or complement). On the other hand, the bacterium could utilize the host epithelium as a breeding ground, as has been demonstrated for *P. aeruginosa* (Fig. 6A) (27) and for *Salmonella* spp. (25), which have been shown to replicate within the endocytic vacuoles of their host epithelial cells. Another reason for this bacterial invasion of host epithelial cells has been alluded to earlier, that is, the invasion of epithelial cells to gain access to the host tissues that reside beyond the epithelium (25, 89). A classic example of this strategy is that employed by *Salmonella enterica* serovar Typhi, which invades the epithelium of the small intestine to gain access to the underlying tissues. After penetrating the host's intestinal epithelium, the bacterium invades the host macrophages, which it then uses as a vehicle to disseminate throughout the host's tissues (Fig. 6B).

This latter case is of particular interest in that it highlights the exploitation by a bacterium of a host endocytic process that was originally intended for other purposes. When serovar Typhi infects the intestinal epithelium,

it does so initially by entering specialized epithelial cells known as M cells. These M cells are dedicated to endocytosing antigens present in the intestinal lumen and transporting them to macrophages and lymphocytes, which actually reside within a cytoplasmic pocket of the M cell. Once transported into a macrophage, the bacterium shuts down the macrophage's normal antigen-processing and presentation machinery and utilizes the macrophage to seed itself into deeper host tissues. The mechanisms by which serovar Typhi avoids and inactivates the phagocytic and processing machinery of the macrophage are beyond the scope of this chapter and will be discussed elsewhere in this text (chapter 7). However, the means by which the bacterium enters its host's epithelia and the subsequent alterations that occur in the bacterium and the host cell are active topics of research that deserve further exploration and are addressed in more detail below.

Turning the Tables—Converting a Host Defense Mechanism into a Host Susceptibility Factor

Although there are many ways in which pathogens circumvent these defenses, in some cases microbes do not merely succeed in evading host defense mechanisms but actually turn the host defense mechanism to their own advantage. This is exemplified by the internalization of the bacterium *P. aeruginosa* by the epithelial cells of its host's airway and cornea. Through comparisons of bacterial internalization by corneal epithelial cells and disease pathology in mice that express either normal CFTR or mutant CFTR, it was concluded that CFTR mediates internalization of *P. aeruginosa* by epithelial cells of the cornea, in the same manner as it does with epithelial cells of the airway (117). However, in the case of corneal infection, internalization of bacteria by epithelial cells correlates with more severe disease, so that mice with mutant CFTR protein (where the epithelial cells were less capable of internalizing the bacteria) are resistant to *P. aeruginosa* eye infection (117). Thus, one

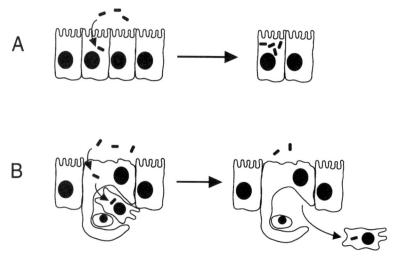

FIGURE 6 Invasion of bacteria into nonprofessional phagocytes allows bacterial proliferation and spread. Diagram A depicts the intracellular replication of bacteria within host epithelium, as has been demonstrated to occur in the case of several microorganisms including *P. aeruginosa* and serovar Typhi. Diagram B illustrates that invasion of bacteria into a nonprofessional phagocytic host can serve as a means of accessing deeper host tissues. The specific example shown in the figure depicts serovar Typhi passing through the intestinal epithelium to reach a host macrophage, which it then infects and uses to disseminate throughout the host's body.

interaction between the host and the bacterium (CFTR-mediated internalization) can benefit the host under certain circumstances (in the airway), but can be detrimental to it under others (in the cornea). The reasons for this difference likely lie in the fate of the bacterium-laden epithelial cell after bacterial internalization. In the case of the airway epithelium, the cells that internalize bacteria are part of an epithelium that is only one cell layer thick. Therefore, bacterium-laden epithelial cells are able to desquamate and are then free to be removed from the airway (Fig. 7A). This is quite different from the situation in the cornea, particularly when the mouse model of corneal scratch infection is used. In this tissue, the epithelium is several cell layers in thickness. The epithelial cells that lie in the deeper, more basal layers are likely the ones that actually internalize bacteria during acute infection. For example, acute corneal infection in humans is frequently associated with

some form of corneal injury, trauma, or irritation. Thus, one prerequisite for acute infection may be damage to the more superficial epithelial layers (Fig. 7B), providing access for the bacteria to deeper epithelial cells. Another finding implicating basal epithelial cells of the cornea in acute infection by *P. aeruginosa* is that when CFTR protein is detected by immunofluorescent staining, the basal epithelial cells are found to express high quantities of CFTR (117). Since there is a direct correlation between CFTR expression level and efficiency of bacterial internalization, it is thought that the basal epithelial cells that express CFTR also internalize the bacteria. Unable to desquamate because of their nonsurface location, subsurface corneal epithelial cells support the replication of internalized bacteria, which are protected from antibody, complement, and phagocytic cells. This allows the level of infecting bacteria to dramatically rise, enhancing the tissue damage and ultimate pa-

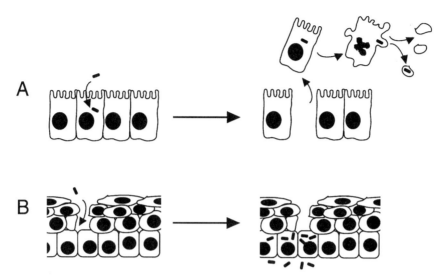

FIGURE 7 Two models intended to explain the differential outcome of CFTR-mediated internalization of *P. aeruginosa* in two different host tissues. (A) In the airway, internalization of *P. aeruginosa* induces apoptosis of bacterium-laden epithelial cells. These epithelial cells detach from the tissue and are expectorated, thus removing the bacteria they contain from the airways. It is possible that apoptotic bodies derived from the bacterium-laden cells are phagocytosed by dendritic cells and presented to helper T cells by the dendritic cell. (B) In the cornea, bacteria that breach the superficial, squamous epithelial cells are internalized by basal epithelial cells. Due to their anatomic location, the bacterium-laden epithelial cells are unable to desquamate and serve instead as a niche for the replication of intracellular bacteria.

thology experienced as part of a *P. aeruginosa* corneal infection. It is possible that CFTR-mediated internalization of *P. aeruginosa* by surface corneal epithelial cells contributes to innate immunity in the eye much as it does in the lung, but exposure of the subsurface epithelial cells by trauma, hypoxia from use of extended-wear contact lenses, or other damage, turns the tables on the host, converting a normal defense mechanism into a means to promote microbial virulence.

Another example of use of a host cell receptor-microbial interaction that can lead to both immunity and disease can be garnered from the recent report that serovar Typhi, like *P. aeruginosa*, uses CFTR protein to enter epithelial cells, although for the former pathogen this would be in the gastrointestinal (GI) tract (82). Cells from CF patients internalized fewer serovar Typhi cells than the same cell lines transfected with a wild-type copy of the CFTR gene. Reagents that blocked the interaction of serovar Typhi with amino acids 108 to 117 in the CFTR protein (the same region of the protein that binds *P. aeruginosa*) reduced bacterial entry into cells and tissues. Importantly, when compared to those in wild-type mice, GI epithelial cells in mice with one mutant ΔF508 CFTR allele had an 86% reduction in internalized serovar Typhi, and homozygous ΔF508 CFTR mice had essentially no internalized serovar Typhi. This led to the proposal that mutant alleles of CFTR, particularly the ΔF508 allele, have been maintained at high levels (4 to 5% in some populations) because of a heterozygote advantage in resistance to serovar Typhi infection and development of typhoid fever. Under this scenario, ingestion of serovar Typhi at levels that do not lead to clinical disease (which requires about 10^5 CFU be orally ingested [46]) leads to clearance from the GI tract due, in part, to epithelial cell ingestion and desquamation. However, when higher doses of serovar Typhi are ingested, the capacity of epithelial cell ingestion to remove bacteria is overwhelmed, and the damage to the M cells and neighboring enterocytes from shedding forms a "hole in the epithelium," through which noningested serovar Typhi cells gain access to the submucosa (47). Thus, an initial host defense mechanism is exploited by a pathogenic microbe to spread to deeper tissues.

Biochemical Responses of Host Cells during Invasion

An appropriate starting point for a discussion of host response to bacterial invasion is host cell alterations that occur at the level of the host cell receptor, which binds the bacterium. This event can be similar to the binding of any other ligand to a receptor. Indeed, it has been shown that many of the early signal transduction events commonly associated with receptor activation (e.g., tyrosine phosphorylation and Ca^{2+} flux [17, 77, 78, 86]) are also seen during invasion of epithelial cells by Salmonella and Shigella spp. These signaling events may largely function in the actual endocytic process by which bacteria are internalized by the host cell. For example, it has been shown that Shigella flexneri induces phosphorylation of cortactin, a cytoskeleton-associated protein that is also a substrate for pp60-src (17). Furthermore, recruitment of pp60-src was found to be localized to the membrane ruffles that formed near the point of bacterial entry into the host cell. However, while several clear links between host cell signaling events and downstream effects have been demonstrated, the importance of these signaling events in bacterial invasion seems to vary between experimental models of infection. Researchers have demonstrated that treatment of epithelial cells with kinase inhibitor prior to infection reduces bacterial invasion (91). However, Rosenshine and coworkers (86) report that although epithelial cell phosphorylation events do occur during invasion of S. enterica serovar Typhimurium, inhibiting these phosphorylation events has no effect on invasion. These

host cell signaling events are essential for the bacterium to recruit host cell components necessary to internalize and disseminate the bacterium. The next section will discuss some of the biochemical changes that occur in the bacterial cell to achieve these ends.

Biochemical Changes in Bacteria during Invasion

The biochemical basis of bacterial invasion into host epithelium has been a topic of intense investigation since the late 1980s. The early discovery of bacterial genetic loci capable of conferring an invasive phenotype upon normally noninvasive bacterial strains (66) set the stage for intense genetic analysis of the bacterial invasive mechanism (31, 32, 92). One of the most influential findings in the field of bacterial invasion was the discovery that encounters with the host epithelium induce the de novo expression of bacterial proteins and that these bacterial proteins are essential for invasion into epithelial cells (26). This finding led to the identification of a novel bacterial mechanism for interacting with the cells of its host, the type III secretion system.

Many bacterial pathogens, including Legionella pneumophila (13, 75), serovar Typhimurium (60), Burkholderia cepacia (7), Listeria monocytogenes (30), and Shigella spp. (93), are capable of multiplying within host cells. These bacterial pathogens can multiply at various intracellular sites, including endocytic vesicles and the cytoplasm. Escape from the endocytic vacuole by S. flexneri (42) and L. monocytogenes (11, 33) is a necessary step for bacterial replication or for spread of infection to other host cells. Using actin polymerization for motility within the cell and for cell-to-cell spread (otherwise called "rocket motility"), both Shigella and Listeria spp. take advantage of the enormous biomechanical activity within the host cell's own cytoskeleton (Fig. 8A). Research from the past half-decade has done much to elucidate the mechanism of this form of motility. Following its invasion into the host epithelial cell (39, 42, 92), the IcsA protein on the outer surface of S. flexneri (filled small cir-

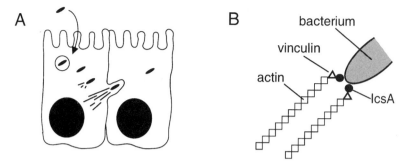

FIGURE 8 Actin motility (also called "rocket motility") of *S. flexneri* after its invasion of a host epithelial cell. (A) After invasion of the epithelial cell, *Shigella* escapes the endocytic vacuole and then forms actin filaments that propel the *Shigella* from its original host epithelial cell to adjacent cells. (B) Formation of actin filaments requires the participation of bacterial proteins such as IcsA, as well as the recruitment of host cytoskeletal elements vinculin and actin. Synthesis of IcsA is asymmetric on the bacterial cell, resulting in the formation of actin filaments on only one pole of the bacterium (the "old" pole). Vinculin binds IcsA, initiating the polymerization of actin.

cles in Fig. 8B) binds host cell vinculin (106), which initiates assembly of actin subunits on the bacterial surface. Since IcsA is preferentially targeted to the "old pole" of the bacterium (105), actin polymerizes at only one end of the bacterial cell. IcsA function is negatively regulated by its own phosphorylation (12) and proteolytic cleavage (97). There is also evidence (57) that proteolytic cleavage of vinculin is essential to initiate these motility complexes. As actin polymerizes at one pole of the bacterial cell, the resulting force propels the bacterium through the host cell cytoplasm, eventually stabbing through it and carrying the bacterium into the adjacent host cell. A similar mechanism of motility is observed in the infection of monocytes and enterocytes by *L. monocytogenes* (67).

CONCLUSIONS

The interaction of a bacterial pathogen with the host epithelium is an event that determines the course of a given infectious event. During this process, biochemical alterations occur in both the bacterial pathogen and the host epithelial cell. For the bacterial cell, these alterations are aimed at recruiting the cellular components of its epithelial host cell to facilitate the bacterium's invasion of (and/or passage through) the epithelial barrier. From the standpoint of the host epithelial cell, the biochemical alterations are part of innate and acquired immune responses to the pathogens, which exist either to limit the bacterium's persistence in the host or to destroy it outright. The defense mechanisms that the host epithelium employs range from simple mechanical removal of the bacterial cells to extremely intricate cytokine networks that orchestrate a response by dedicated immune effector cells. These defense mechanisms, however, do not come without a heavy price. Under some circumstances, the host response to bacterial infection increases host susceptibility to infection, either by providing bacteria with intracellular niches that the bacterium can use to evade host defenses or by damaging nearby host tissue, thus creating access points for bacterial dissemination. A fuller understanding of the relative importance of these mechanisms and of the interactions that occur between them will greatly assist future strategies for prophylactic or therapeutic intervention in bacterial infectious disease.

REFERENCES

1. Agerberth, B., J. Grunewald, E. Castanos-Velez, B. Olsson, H. Jornvall, H. Wigzell,

A. Eklund, and G. H. Gudmundsson. 1999. Antibacterial components in bronchoalveolar lavage fluid from healthy individuals and sarcoidosis patients. *Am. J. Respir. Crit. Care Med.* **160:**283–290.

2. Beck-Schimmer, B., R. C. Schimmer, R. L. Warner, H. Schmal, G. Nordblom, et al. 1997. Expression of lung vascular and airway ICAM-1 after exposure to bacterial lipopolysaccharide. *Am. J. Respir. Cell Mol. Biol.* **17:**344–352.

3. Brandtzaeg, P., T. S. Halstensen, H. S. Huitfeldt, P. Krajci, D. Kvale, H. Scott, and P. S. Thrane. 1992. Epithelial expression of HLA, secretory component (poly-Ig receptor), and adhesion molecules in the human alimentary tract. *Ann. N.Y. Acad. Sci.* **664:**157–179.

4. Brogden, K. A., M. R. Ackermann, P. B. McCray, Jr., and K. M. Huttner. 1999. Differences in the concentrations of small, anionic, antimicrobial peptides in bronchoalveolar lavage fluid and in respiratory epithelia of patients with and without cystic fibrosis. *Infect. Immun.* **67:**4256–4259.

5. Bryan, R., D. Kube, A. Perez, P. Davis, and A. Prince. 1998. Overproduction of the CFTR R domain leads to increased levels of asialoGM1 and increased Pseudomonas aeruginosa binding by epithelial cells. *Am. J. Respir. Cell Mol. Biol.* **19:**269–277.

6. Buommino, E., F. Morelli, S. Metafora, F. Rossano, B. Perfetto, A. Baroni, and M. A. Tufano. 1999. Porin from *Pseudomonas aeruginosa* induces apoptosis in an epithelial cell line derived from rat seminal vesicles. *Infect. Immun.* **67:**4794–4800.

7. Burns, J. L., M. Jonas, E. Y. Chi, D. K. Clark, A. Berger, and A. Griffith. 1996. Invasion of respiratory epithelial cells by *Burkholderia* (*Pseudomonas*) *cepacia. Infect. Immun.* **64:**4054–4059.

8. Christensen, P. J., S. Kim, R. H. Simon, G. B. Toews, and R. D. Paine. 1993. Differentiation-related expression of ICAM-1 by rat alveolar epithelial cells. *Am. J. Respir. Cell Mol. Biol.* **8:**9–15.

9. Clark, D. A., P. J. Lamey, R. F. Jarrett, and D. E. Onions. 1994. A model to study viral and cytokine involvement in Sjogren's syndrome. *Autoimmunity* **18:**7–14.

10. Comolli, J. C., L. L. Waite, K. E. Mostov, and J. N. Engel. 1999. Pili binding to asialo-GM1 on epithelial cells can mediate cytotoxicity or bacterial internalization by *Pseudomonas aeruginosa. Infect. Immun.* **67:**3207–3214.

11. Conte, M. P., G. Petrone, C. Longhi, P. Valenti, R. Morelli, F. Superti, and L. Se-ganti. 1996. The effects of inhibitors of vacuolar acidification on the release of *Listeria monocytogenes* from phagosomes of Caco-2 cells. *J. Med. Microbiol.* **44:**418–424.

12. d'Hauteville, H., and P. J. Sansonetti. 1992. Phosphorylation of IcsA by cAMP-dependent protein kinase and its effect on intracellular spread of Shigella flexneri. *Mol. Microbiol.* **6:**833–841.

13. Daisy, J. A., C. E. Benson, J. McKitrick, and H. M. Friedman. 1981. Intracellular replication of Legionella pneumophila. *J. Infect. Dis.* **143:**460–464.

14. Dang, L. H., M. T. Michalek, F. Takei, B. Benaceraff, and K. L. Rock. 1990. Role of ICAM-1 in antigen presentation demonstrated by ICAM-1 defective mutants. *J. Immunol.* **144:**4082–4091.

15. de Bentzmann, S., P. Roger, and E. Puchelle. 1996. Pseudomonas aeruginosa adherence to remodelling respiratory epithelium. *Eur. Respir. J.* **9:**2145–2150.

16. De Panfilis, G., G. C. Manara, C. Ferrari, C. Torresani, and A. Lonati. 1992. Adhesion molecules on the plasma membrane of epidermal cells. IV. Immunolocalization of the intercellular adhesion molecule-1 (ICAM-1, CD54) on the cell surface of a small subpopulation of keratinocytes freshly isolated from normal human epidermis. *Reg. Immunol.* **4:**119–129.

17. Dehio, C., M. C. Prevost, and P. J. Sansonetti. 1995. Invasion of epithelial cells by Shigella flexneri induces tyrosine phosphorylation of cortactin by a pp60c-src-mediated signalling pathway. *EMBO J.* **14:**2471–2482.

18. Denning, S. M., D. T. Tuck, L. W. Vollger, T. A. Springer, K. H. Singer, and B. F. Haynes. 1987. Monoclonal antibodies to CD2 and lymphocyte function-associated antigen 3 inhibit human thymic epithelial cell-dependent mature thymocyte activation. *J. Immunol.* **139:**2573–2578.

19. Eckle, I., G. Kolb, and K. Havemann. 1991. Inhibition of neutrophil chemotaxis by elastase-generated IgG fragments. *Scand. J. Immunol.* **34:**359–364.

20. Eckmann, L., H. C. Jung, C. Schurer-Maly, A. Panja, E. Morzycka-Wroblewska, and M. F. Kagnoff. 1993. Differential cytokine expression by human intestinal epithelial cell lines: regulated expression of interleukin 8 [comment]. *Gastroenterology* **105:**1689–1697.

21. Elner, S. G., R. M. Strieter, V. M. Elner, B. J. Rollins, M. A. Del Monte, and S. L. Kunkel. 1991. Monocyte chemotactic protein gene expression by cytokine-treated human retinal pigment epithelial cells. *Lab. Invest.* **64:**819–825.

22. Elner, V. M., R. M. Strieter, S. G. Elner, M. Baggiolini, I. Lindley, and S. L. Kunkel. 1990. Neutrophil chemotactic factor (IL-8) gene expression by cytokine-treated retinal pigment epithelial cells. *Am. J. Pathol.* **136:**745–750.

23. Eriksson, K., E. Ahlfors, A. George-Chandy, D. Kaiserlian, and C. Czerkinsky. 1996. Antigen presentation in the murine oral epithelium. *Immunology* **88:**147–152.

24. Feldman, M., R. Bryan, S. Rajan, L. Scheffler, S. Brunnert, H. Tang, and A. Prince. 1998. Role of flagella in pathogenesis of *Pseudomonas aeruginosa* pulmonary infection. *Infect. Immun.* **66:**43–51.

25. Finlay, B. B., J. Fry, E. P. Rock, and S. Falkow. 1989. Passage of Salmonella through polarized epithelial cells: role of the host and bacterium. *J. Cell Sci. Suppl.* **11:**99–107.

26. Finlay, B. B., F. Heffron, and S. Falkow. 1989. Epithelial cell surfaces induce Salmonella proteins required for bacterial adherence and invasion. *Science* **243:**940–943.

27. Fleiszig, S. M., T. S. Zaidi, and G. B. Pier. 1995. *Pseudomonas aeruginosa* invasion of and multiplication within corneal epithelial cells in vitro. *Infect. Immun.* **63:**4072–4077.

28. Franzetti, F., M. Cernuschi, R. Esposito, and M. Moroni. 1992. Pseudomonas infections in patients with AIDS and AIDS-related complex. *J. Intern. Med.* **231:**437–443.

29. Fujiwara, Y., T. Arakawa, T. Fukuda, E. Sasaki, K. Nakagawa, K. Fujiwara, K. Higuchi, K. Kobayashi, and A. Tarnawski. 1997. Interleukin-8 stimulates leukocyte migration across a monolayer of cultured rabbit gastric epithelial cells. Effect associated with the impairment of gastric epithelial barrier function. *Dig. Dis. Sci.* **42:**1210–1215.

30. Gaillard, J. L., P. Berche, J. Mounier, S. Richard, and P. Sansonetti. 1987. In vitro model of penetration and intracellular growth of *Listeria monocytogenes* in the human enterocyte-like cell line Caco-2. *Infect. Immun.* **55:**2822–2829.

31. Galan, J. E., and R. D. Curtiss. 1990. Expression of *Salmonella typhimurium* genes required for invasion is regulated by changes in DNA supercoiling. *Infect. Immun.* **58:**1879–1885.

32. Galan, J. E., C. Ginocchio, and P. Costeas. 1992. Molecular and functional characterization of the *Salmonella* invasion gene *invA*: homology of InvA to members of a new protein family. *J. Bacteriol.* **174:**4338–4349.

33. Goldfine, H., C. Knob, D. Alford, and J. Bentz. 1995. Membrane permeabilization by Listeria monocytogenes phosphatidylinositol-specific phospholipase C is independent of phospholipid hydrolysis and cooperative with listeriolysin O. *Proc. Natl. Acad. Sci. USA* **92:**2979–2983. (Retraction, *Proc. Natl. Acad. Sci. USA*, 1997, **94:**2772.)

34. Goldman, M. J., G. M. Anderson, E. D. Stolzenberg, U. P. Kari, M. Zasloff, and J. M. Wilson. 1997. Human beta-defensin-1 is a salt-sensitive antibiotic in lung that is inactivated in cystic fibrosis. *Cell* **88:**553–560.

35. Govan, J. R., and V. Deretic. 1996. Microbial pathogenesis in cystic fibrosis: mucoid *Pseudomonas aeruginosa* and *Burkholderia cepacia*. *Microbiol. Rev.* **60:**539–574.

36. Govan, J. R., and J. W. Nelson. 1992. Microbiology of lung infection in cystic fibrosis. *Br. Med. Bull.* **48:**912–930.

37. Greenfield, E. A., K. A. Nguyen, and V. K. Kuchroo. 1998. CD28/B7 costimulation: a review. *Crit. Rev. Immunol.* **18:**389–418.

38. Gupta, S. K., R. S. Berk, S. Masinick, and L. D. Hazlett. 1994. Pili and lipopolysaccharide of *Pseudomonas aeruginosa* bind to the glycolipid asialo GM1. *Infect. Immun.* **62:**4572–4579.

39. Hale, T. L. 1986. Invasion of epithelial cells by shigellae. *Ann. Inst. Pasteur Microbiol.* **137A:**311–314.

40. Hauser, A. R., and J. N. Engel. 1999. *Pseudomonas aeruginosa* induces type-III-secretion-mediated apoptosis of macrophages and epithelial cells. *Infect. Immun.* **67:**5530–5537.

41. Hershberg, R. M., D. H. Cho, A. Youakim, M. B. Bradley, J. S. Lee, P. E. Framson, and G. T. Nepom. 1998. Highly polarized HLA class II antigen processing and presentation by human intestinal epithelial cells. *J. Clin. Invest.* **102:**792–803.

42. High, N., J. Mounier, M. C. Prevost, and P. J. Sansonetti. 1992. IpaB of Shigella flexneri causes entry into epithelial cells and escape from the phagocytic vacuole. *EMBO J.* **11:**1991–1999.

43. Housseau, F., N. Rouas-Freiss, M. Roy, J. M. Bidart, J. G. Guillet, and D. Bellet. 1997. Antigen-presenting function of murine gonadal epithelial cell lines. *Cell. Immunol.* **177:**93–101.

44. Htin, A. 1990. T lymphocyte motility toward IL-1 in patients with interstitial lung diseases. *Bull. Chest Dis. Res. Inst. Kyoto Univ.* **23:**38–47.

45. Imai, Y., M. Yamakawa, and T. Kasajima. 1998. The lymphocyte-dendritic cell system. *Histol. Histopathol.* **13:**469–510.

46. Ivanoff, B., M. M. Levine, and P. H. Lambert. 1994. Vaccination against typhoid fever: present status. *Bull. W. H. O.* **72:**957–971.

47. Jones, B. D., N. Ghori, and S. Falkow. 1994. Salmonella typhimurium initiates murine infection by penetrating and destroying the specialized

epithelial M cells of the Peyer's patches [see comments]. *J. Exp. Med.* **180**:15–23.

48. **Jones, N. L., A. S. Day, H. A. Jennings, and P. M. Sherman.** 1999. *Helicobacter pylori* induces gastric epithelial cell apoptosis in association with increased Fas receptor expression. *Infect. Immun.* **67**:4237–4242.

49. **Jorens, P. G., J. B. Richman-Eisenstat, B. P. Housset, P. P. Massion, I. Ueki, and J. A. Nadel.** 1994. Pseudomonas-induced neutrophil recruitment in the dog airway in vivo is mediated in part by IL-8 and inhibited by a leumedin. *Eur. Respir. J.* **7**:1925–1931.

50. **Kaiserlian, D., D. Rigal, J. Abello, and J. P. Revillard.** 1991. Expression, function and regulation of the intercellular adhesion molecule-1 (ICAM-1) on human intestinal epithelial cell lines. *Eur. J. Immunol.* **21**:2415–2421.

51. **Kharazmi, A., H. Nielsen, and K. Bendtzen.** 1988. Modulation of human neutrophil and monocyte chemotaxis and superoxide responses by recombinant TNF-alpha and GM-CSF. *Immunobiology* **177**:363–370.

52. **Kielhofner, M., R. L. Atmar, R. J. Hamill, and D. M. Musher.** 1992. Life-threatening Pseudomonas aeruginosa infections in patients with human immunodeficiency virus infection. *Clin. Infect. Dis.* **14**:403–411.

53. **Kim, J. M., L. Eckmann, T. C. Savidge, D. C. Lowe, T. Witthoft, and M. F. Kagnoff.** 1998. Apoptosis of human intestinal epithelial cells after bacterial invasion. *J. Clin. Invest.* **102**:1815–1823.

54. **Konturek, P. C., P. Pierzchalski, S. J. Konturek, H. Meixner, G. Faller, T. Kirchner, and E. G. Hahn.** 1999. Helicobacter pylori induces apoptosis in gastric mucosa through an upregulation of Bax expression in humans. *Scand. J. Gastroenterol.* **34**:375–383.

55. **Krivan, H. C., D. D. Roberts, and V. Ginsburg.** 1988. Many pulmonary pathogenic bacteria bind specifically to the carbohydrate sequence GalNAc beta 1-4Gal found in some glycolipids. *Proc. Natl. Acad. Sci. USA* **85**:6157–6161.

56. **Kunkel, S. L., T. Standiford, K. Kasahara, and R. M. Strieter.** 1991. Interleukin-8 (IL-8): the major neutrophil chemotactic factor in the lung. *Exp. Lung Res.* **17**:17–23.

57. **Laine, R. O., W. Zeile, F. Kang, D. L. Purich, and F. S. Southwick.** 1997. Vinculin proteolysis unmasks an ActA homolog for actin-based Shigella motility. *J. Cell Biol.* **138**:1255–1264.

58. **Lam, J., R. Chan, K. Lam, and J. W. Costerton.** 1980. Production of mucoid microcolonies by *Pseudomonas aeruginosa* within infected

lungs in cystic fibrosis. *Infect. Immun.* **28**:546–556.

59. **Lee, K. K., H. B. Sheth, W. Y. Wong, R. Sherburne, W. Paranchych, R. S. Hodges, C. A. Lingwood, H. Krivan, and R. T. Irvin.** 1994. The binding of Pseudomonas aeruginosa pili to glycosphingolipids is a tip-associated event involving the C-terminal region of the structural pilin subunit. *Mol. Microbiol.* **11**:705–713.

60. **Leung, K. Y., and B. B. Finlay.** 1991. Intracellular replication is essential for the virulence of Salmonella typhimurium. *Proc. Natl. Acad. Sci. USA* **88**:11470–11474.

61. **Li, X. C., A. M. Jevnikar, and D. R. Grant.** 1997. Expression of functional ICAM-1 and VCAM-1 adhesion molecules by an immortalized epithelial cell clone derived from the small intestine. *Cell Immunol.* **175**:58–66.

62. **Maekawa, T., Y. Kinoshita, Y. Matsushima, A. Okada, H. Fukui, et al.** 1997. Helicobacter pylori induces proinflammatory cytokines and major histocompatibility complex class II antigen in mouse gastric epithelial cells. *J. Lab. Clin. Med.* **130**:442–449.

63. **Medzhitov, R., and C. A. Janeway, Jr.** 1998. Innate immune recognition and control of adaptive immune responses. *Semin. Immunol.* **10**:351–353.

64. **Mellman, I., S. J. Turley, and R. M. Steinman.** 1998. Antigen processing for amateurs and professionals. *Trends Cell Biol.* **8**:231–237.

65. **Meyer, K. C., and J. Zimmerman.** 1993. Neutrophil mediators, Pseudomonas, and pulmonary dysfunction in cystic fibrosis [see comments]. *J. Lab. Clin. Med.* **121**:654–661.

66. **Miller, V. L., and S. Falkow.** 1988. Evidence for two genetic loci in *Yersinia enterocolitica* that can promote invasion of epithelial cells. *Infect. Immun.* **56**:1242–1248.

67. **Mounier, J., A. Ryter, M. Coquis-Rondon, and P. J. Sansonetti.** 1990. Intracellular and cell-to-cell spread of *Listeria monocytogenes* involves interaction with F-actin in the enterocytelike cell line Caco-2. *Infect. Immun.* **58**:1048–1058.

68. **Mulvey, M. A., Y. S. Lopez-Boado, C. L. Wilson, R. Roth, W. C. Parks, J. Heuser, and S. J. Hultgren.** 1998. Induction and evasion of host defenses by type 1-piliated uropathogenic Escherichia coli. *Science* **282**:1494–1497. [Erratum, *Science*, 1999, **283**:795.]

69. **Nadel, J. A., B. Davis, and R. J. Phipps.** 1979. Control of mucus secretion and ion transport in airways. *Annu. Rev. Physiol.* **41**:369–381.

70. **Nagineni, C. N., R. K. Kutty, B. Detrick, and J. J. Hooks.** 1996. Inflammatory cytokines

induce intercellular adhesion molecule-1 (ICAM-1) mRNA synthesis and protein secretion by human retinal pigment epithelial cell cultures. *Cytokine* **8**:622–630.

71. **Nakagawa, H., K. Watanabe, and K. Sato.** 1988. Inhibitory action of synthetic proteinase inhibitors and substrates on the chemotaxis of rat polymorphonuclear leukocytes in vitro. *J. Pharmacobiodyn.* **11**:674–678.

72. **Nash, S., J. Stafford, and J. L. Madara.** 1987. Effects of polymorphonuclear leukocyte transmigration on the barrier function of cultured intestinal epithelial monolayers. *J. Clin. Invest.* **80:** 1104–1113.

73. **Newell, D. G., and A. Pearson.** 1984. The invasion of epithelial cell lines and the intestinal epithelium of infant mice by Campylobacter jejuni/coli. *J. Diarrhoeal Dis. Res.* **2**:19–26.

74. **Niesel, D. W., C. B. Hess, Y. J. Cho, K. D. Klimpel, and G. R. Klimpel.** 1986. Natural and recombinant interferons inhibit epithelial cell invasion by *Shigella* spp. *Infect. Immun.* **52**:828–833.

75. **Oldham, L. J., and F. G. Rodgers.** 1985. Adhesion, penetration and intracellular replication of Legionella pneumophila: an in vitro model of pathogenesis. *J. Gen. Microbiol.* **131**:697–706.

76. **Osusky, R., R. J. Dorio, Y. K. Arora, S. J. Ryan, and S. M. Walker.** 1997. MHC class II positive retinal pigment epithelial (RPE) cells can function as antigen-presenting cells for microbial superantigen. *Ocul. Immunol. Inflamm.* **5**:43–50.

77. **Pace, J., M. J. Hayman, and J. E. Galan.** 1993. Signal transduction and invasion of epithelial cells by S. typhimurium. *Cell* **72**:505–514.

78. **Pace, J. L., and J. E. Galan.** 1994. Measurement of free intracellular calcium levels in epithelial cells as consequence of bacterial invasion. *Methods Enzymol.* **236**:482–490.

79. **Panjwani, N., T. S. Zaidi, J. E. Gigstad, F. B. Jungalwala, M. Barza, and J. Baum.** 1990. Binding of Pseudomonas aeruginosa to neutral glycosphingolipids of rabbit corneal epithelium. *Infect. Immun.* **58**:114–118.

80. **Patel, J. A., M. Kunimoto, T. C. Sim, R. Garofalo, T. Eliott, et al.** 1995. Interleukin-1 alpha mediates the enhanced expression of intercellular adhesion molecule-1 in pulmonary epithelial cells infected with respiratory syncytial virus. *Am. J. Respir. Cell Mol. Biol.* **13**:602–609.

81. **Perdomo, J. J., P. Gounon, and P. J. Sansonetti.** 1994. Polymorphonuclear leukocyte transmigration promotes invasion of colonic epithelial monolayer by Shigella flexneri. *J. Clin. Invest.* **93**:633–643.

82. **Pier, G. B., M. Grout, T. Zaidi, G. Meluleni, S. S. Mueschenborn, G. Banting, R.** Ratcliff, M. J. Evans, and W. H. Colledge. 1998. Salmonella typhi uses CFTR to enter intestinal epithelial cells. *Nature* **393**:79–82.

83. **Pier, G. B., M. Grout, and T. S. Zaidi.** 1997. Cystic fibrosis transmembrane conductance regulator is an epithelial cell receptor for clearance of Pseudomonas aeruginosa from the lung. *Proc. Natl. Acad. Sci. USA* **94**:12088–12093.

84. **Pier, G. B., M. Grout, T. S. Zaidi, and J. B. Goldberg.** 1996. How mutant CFTR may contribute to Pseudomonas aeruginosa infection in cystic fibrosis. *Am. J. Respir. Crit. Care Med.* **154:** S175–S182.

85. **Pier, G. B., M. Grout, T. S. Zaidi, J. C. Olsen, L. G. Johnson, J. R. Yankaskas, and J. B. Goldberg.** 1996. Role of mutant CFTR in hypersusceptibility of cystic fibrosis patients to lung infections. *Science* **271**:64–67.

86. **Rosenshine, I., S. Ruschkowski, V. Foubister, and B. B. Finlay.** 1994. *Salmonella typhimurium* invasion of epithelial cells: role of induced host cell tyrosine protein phosphorylation. *Infect. Immun.* **62**:4969–4974.

87. **Rovere, P., A. A. Manfredi, C. Vallinoto, V. S. Zimmermann, U. Fascio, et al.** 1998. Dendritic cells preferentially internalize apoptotic cells opsonized by anti-beta2-glycoprotein I antibodies. *J. Autoimmun.* **11**:403–411.

88. **Rovere, P., C. Vallinoto, A. Bondanza, M. C. Crosti, M. Rescigno, P. Ricciardi-Castagnoli, C. Rugarli, and A. A. Manfredi.** 1998. Bystander apoptosis triggers dendritic cell maturation and antigen-presenting function. *J. Immunol.* **161**:4467–4471.

89. **Rubens, C. E., S. Smith, M. Hulse, E. Y. Chi, and G. van Belle.** 1992. Respiratory epithelial cell invasion by group B streptococci. *Infect. Immun.* **60**:5157–5163.

90. **Saiman, L., and A. Prince.** 1993. Pseudomonas aeruginosa pili bind to asialoGM1 which is increased on the surface of cystic fibrosis epithelial cells. *J. Clin. Invest.* **92**:1875–1880.

91. **Sandros, J., P. N. Madianos, and P. N. Papapanou.** 1996. Cellular events concurrent with Porphyromonas gingivalis invasion of oral epithelium in vitro. *Eur. J. Oral Sci.* **104**:363–371.

92. **Sansonetti, P. J.** 1991. Genetic and molecular basis of epithelial cell invasion by Shigella species. *Rev. Infect. Dis.* **13**(Suppl 4):S285–S292.

93. **Sansonetti, P. J., A. Ryter, P. Clerc, A. T. Maurelli, and J. Mounier.** 1986. Multiplication of *Shigella flexneri* within HeLa cells: lysis of the phagocytic vacuole and plasmid-mediated contact hemolysis. *Infect. Immun.* **51**:461–469.

94. **Sayers, T. J., T. A. Wiltrout, C. A. Bull, A. C. Denn, A. M. Pilaro, and B. Lokesh.** 1988. Effect of cytokines on polymorphonuclear

neutrophil infiltration in the mouse. Prostaglan-din- and leukotriene-independent induction of infiltration by IL-1 and tumor necrosis factor. *J. Immunol.* **141:**1670–1677.

95. **Schnapp, D., and A. Harris.** 1998. Antibacterial peptides in bronchoalveolar lavage fluid. *Am. J. Respir. Cell Mol. Biol.* **19:**352–356.

96. **Schurr, M. J., D. W. Martin, M. H. Mudd, N. S. Hibler, J. C. Boucher, and V. Deretic.** 1993. The algD promoter: regulation of alginate production by Pseudomonas aeruginosa in cystic fibrosis. *Cell Mol. Biol. Res.* **39:**371–376.

97. **Shere, K. D., S. Sallustio, A. Manessis, T. G. D'Aversa, and M. B. Goldberg.** 1997. Disruption of IcsP, the major Shigella protease that cleaves IcsA, accelerates actin-based motility. *Mol. Microbiol.* **25:**451–462.

98. **Shimizu, A., A. Takeuchi, H. Ohto, T. Hashimoto, and T. Miyamoto.** 1988. Inhibition of neutrophil chemotaxis by a monoclonal antibody (TM316). *Scand. J. Immunol.* **28:** 675–685.

99. **Shimizu, T., C. X. Cao, R. G. Shao, and Y. Pommier.** 1998. Lamin B phosphorylation by protein kinase calpha and proteolysis during apoptosis in human leukemia HL60 cells. *J. Biol. Chem.* **273:**8669–8674.

100. **Siu, G., S. M. Hedrick, and A. A. Brian.** 1989. Isolation of the murine intercellular adhesion molecule 1 (ICAM-1) gene. ICAM-1 enhances antigen-specific T cell activation. *J. Immunol.* **143:**3813–3820.

101. **Smart, S. J., and T. B. Casale.** 1994. Pulmonary epithelial cells facilitate TNF-alpha-induced neutrophil chemotaxis. A role for cytokine networking. *J. Immunol.* **152:**4087–4094.

102. **Smith, J. J., S. M. Travis, E. P. Greenberg, and M. J. Welsh.** 1996. Cystic fibrosis airway epithelia fail to kill bacteria because of abnormal airway surface fluid. *Cell* **85:**229–236. [Erratum, *Cell,* 1996, **87:**following 355.]

103. **Standiford, T. J., S. L. Kunkel, M. A. Basha, S. W. Chensue, J. P. Lynch, G. B. Toews, J. Westwick, and R. M. Strieter.** 1990. Interleukin-8 gene expression by a pulmonary epithelial cell line. A model for cytokine networks in the lung. *J. Clin. Invest.* **86:**1945–1953.

104. **Stein, M. E., and M. J. Stadecker.** 1987. Characterization and antigen-presenting function of a murine thyroid-derived epithelial cell line. *J. Immunol.* **139:**1786–1791.

105. **Steinhauer, J., R. Agha, T. Pham, A. W. Varga, and M. B. Goldberg.** 1999. The unipolar Shigella surface protein IcsA is targeted di-rectly to the bacterial old pole: IcsP cleavage of IcsA occurs over the entire bacterial surface. *Mol. Microbiol.* **32:**367–377.

106. **Suzuki, T., S. Saga, and C. Sasakawa.** 1996. Functional analysis of Shigella VirG domains essential for interaction with vinculin and actin-based motility. *J. Biol. Chem.* **271:**21878–21885.

107. **Svanborg, C., W. Agace, S. Hedges, H. Linder, and M. Svensson.** 1993. Bacterial adherence and epithelial cell cytokine production. *Zentralbl. Bakteriol.* **278:**359–364.

108. **Takahashi, I., and H. Kiyono.** 1999. Gut as the largest immunologic tissue. *J. Parenter. Enteral Nutr.,* in press.

109. **Takaya, M., Y. Ichikawa, H. Shimizu, M. Uchiyama, J. Moriuchi, and S. Arimori.** 1990. Expression of MHC class II antigens and other T cell activation antigens on T cells and salivary duct epithelial cells in the salivary gland of cases of Sjogren's syndrome. *Tokai J. Exp. Clin. Med.* **15:**27–33.

110. **Taylor, J. L., and W. J. O'Brien.** 1985. Interferon production and sensitivity of rabbit corneal epithelial and stromal cells. *Invest. Ophthalmol. Vis. Sci.* **26:**1502–1508.

111. **Tosi, M. F., J. M. Stark, C. W. Smith, A. Hamedani, D. C. Gruenert, and M. D. Infeld.** 1992. Induction of ICAM-1 expression on human airway epithelial cells by inflammatory cytokines: effects on neutrophil-epithelial cell adhesion. *Am. J. Respir. Cell Mol. Biol.* **7:**214–221.

112. **Tuteja, R., and N. Tuteja.** 1998. Nucleolin: a multifunctional major nucleolar phosphoprotein. *Crit. Rev. Biochem. Mol. Biol.* **33:**407–436.

113. **Vejlsgaard, G. L., E. Ralfkiaer, C. Avnstorp, M. Czajkowski, S. D. Marlin, and R. Rothlein.** 1989. Kinetics and characterization of intercellular adhesion molecule-1 (ICAM-1) expression on keratinocytes in various inflammatory skin lesions and malignant cutaneous lymphomas. *J. Am. Acad. Dermatol.* **20:**782–790.

114. **Webster, P., L. Vanacore, A. C. Nairn, and C. R. Marino.** 1994. Subcellular localization of CFTR to endosomes in a ductal epithelium. *Am. J. Physiol.* **267:**C340–C348.

115. **Yard, B. A., M. R. Daha, M. Kooymans-Couthino, J. A. Bruijn, M. E. Paape, E. Schrama, L A. van Es, and F. J. van der Woude.** 1992. IL-1 alpha stimulated TNF alpha production by cultured human proximal tubular epithelial cells. *Kidney Int.* **42:**383–389.

116. **Ye, G., C. Barrera, X. Fan, W. K. Gourley, S. E. Crowe, P. B. Ernst, and V. E. Reyes.** 1997. Expression of B7-1 and B7-2 costimulatory molecules by human gastric epithelial cells:

potential role in CD4+ T cell activation during Helicobacter pylori infection. *J. Clin. Invest.* **99:** 1628–1636.

117. **Zaidi, T. S., J. Lyczak, M. Preston, and G. B. Pier.** 1999. Cystic fibrosis transmembrane conductance regulator-mediated corneal epithelial cell ingestion of *Pseudomonas aeruginosa* is a key component in the pathogenesis of ex-perimental murine keratitis. *Infect. Immun.* **67:** 1481–1492.

118. **Zar, H., L. Saiman, L. Quittell, and A. Prince.** 1995. Binding of Pseudomonas aeruginosa to respiratory epithelial cells from patients with various mutations in the cystic fibrosis transmembrane regulator. *J. Pediatr.* **126:**230–233.

ECOLOGICAL ASPECTS OF HOST COLONIZATION: COAGGREGATION, OSMOADAPTATION, AND ACID TOLERANCE OR RESISTANCE

Thad Stanton

4

To successfully colonize a host, a bacterium must reach a suitable microhabitat on or within the host, establish dividing cell populations, and persist. A pathogenic bacterium has the additional task of damaging host tissues inadvertently or in the course of accomplishing the other steps. Each colonization step requires that characteristics of the bacterium overcome (or at least balance) environmental forces working against the bacterium. Through evolution, adaptations have developed that enable a bacterium to take advantage of or contend with factors in its environment.

Colonization is the successful outcome of bacterial interactions with the living and nonliving components of the environment. For the host-associated bacterium, the living components of its environment are host tissues and frequently, but not always, other colonizing microorganisms. The nonliving components can include fluid flow, pH, osmolality, oxygen/redox potential, inert surfaces/viscous matrices (mucus), essential substrates or nutrients, toxicants, and bacteriophages.

This chapter describes three examples of adaptations that are important for bacteria to colonize human hosts. Coaggregation refers to specific physical associations that occur between bacterial species colonizing the human mouth. Acid resistance and osmoadaptation are features of *Helicobacter pylori* and *Escherichia coli*, bacteria that colonize, respectively, the stomach and lower gastrointestinal tract. The chapter provides a short preview of each topic, not a review. Several excellent accounts of coaggregation (50, 52, 100), osmoadaptation (17, 45, 84, 101), and acid resistance and tolerance (31, 41, 67) provide more extensive coverage for the interested reader.

COAGGREGATION

A complex microbial community exists within the human gingival crevice, the space below the gum line between tooth surface and epithelial tissue. The tooth surface provides a stable physical support for bacterial accumulation and plaque formation. Crevicular fluid, a biochemical equivalent of blood serum, flows steadily from host tissues over the tooth surface and provides nutrients for microbial growth.

In the healthy gingival crevice, the growth of plaque bacteria is held in check by internal and external factors (e.g., rinsing action and

Thad Stanton   Zoonotic Diseases Research Unit, National Animal Disease Center, Agricultural Research Service, U.S. Department of Agriculture, Ames, IA 50010.

Virulence Mechanisms of Bacterial Pathogens, 3rd ed., Edited by K. A. Brogden et al.
©2000 ASM Press, Washington, D.C.

protective immunoglobins provided by the crevicular fluid and dental floss). In the diseased state, the gingival crevice becomes a periodontal pocket. Plaque accumulates and the resident bacterial populations increase. Most significantly, the species composition of the plaque microbiota changes, favoring those species that are periodontopathogenic, especially gram-negative anaerobes. The plaque-coated tooth becomes a staging area for bacterial chemical assaults and invasions of nearby tissues. A combination of bacterial activities and host inflammatory responses leads to death of tissues around the tooth, destruction of collagen connecting tooth to bone, loss of bone, loss of tooth, and various forms of periodontal disease.

Periodontal disease appears as a progressive, mixed bacterial infection. Hundreds of bacterial species have been isolated from subgingival plaque of patients with severe periodontal disease (40, 100). Of these, 10 or more species have characteristics that fit a periodontopathogenic profile. Species implicated as periodontal pathogens are *Actinobacillus actinomycetemcomitans, Porphyromonas gingivalis, Fusobacterium nucleatum, Bacteroides forsythus, Prevotella intermedia, Eikenella corrodens, Peptostreptococcus micros, Streptococcus intermedius, Selenomonas* sp., *Campylobacter rectus, Eubacterium* sp., and various known and unknown spirochetes (40). They are late colonizers of plaque. They increase in numbers and are consistently detectable at periodontal disease sites. They display tissue-damaging activities (such as tissue invasiveness and hemolytic, proteolytic, or cytolytic capacities) and immunologically reactive properties (they can induce antibodies and impair immune function).

Identification of periodontopathogens and their pathogenic properties is the basis for understanding and controlling periodontal disease. One commonly shared characteristic of human oral bacteria, especially of those colonizing the tooth surface, is coaggregation. Certain combinations of oral strains, when mixed together in suspension, will form cell pairs and multicellular aggregates (Fig. 1). Al-

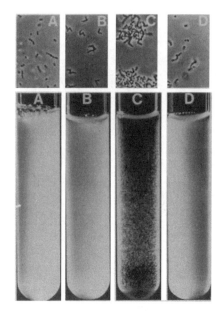

FIGURE 1 Coaggregation between oral bacteria species, *Streptococcus sanguis* and *Actinomyces viscosus*. At the top are phase-contrast micrographs of cells taken from tubes containing suspensions. (A) *S. sanguis* cells. (B) *A. viscosus* cells. (C) Mixture of *S. sanguis* and *A. viscosus* cells. (D) Mixture of cells to which 100 mM lactose (final concentration) has been added. Aggregates of mixed-cell types form (C) when *S. sanguis* and *A. viscosus* cell suspensions are combined and are visible both macroscopically (bottom) and microscopically (top). Lactose inhibition of aggregation (D) indicates involvement of that sugar in the receptor-adhesin interactions of the two cell types. Photograph from (50) and used with the permission of the author.

though the adhering cells belong to different strains, species, or genera (50, 52), the cell pairings are not random. Cell-to-cell recognition between specific partners takes place. The bases for this recognition are adhesins and receptors on the bacterial surfaces (100).

Bacterial coaggregations have also been reported, but much less frequently, among host-associated bacteria from nonoral environments, e.g., from urinary tract infections (85) and from digestive tracts of animals and insects (7, 96). This infrequent occurrence could be due to a lack of a solid support (tooth surface) for bacterial colonization in those environments, the presence of alternative physical supports (e.g., mucous layer over intestinal epithelium), or insufficient investigations of microbial aggregation in those environments.

Treponema denticola is the best-characterized oral spirochete that fits the periodontopathogen profile (29, 39, 40). There are also unidentified and yet-to-be-cultivated spirochetes in the human gingival crevice (11, 19, 68, 78), and some of these unknown species are likely to be associated with disease (11, 68, 87).

T. denticola cells aggregate with or bind to cells of *F. nucleatum, P. gingivalis,* and *B. forsythus* (37, 54, 105). The spirochete does not coaggregate with nonoral fusobacteria or with 45 other strains of oral bacteria representing 9 genera that are early colonizers of plaque (54, 100). *F. nucleatum* strains are bacterial bridges within dental plaque (Fig. 2), forming coag-

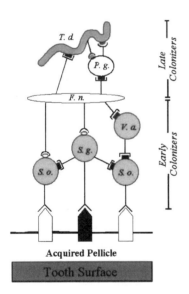

FIGURE 2 Schematic representation of physical interactions between selected oral bacteria in plaque. Interactions are based on in vitro coaggregations of *Streptococcus oralis* (S.o.), *S. gordonii* (S.g.), *Veillonella atypica* (V.a.), *Fusobacterium nucleatum* (F.n.), *Porphyromonas gingivalis* (P.g.), and *Treponema denticola* (T.d.). Adhesins are depicted as stemlike appendages with branched ends, and receptors are triangular, rectangular, or semicircular projections from the bacterial surface and tooth pellicle. Adapted from Kolenbrander and London (52).

gregation partnerships with many other species and linking early and late plaque-colonizing species that do not coaggregate (51, 52). *Fusobacterium* strains display an assortment of adhesins and receptors for other bacteria and for host cells (100).

Coaggregation between *T. denticola* and *F. nucleatum* cells is inhibited if *F. nucleatum* cells are preheated (85°C, 30 min). Preheating *T. denticola* cells alone does not affect cell-to-cell attachment (54). On this basis, the *F. nucleatum* cell surface is thought to display a heat-sensitive proteinaceous adhesin that interacts with a receptor on the *T. denticola* cell surface. D-Galactosamine and lactose inhibition of *T. denticola*–*F. nucleatum* coaggregation suggests that the *F. nucleatum* adhesin is a lectinlike protein interacting with a carbohydrate-containing receptor on *T. denticola* cells (54).

Coaggregation between *T. denticola* and *P. gingivalis* cells is not affected by heating cells of either species but is partially inhibited by heating cells of both species (37, 105). Thus, the bacteria participate in a bimodal (or multimodal) binding, with each species expressing at least one adhesin and receptor complementary to counterparts on the other species (Fig. 2). Rhamnose, fucose, arginine, and galactosamine partially inhibit both binding and coaggregation, also indicating that multiple receptors are involved (37, 105). Although fimbriae mediate many intergeneric aggregations, the failure of purified *P. gingivalis* fimbriae to block *T. denticola*–*P. gingivalis* aggregation suggests that those appendages are not involved in this interaction (105).

Several observations suggest that the intergeneric coaggregations of bacterial species in the test tube represent physical associations that can occur within the host (50). One line of evidence is that coaggregation partners appear at the same time and place within oral microhabitats. For example, *T. denticola*, a late colonizer of plaque, coaggregates specifically with other late colonizers of plaque, such as *B. forsythus* and *P. gingivalis*, and not with early colonizers (54). Spirochetes reactive with monoclonal antibodies to *T. denticola* are present only in those subgingival plaque samples in which *P. gingivalis* cells are also present (90). Saito and colleagues used specific antisera and immunoelectron microscopy to identify *T. denticola* and *P. gingivalis* in ultrathin sections of periodontal subgingival plaque (47). *T. denticola* cells were observed near *P. gingivalis* cells in deeper regions of plaque samples.

More direct evidence for the in vivo occurrence of coaggregation comes from studies of oral cocci. Two early colonizing bacteria, the filamentous rod *Capnocytophaga matruchoti* and the gram-positive coccus *Streptococcus sanguis*, form conspicuous cell aggregates resembling corncobs in vitro. By direct microscopy, the distinct corncob formations of these species can be detected both in plaque and in the test tube (55, 69, 70).

McBride and van der Hoeven demonstrated the importance of coaggregation in colonization of a host by oral cocci (63). *Veillonella parvula* (*alcalescens*) cells colonized tooth surfaces of gnotobiotic rats within 2 h after inoculation if cells of a coaggregating strain of *Streptococcus mutans* were present. If the animals were not colonized by the coaggregating strain or were colonized by an *S. mutans* strain that did not coaggregate, *V. parvula* colonization was reduced 100- to 1,000-fold.

Microorganisms involved in intergeneric associations are likely to collaborate in other ways (Table 1). Some examples are based on experimental evidence. Some are predictable and others, such as horizontal gene transfer, are feasible on the basis of recent studies of low-frequency gene transfer in a biofilm community (12).

The aggregation of *Veillonella* cells with lactate-producing *Streptococcus* species can provide metabolic benefits to both partners (98). *Veillonella* species benefit by using lactate as a growth substrate. The removal of lactate from their external environment could stimulate glycolysis and increase growth of streptococcal cells.

An elegant study demonstrating coaggregation-mediated interactions between oral bacteria exposed to oxygen was recently provided by Bradshaw and coworkers (5). The obligately anaerobic bacteria *P. gingivalis* and *Prevotella nigrescens* were cocultured with seven oxygen-consuming species, aerobic and facultatively anaerobic bacteria. The anaerobes and oxygen-consuming species did not coaggregate well with each other, but all nine species coaggregated strongly with *F. nucleatum*. Consequently, *F. nucleatum* cells could be used as a bridge to coaggregate anaerobic and oxygen-consuming bacteria. The anaerobes could not grow in aerated cultures by themselves and grew poorly in aerated cultures containing oxygen-consuming bacteria but no *F. nucleatum* cells. Adding *F. nucleatum* to aerated chemostat cultures of all nine species resulted in 100- to 1,000-fold increases in *P. gingivalis* and *P. nigrescens* population densities. In anaerobic (control) cultures containing the same species, the growth of the anaerobes was unaffected by the presence of *F. nucleatum*. Results of these studies illustrate the impor-

TABLE 1 Possible coaggregation-mediated interactions and outcomes

Possible outcome	Reference(s)
1. Persistence in a microenvironment. Primary colonizers provide support for secondary colonizers	53, 63, 88, 96
2. Physiological cooperation	38, 65, 74, 94
3. Protection from environmental assaults	4, 5, 50
4. Movement of nonmotile cells by motile cells	6
5. Virulence enhancement due to physiological cooperation, protection from immune assault, and synergistic pathogenic mechanisms	3, 46, 85, 107
6. Concentrated antagonism/defenses against competing microbes	36, 76, 86
7. Horizontal gene transfer	12

tance of coaggregation-based interactions between anaerobic bacteria, *F. nucleatum*, and oxygen-consuming bacteria in providing protection for the anaerobes in a hostile (aerated) environment.

Our knowledge of microorganisms is overwhelmingly biased toward pure cultures of suspended cells, a microbial situation that seldom exists in nature (14). Studies of the coaggregation of oral bacteria provide a foundation for understanding the physical structure and exploring the dynamic interactions of a bacterial community. The observation that cells of two different microbial species from the same habitat are able to aggregate is a reasonable predictor that other symbiotic associations exist between the pair. Mechanisms of intermicrobial binding in coaggregation and regulation of those mechanisms are now being intensely examined. Additional research will undoubtedly discover the sequels of bacterial attachment—the genetic, physiological, and pathogenic interactions facilitated by coaggregation.

OSMOADAPTATION

A preprogrammed, coordinated series of biochemical events takes place within minutes after *E. coli* bacteria are switched between environments of different osmolalities. Approximate osmolalities (OsM = moles per kilogram) of various liquids are as follows: bacterial minimum culture media and environmental water samples, 0.15 to 0.25 OsM; Luria-Bertani culture broth, 0.4 OsM; LB broth plus 0.3 M NaCl, 1.0 OsM; human colonic contents, 0.3 to 0.6 OsM (102); human plasma, 0.3 OsM; normal human urine, 0.5 to 0.8 OsM; cell cytoplasm of *E. coli* in minimal medium, 0.3 OsM (92). A hyperosmotic stress can occur when bacteria are transported by ingestion from a contaminated drinking water source to a site in the human lower bowel or when bacterial cells infecting the urinary tract are exposed to fluctuating urine osmolalities. Under experimental conditions, a hyperosmotic stress occurs when *E. coli* cells cultured in dilute minimal medium are exposed to me-

dia containing 0.3 to 0.5M NaCl. The osmotic shock stimulates the synthesis of about 40 proteins and results in decreased synthesis of about 30 other proteins as the cells adjust to the new external environment (17). The induced proteins are involved in maintaining internal osmotic conditions that are optimal for cell functions. Among the proteins regulating osmoadaptation is σ^S (Kat^F), an alternate sigma factor of RNA polymerase involved in transcriptional control of stress response proteins induced during the transition to stationary growth phase (60). Osmoadaptation is a widely distributed, if not universal, stress response of Eubacteria, Archaea, and Eukarya (33, 45, 62).

Why is osmoadaptation important for bacterial growth? Bacterial cells are pressurized. They have an internal turgor pressure, which is the outward pressure exerted by the cytoplasmic membrane on the cell wall (17). The pressure is an estimated 3 to 10 $\times$ 10^5 Pa (3.5 atm for *E. coli*) for gram-negative bacteria. Against gram-positive cell walls, the pressure is 2 to 10 times greater (99). Turgor pressure inside *Bacillus subtilis* cells is reported by Kempf and Bremer to be 10 times greater than the air pressure of an automobile tire (45).

Turgor pressure arises both from the ability of bacterial cells to actively transport and concentrate environmental nutrients in their cytoplasm and from the elastic peptidoglycan cell wall surrounding the cytoplasm. The cell cytoplasm exists at a higher osmolality than the environment. A strong cell wall was a necessary adaptation in bacterial evolution to maintain this elevated osmolality and prevent cell swelling and bursting due to the influx of water (49). When external osmotic conditions change, bacteria maintain turgor pressure by adjusting the concentrations of osmotically active components in their cytoplasm. Adjustments are made by regulating the uptake and synthesis of osmotic solutes, both ions and compatible solutes (Fig. 3, Table 2).

Within seconds after hyperosmotic stress, or "osmotic upshock," water in the cytoplasm of *E. coli* cells passively flows outward through

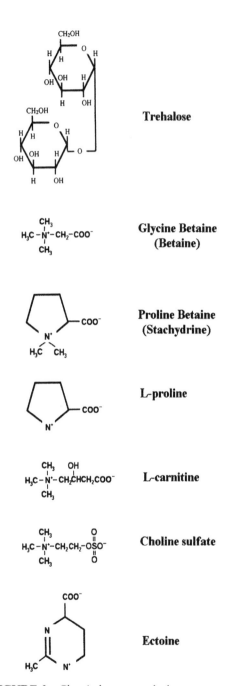

Trehalose

**Glycine Betaine
(Betaine)**

**Proline Betaine
(Stachydrine)**

L-proline

L-carnitine

Choline sulfate

Ectoine

FIGURE 3 Chemical compounds that serve as osmoprotectants for *E. coli.*

membrane channels or "aquaporins" (8), raising the intracellular osmolality by increasing the cytoplasmic solute concentrations (45). Additionally, cell transport mechanisms (Table 2) are stimulated to take up K^+ from the environment.

Potassium accumulation is a rapid osmotic balancing response and is a logical signal for turning on slower osmoadaptation responses. However, the accumulation and persistence of high levels of intracellular K^+ can inhibit cell growth in several ways. To maintain the electrical gradient across the cell membrane, protons (H^+) are exchanged for K^+ ions entering the cell. The outflow of H^+ increases the intracellular pH, affecting the electrochemical potential across the cytoplasmic membrane (17). Although K^+ is an essential nutrient for bacteria, high concentrations impair protein activities, denaturing macromolecules (45) and tying up free water essential for nucleic acid-protein interactions (84). To counteract high concentrations of K^+, *E. coli* cells accumulate, by synthesis (Table 1), a second ionic solute, glutamate. This acidic amino acid is an obligatory counterion to K^+ and a buffer for the cell against rising pH due to H^+ efflux (17, 21, 103).

In a second phase of responding to osmotic upshock, which is essential at external osmolalities >0.25 OsM, the bacteria increase their intracellular levels of compatible solutes to raise intracellular osmolality. Compatible solutes (Fig. 3) are highly soluble compounds, are not charged at physiological pH, and can reach high (molar) concentrations inside the cell without affecting vital biochemical processes (hence the term compatible). Trehalose, a disaccharide containing two glucose molecules, can account for as much as 20% of the osmolality of *E. coli* cells cultured in high-osmolality media (17). The osmoprotectants found in *E. coli*, as well as related chemical compounds, are widely distributed among prokaryotic and eukaryotic cells (33, 104). Different osmoprotectants are available in plant tissues ingested by the host, in animal urine and tissues, and as metabolic products of

TABLE 2 *E. coli* proteins directly involved in the accumulation of osmoprotectants

Osmoprotectant	Related protein and function
K$^+$	Trk, Kdp, Kup systems (K$^+$ transport; KdpD and E form a two-component system with capabilities as a turgor sensor[a])
Glutamate	GltB/D, GlnE (enzymes for alternative routes of glutamate synthesis)
Trehalose	OtsA, OtsB (trehalose-6-PO$_4$ synthase and trehalose-6-PO$_4$ phosphatase, enzymes for trehalose synthesis)
	TreA, TreF (trehalase enzymes for metabolism and recycling trehalose)
Glycine betaine	BetT (cytoplasmic membrane protein for choline transport)
	BetA, BetB (enzymes for converting choline to glycine betaine)
Proline, proline betaine, choline, carnitine, glycine betaine, others	ProP, ProU system (cytoplasmic membrane proteins for transport of various compatible solutes; ProU is induced up to 100-fold in response to increases in medium osmolality)

[a] Wood has recently provided a thorough review of bacterial osmosensors (101).

other bacteria and thus are generally available for bacteria that are colonizing a host.

In 30 min to several hours following hyperosmotic stress, compatible solute levels increase in *E. coli* cells. Proline and betaine, when available, are taken up from the environment or are synthesized. Trehalose used as an osmoprotectant must be synthesized de novo by *E. coli* cells (Table 1) (93). Accumulation of trehalose provides the bacterium with osmoprotection sufficient to withstand osmotic stress generated by adding 0.45 M NaCl to a minimum culture broth (93).

In the final events of osmoadaptation to hyperosmotic stress, cells continue to accumulate proline and betaine by transport or by metabolism; concentrations of cytoplasmic K$^+$, glutamate, and trehalose decrease, and there is a large increase in free cytoplasmic water (83). Optimum growth rates of the bacteria are restored.

Despite extensive knowledge of the biochemical events, metabolic pathways, genes, and chemicals involved in osmoadaptation, understanding of osmoadaptive behaviors of bacteria in their natural environments and the significance of osmoadaptation for diverse host associated species is limited. This is unfortunate, since the importance of osmoadaptation in bacterial ecology is reflected in its universal adoption by life forms. Further, it has been proposed that chemical analogs of osmoprotectants, specifically betaine, could be toxic for bacteria and thus useful as antibiotics against certain infections (10, 79).

There is evidence that osmoadaptation plays a role in bacterial virulence, both through a linkage to expression of virulence genes and by affecting bacterial growth in vivo. Galan and Curtiss reported an eightfold increase in transcription of *invA*, a gene essential for *Salmonella enterica* serovar Typhimurium invasion of tissue culture cells, after osmotic stress of bacterial cells (32). They also determined that wild-type bacterial cells became more invasive in tissue culture assays after they were cultured in broth to which 0.3 M NaCl had been added to increase the osmolality. Recent studies by Culham and colleagues provided the first direct evidence that osmoadaptation is important in bacterial colonization of a host (18). These investigators deleted the *proP* gene (Table 2) from an *E. coli* strain that had been isolated from a case of human pyelonephritis. Compared to its isogenic parent strain, the *proP* mutant grew more slowly in human urine samples (0.8 to 0.95 OsM) and displayed a 100-fold reduced ability to colonize the urinary bladder of mice.

ACID TOLERANCE AND RESISTANCE

The human stomach is an early defense against intestinal disease. Hydrochloric acid in amounts of 1 to 2 liters per day is secreted into the stomach contents by gastric glands. Bacteria, ingested or carried downstream from the mouth and esophagus, are mixed with the gastric juices of pH 1.0 to 2.0 and most are killed. (The pH of the fasting stomach is 2.0.) Human pathogens, on their way to colonizing the small intestine and colon, must first survive a passage of up to 2 h through the stomach (34). The more acid insensitive the bacteria entering the stomach are, the more likely they will survive passage. If gastric acidity is reduced by antacids, medical therapy, or disease, bacteria in the stomach increase in number (22, 106). In the case of acid-sensitive intestinal pathogens, such as *Vibrio cholerae*, the number of bacteria necessary to cause disease can be reduced 10,000-fold when gastric pH values are elevated by $NaHCO_3$ treatment (9).

Enteric species have evolved protective mechanisms against acid killing that improve their success in colonizing the intestinal tract (41). *E. coli* and *Salmonella* cells can become acid tolerant (a term used to indicate that they have an inducible insensitivity to acid). A nonlethal exposure to mild acid conditions will transiently protect these species against lethal, more severe acid exposures. The mild acid shock activates synthesis of a network of more than 50 proteins that serve to maintain intracellular pH and protect cytoplasmic macromolecules (31). Several acid survival systems induced or selected by low pH have been identified for *E. coli* (44, 58).

Acid tolerance could play a role in foodborne diseases that involve fecal–oral transmission between animals and humans. In a recent study by Diez-Gonzalez and coworkers, the number of acid tolerant (nonpathogenic) *E. coli* cells in the colons of cattle were counted (20). When the cattle were fed mostly grain (>80% of diet), the pH of the colon contents averaged 5.9 and the population levels of acid-tolerant *E. coli* (i.e., those surviving for 1 h in culture broth at pH 2.0) reached 2.5×10^5 cells/g of contents. The colon contents of grass-fed cattle had a pH of 7.1 and did not contain detectable acid-tolerant *E. coli* cells. An explanation for these results is that a grain diet raises the level of carbohydrate (starch) fermentation in the colon, both stimulating *E. coli* population growth and raising concentrations of volatile fatty acids that induce acid tolerance in the *E. coli* cells.

E. coli O157:H7, a toxin producing enterohemorrhagic pathogen of humans, resides in the cattle intestinal tract and can be transmitted in feces-contaminated, undercooked meat. Isolates of that strain become acid tolerant after growth in media with pH values of 4.8 to 5.0 (20, 59). Diez-Gonzalez and coworkers suggest that *E. coli* O157:H7 cells exiting the colons (pH of 5.5 to 5.9) of grain-fed cattle may have an induced tolerance for acid and thus an enhanced ability to survive passage through the human stomach and cause disease. The feeding of hay may raise the colon pH and impair the transmission of this pathogen. Additional studies are needed to examine various aspects of these observations and to test this interesting, but controversial, hypothesis. In contrast to the findings of Diez-Gonzalez et al. (20), Hovde et al. recently reported, for grain-fed cattle, only slight increases (1.7-fold) in the numbers of acid-tolerant *E. coli* and decreased shedding of *E. coli* O157:H7 cells from animals on a grain diet (43). Transmission is likely a multifactor process involving numerous bacterial traits.

As the mechanisms and genes involved in acid tolerance become better known, its significance in virulence will become more amenable to investigation through use of strains with specific mutations. Exposure to acidic environments outside the host or during passage through the stomach may signal the induction of bacterial virulence mechanisms. Regulation by pH is a feature of a variety of bacterial virulence-associated genes (41). O'Driscoll and coworkers isolated spontaneous *Listeria monocytogenes* mutant strains with increased acid tolerance (75). The mutant strains were more virulent than the wild-type strain in an experimental mouse model, reaching higher population levels in the spleens of the animals. Both effects are likely due to increased survival of the acid-tolerant mutants within macrophages.

If awards were given for host-associated bacteria most likely to succeed under acid stress, the honor would arguably belong to *Helicobacter* species. Members of that genus are naturally acid resistant in their native environments, not only surviving but thriving within the midst of acidity. *H. pylori* and a handful of other *Helicobacter* species, including *H. mustelae* (ferrets) and *H. felis* (cats), inhabit the stomach.

H. pylori is one of the most common bacterial pathogens in the world, colonizing up to 50% of human populations (16). Damage to gastric and duodenal mucosal tissues by *H. pylori* cells can result in gastric and duodenal ulcers and has been implicated as an early event in the development of some gastric cancers (56, 64, 73, 77).

Various cell traits, among them adhesin, cytotoxin, neutrophil activation factor, and a type IV secretion system, are important for *H. pylori* to colonize and damage the gastric mucosa (16, 56). Two characteristics specifically enable *H. pylori* cells to avoid and tolerate acid exposure in the stomach—flagellar-driven motility and the enzyme urease.

H. pylori cells favor mucosal surfaces of the nonacid secreting, antral portion of the stomach, near the pyloric sphincter. Mucus and bicarbonate secreted by the antral epithelium are a protective buffer zone for the epithelial cells. As a result, a pH gradient ranging from 3.5 in the stomach lumen to 5.4 at the antral mucosal surface is created (82). This mucus-antacid blanket also provides physical support and acid protection for *H. pylori* cells. The active motility of *Helicobacter* cells in mucus, perhaps directed by chemotaxis for urea and bicarbonate in mucosal secretions (72), enables bacteria to move into that buffer zone and colonize underlying epithelial cell surfaces.

Urea is a major end product of nitrogen metabolism by mammals. Diffusing from the blood through mucosal tissues and into the gastrointestinal tract, urea concentrations in the stomach range from 1 to 14 mM (48, and B. J. Marshall and S. R. Langton, Letter, *Lancet* i:965–966, 1986). By metabolizing urea, *H. pylori* is naturally acid resistant. *H. pylori* cells have high levels of urease activity. The enzyme represents 5 to 10% of the total cell protein of cultured cells and has a high affinity ($K_m = 0.5$ to 0.8) for urea (2, 23, 66, 67). The purified urease is a 300 to 600 kDa, Ni^{2+}-containing enzyme with 62- and 30-kDa subunits (23, 28, 66). The enzyme is both in the cytoplasm and associated with the surfaces of cultured bacteria (24, 42, 81) and of bacteria that colonize humans (25). As much as 50 to 90% of *H. pylori* urease activity has been reported to be surface-associated (24, 81), and two mechanisms have been proposed for getting urease out of *H. pylori* cells (81, 97).

The hydrolysis of urea by urease produces ammonia, which reacts with water to form NH_4OH, a strong base (Fig. 4). An early hypothesis had the surface-associated urease surrounding *H. pylori* cells with a cloud of acid-neutralizing ammonia (35). When urea is added to *H. pylori* cell suspensions at low pH values (down to 2.0), bacterial survival is enhanced 10^3–10^6-fold compared with that in suspensions without urea, and survival is associated with rapid alkalinization of the suspensions by urease (13, 30, 61). *H. pylori* mutant strains deficient in urease activity are more susceptible to acid killing (13, 80, 95).

Several observations suggest urease is a virulence-associated property essential for colonization of a host by *H. pylori* and other *Helicobacter* species. If urease activity is impaired biochemically (acetohydroxamic acid) or genetically, the ability of human strains of *H.*

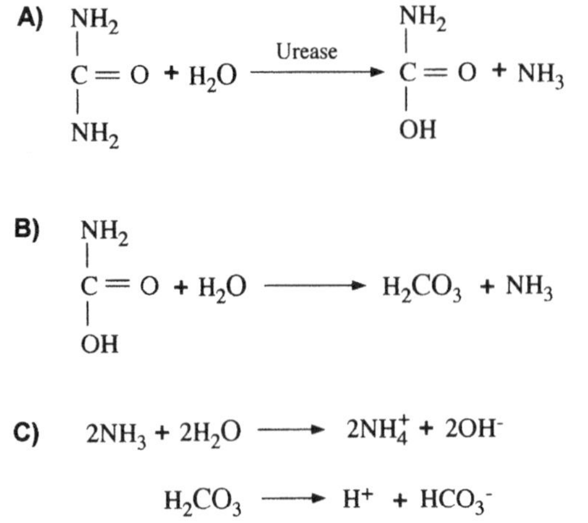

A)

$$\underset{NH_2}{\overset{NH_2}{C}}{=}O + H_2O \xrightarrow{\text{Urease}} \underset{OH}{\overset{NH_2}{C}}{=}O + NH_3$$

B)

$$\underset{OH}{\overset{NH_2}{C}}{=}O + H_2O \longrightarrow H_2CO_3 + NH_3$$

C)

$$2NH_3 + 2H_2O \longrightarrow 2NH_4^+ + 2OH^-$$

$$H_2CO_3 \longrightarrow H^+ + HCO_3^-$$

FIGURE 4 Hydrolysis of urea by urease (urea aminohydrolase, EC 3.5.1.5). Urease hydrolyzes urea to ammonia and carbamate. Carbamate spontaneously converts to NH_3 and carbonic acid. NH_3 equilibrates with water to form ammonium hydroxide, which generates hydroxyls (raises pH). Urease activity is the basis for a noninvasive test for *H. pylori* infections. A patient ingests ^{14}C-labeled urea. If the patient's stomach is colonized by *H. pylori*, the strong ureolytic activity of the bacteria releases $^{14}CO_2$, which is breathed out and can be measured by radioactivity detectors.

pylori cells to establish colonizing populations in experimental animals is reduced (26, 27, 95). Studies of *H. mustelae*, a natural gastric pathogen of ferrets, support a role for urease in that infection (1, 91). A urease-negative mutant strain of *H. mustelae* was constructed by transposon mutagenesis and allelic exchange, inactivating the *ureB* gene of the mutant strain (91). In phosphate-buffered saline at pH 7.4, cells of both the mutant strain and the wild-type parent strain survived equally well, either with or without the addition of 5 mM urea. In buffer at pH 1.5 with urea added, the nonureolytic mutant cell populations were 10^5 to 10^6 times more sensitive to acid killing. In animal challenge studies (1), two ferrets given the wild-type strain both were colonized at 6 months and exhibited inflammation of gastric tissues after inoculation. By contrast, none of four ferrets given the isogenic mutant strain were detectably colonized at 6 months after inoculation based on culture or histopathology.

Urease activity may also contribute to *H. pylori* colonization in ways not directly related to acid protection. Indirectly, through urease protection from acid, *H. pylori* cells may have freedom from competition by acid-sensitive bacteria and have protection from host-immune mechanisms (e.g., phagocytosis) that are impaired by low pH. Alternatively, the enzyme could provide nitrogen for *H. pylori* cells, as it does for other gastrointestinal bacteria (15). Concentrated external urease might generate toxic levels of NH_3^+ and OH^-, which could impair mucosal cell physiology and damage gastric tissues (24, 89).

An alternative role for urease, in addition to acid resistance, has been supported by the observations of Eaton and Krakowka (27). These investigators found that a urease-deficient mutant strain of *H. pylori*, unlike its ureolytic parent strain, could not colonize the stomachs (pH 2) of gnotobiotic piglets, consistent with an acid-sensitive phenotype of the strain. The mutant strain poorly colonized stomachs of pigs that had been treated with drugs to make the stomach achlorhydric (pH 7), suggesting urease may serve additional functions for *H. pylori* cells postcolonization. The availability of urease-negative isogenic mutants will likely lead to more in-depth studies of the physiological effects of urease on *H. pylori* cell growth.

Immunological evidence supports a virulence-associated role of urease and also demonstrates the efficacy of immunization for preventing *H. pylori* colonization. In a recent vaccine trial (71), mice given oral doses of recombinant urease plus mucosal adjuvant were protected from colonization by a challenge inoculum of *H. felis*. Mice were inoculated with bacteria at different times after immunization. A urease assay was used to assess stomach colonization by the bacteria. Protection (no detectable urease activity) ranged from 63 to 90% for the different groups up to a year after immunization. In a separate study (57), the therapeutic efficacy of vaccination was tested by using rhesus monkeys, a species which, like humans, is naturally colonized by *H. pylori*. After 6 doses and 8 weeks, there was no measurable effect on colonization, as assessed by biopsy cultures of the bacterium. The monkeys were treated with antibiotics to eliminate *H. pylori*, were given a seventh immunization, and were orally challenged with bacterial cells 16 weeks later. Based on necropsy histopathological examinations made 23 weeks later, colonization by the pathogen was significantly reduced in the urease-vaccinated group. For the treatment of *H. pylori*-induced gastric disorders, urease inhibitors would be most effective when used prophylactically to prevent bacteria from colonizing the stomach. Urease inhibition is less effective for eliminating *H. pylori* cells from their protective microhabitat over the mucosal epithelium.

CONCLUSIONS

Inflicting damage on host tissues and evading immune surveillance are not only events, but defining properties in the life of a pathogenic bacterium. Equally important are characteristics essential for a pathogen to grow within a host and to access new hosts. Coaggregation, osmoadaptation, and acid tolerance and resistance are not essential for routine growth of bacteria in culture. Nevertheless, these properties enable certain bacteria to colonize host microhabitats, and for pathogenic bacteria, they are virulence-associated traits. A broader awareness of the mechanisms that enable pathogenic bacteria to contend with or take advantage of environmental components will likely lead (as in the use of the *H. pylori* urease for vaccination) to novel strategies for disease prevention and control.

REFERENCES

1. **Andrutis, K. A., J. G. Fox, D. B. Schauer, R. P. Marini, J. C. Murphy, L. L. Yan, and J. V. Solnick.** 1995. Inability of an isogenic urease-negative mutant strain of *Helicobacter mustelae* to colonize the ferret stomach. *Infect. Immun.* **63:** 3722–3725.
2. **Bauerfeind, P., R. Garner, B. E. Dunn, and H. L. Mobley.** 1997. Synthesis and activity of *Helicobacter pylori* urease and catalase at low pH. *Gut* **40:**25–30.
3. **Blake, M., O. D. Rotstein, M. Llano, M. J. Girotti, and G. Reid.** 1989. Aggregation by *fra-*

gilis and non-*fragilis Bacteroides* strains in vitro. *J. Med. Microbiol.* **28:**9–14.
4. **Bossier, P., and W. Verstraete.** 1996. *Comamonas testosteroni* colony phenotype influences exopolysaccharide production and coaggregation with yeast cells. *Appl. Environ. Microbiol.* **62:** 2687–2691.
5. **Bradshaw, D. J., P. D. Marsh, G. K. Watson, and C. Allison.** 1998. Role of *Fusobacterium nucleatum* and coaggregation anaerobe survival in planktonic and biofilm oral microbial communities during aeration. *Infect. Immun.* **66:** 4729–4732.
6. **Breznak, J. A.** 1973. Biology of nonpathogenic, host-associated spirochetes. *CRC Crit. Rev. Microbiol.* **2:**457–489.
7. **Breznak, J. A., and H. S. Pankratz.** 1977. In situ morphology of the gut microbiota of wood-eating termites [*Reticulitermes flavipes* (Kollar) and *Coptotermes formosanus* Shiraki]. *Appl. Environ. Microbiol.* **33:**406–426.
8. **Calamita, G., W. R. Bishai, G. M. Preston, W. B. Guggino, and P. Agre.** 1995. Molecular cloning and characterization of AqpZ, a water channel from *Escherichia coli*. *J. Biol. Chem.* **270:**29063–29066.
9. **Cash, R. A., S. I. Music, J. P. Libonati, M. J. J. Snyder, R. P. Wenzel, and R. B. Hornick.** 1974. Response of man to infection with *Vibrio cholerae*. I. Clinical, serologic, and bacteriologic responses to a known inoculum. *J. Infect. Dis.* **129:**45–52.
10. **Chambers, S. T., C. M. Kunin, D. Miller, and A. Hamada.** 1987. Dimethylthetin can substitute for glycine betaine as an osmoprotectant molecule for *Escherichia coli*. *J. Bacteriol.* **169:** 4845–4847.
11. **Choi, B. K., B. J. Paster, F. E. Dewhirst, and U. B. Gobel.** 1994. Diversity of cultivable and uncultivable oral spirochetes from a patient with severe destructive periodontitis. *Infect. Immun.* **62:**1889–1895.
12. **Christensen, B. B., C. Sternberg, J. B. Andersen, L. Eberl, S. Moller, M. Givskov, and S. Molin.** 1998. Establishment of new genetic traits in a microbial biofilm community. *Appl. Environ. Microbiol.* **64:**2247–2255.
13. **Clyne, M., A. Labigne, and B. Drumm.** 1995. *Helicobacter pylori* requires an acidic environment to survive in the presence of urea. *Infect. Immun.* **63:**1669–1673.
14. **Costerton, J. W., Z. Lewandowski, D. E. Caldwell, D. R. Korber, and H. M. Lappin-Scott.** 1995. Microbial biofilms. *Annu. Rev. Microbiol.* **49:**711–745.
15. **Cotta, M. A., and J. B. Russell.** 1997. Digestion of nitrogen in the rumen: a model for metabolism of nitrogen compounds in gastro-

intestinal environments, p. 380–423. *In* R. I. Mackie and B. A. White (ed.), *Gastrointestinal Microbiology*, vol. I. *Gastrointestinal Ecosystems and Fermentations*. Chapman & Hall, Ltd., New York.

16. **Covacci, A., J. L. Telford, G. Del Giudice, J. Parsonnet, and R. Rappuoli.** 1999. *Helicobacter pylori* virulence and genetic geography. *Science* **284:**1328–1333.

17. **Csonka, L. N., and W. Epstein.** 1996. Osmoregulation, p. 1210–1223. *In* F. C. Neidhardt (ed.), *Escherichia coli and Salmonella: Cellular and Molecular Biology*, 2nd ed., vol. 1. ASM Press, Washington, D.C.

18. **Culham, D. E., C. Dalgado, C. L. Gyles, D. Mamelak, S. Maclellan, and J. M. Wood.** 1998. Osmoregulatory transporter ProP influences colonization of the urinary tract by *Escherichia coli*. *Microbiology* (*UK*) **144:**91–102.

19. **Dahle, U. R., I. Olsen, L. Tronstad, and D. A. Caugant.** 1995. Population genetic analysis of oral treponemes by multilocus enzyme electrophoresis. *Oral Microbiol. Immunol.* **10:**265–270.

20. **Diez-Gonzalez, F., T. R. Callaway, M. G. Kizoulis, and J. B. Russell.** 1998. Grain feeding and the dissemination of acid-resistant *Escherichia coli* from cattle. *Science* **281:**1666–1668.

21. **Dinnbier, U., E. Limpinsel, R. Schmid, and E. P. Bakker.** 1988. Transient accumulation of potassium glutamate and its replacement by trehalose during adaptation of growing cells of *Escherichia coli* K-12 to elevated sodium chloride concentrations. *Arch. Microbiol.* **150:**348–357.

22. **Drasar, B. S., M. Shiner, and G. M. McLeod.** 1969. Studies on the intestinal flora. I. The bacterial flora of the gastrointestinal tract in healthy and achlorhydric persons. *Gastroenterology* **56:**71–79.

23. **Dunn, B. E., G. P. Campbell, G. I. Perez-Perez, and M. J. Blaser.** 1990. Purification and characterization of urease from *Helicobacter pylori*. *J. Biol. Chem.* **265:**9464–9469.

24. **Dunn, B. E., C. C. Sung, N. S. Taylor, and J. G. Fox.** 1991. Purification and characterization of *Helicobacter mustelae* urease. *Infect. Immun.* **59:**3343–3345.

25. **Dunn, B. E., N. B. Vakil, B. G. Schneider, M. M. Miller, J. B. Zitzer, T. Peutz, and S. H. Phadnis.** 1997. Localization of *Helicobacter pylori* urease and heat shock protein in human gastric biopsies. *Infect. Immun.* **65:**1181–1188.

26. **Eaton, K. A., C. L. Brooks, D. R. Morgan, and S. Krakowka.** 1991. Essential role of urease in pathogenesis of gastritis induced by *Helicobacter pylori* in gnotobiotic piglets. *Infect. Immun.* **59:**2470–2475.

27. **Eaton, K. A., and S. Krakowka.** 1994. Effect of gastric pH on urease-dependent colonization of gnotobiotic piglets by *Helicobacter pylori*. *Infect. Immun.* **62:**3604–3607.

28. **Evans, D. J., Jr., D. G. Evans, S. S. Kirkpatrick, and D. Y. Graham.** 1991. Characterization of the *Helicobacter pylori* urease and purification of its subunits. *Microb. Pathog.* **10:**15–26.

29. **Fenno, J. C., and B. C. Mcbride.** 1998. Virulence factors of oral treponemes. *Anaerobe* **4:**1–17.

30. **Ferrero, R. L., and A. Lee.** 1991. The importance of urease in acid protection for the gastric-colonising bacteria *Helicobacter pylori* and *Helicobacter felis* sp. nov. *Microb. Ecol. Health Dis.* **4:**121–134.

31. **Foster, J. W., and M. Moreno.** 1999. Inducible acid tolerance mechanisms in enteric bacteria, p. 55–69. *In Bacterial Responses to pH. Novartis Foundation Symposium 221*. Wiley, Chichester.

32. **Galan, J. E., and R. D. Curtiss.** 1990. Expression of *Salmonella typhimurium* genes required for invasion is regulated by changes in DNA supercoiling. *Infect. Immun.* **58:**1879–1885.

33. **Galinski, E. A., and H. G. Trüper.** 1994. Microbial behaviour in salt-stressed ecosystems. *FEMS Microbiol. Rev.* **15:**95–108.

34. **Giannella, R. A., S. A. Broitman, and N. Zamcheck.** 1972. Gastric acid barrier to ingested microorganisms in man: studies in vivo and in vitro. *Gut* **13:**251–256.

35. **Goodwin, C. S., J. A. Armstrong, and B. J. Marshall.** 1986. *Campylobacter pyloridis*, gastritis, and peptic ulceration. *J. Clin. Pathol.* **39:**353–365.

36. **Grenier, D.** 1996. Antagonistic effect of oral bacteria towards *Treponema denticola*. *J. Clin. Microbiol.* **34:**1249–1252.

37. **Grenier, D.** 1992. Demonstration of a bimodal coaggregation reaction between *Porphyromonas gingivalis* and *Treponema denticola*. *Oral Microbiol. Immunol.* **7:**280–284.

38. **Grenier, D., and D. Mayrand.** 1986. Nutritional relationships between oral bacteria. *Infect. Immun.* **53:**616–620.

39. **Haffajee, A. D., M. A. Cugini, A. Tanner, R. P. Pollack, C. Smith, R. L. Kent, Jr., and S. S. Socransky.** 1998. Subgingival microbiota in healthy, well-maintained elder and periodontitis subjects. *J. Clin. Periodontol.* **25:**346–353.

40. **Haffajee, A. D., and S. S. Socransky.** 1994. Microbial etiological agents of destructive periodontal diseases. *Periodontal. 2000* **5:**78–111.

41. **Hall, H. K., K. L. Karem, and J. W. Foster.** 1995. Molecular responses of microbes to envi-

ronmental pH stress. *Adv. Microb. Physiol.* **37:** 229–272.

42. **Hawtin, P. R., A. R. Stacey, and D. G. Newell.** 1990. Investigation of the structure and localization of the urease of *Helicobacter pylori* using monoclonal antibodies. *J. Gen. Microbiol.* **136:** 1995–2000.

43. **Hovde, C. J., P. R. Austin, K. A. Cloud, C. J. Williams, and C. W. Hunt.** 1999. Effect of cattle diet on *Escherichia coli* O157:H7 acid resistance. *Appl. Environ. Microbiol.* **65:**3233–3235.

44. **Jordan, K. N., L. Oxford, and C. P. O'Byrne.** 1999. Survival of low-pH stress by *Escherichia coli* O157:H7: correlation between alterations in the cell envelope and increased acid tolerance. *Appl. Environ. Microbiol.* **65:**3048–3055.

45. **Kempf, B., and E. Bremer.** 1998. Uptake and synthesis of compatible solutes as microbial stress responses to high-osmolality environments. *Arch. Microbiol.* **170:**319–330.

46. **Kesavalu, L., S. C. Holt, and J. L. Ebersole.** 1998. Virulence of a polymicrobic complex, *Treponema denticola* and *Porphyromonas gingivalis*, in a murine model. *Oral Microbiol. Immunol.* **13:** 373–377.

47. **Kigure, T., A. Saito, K. Seida, S. Yamada, K. Ishihara, and K. Okuda.** 1995. Distribution of *Porphyromonas gingivalis* and *Treponema denticola* in human subgingival plaque at different periodontal pocket depths examined by immunohistochemical methods. *J. Periodontal Res.* **30:**332–341.

48. **Kim, H., C. Park, W. I. Jang, K. H. Lee, S. O. Kwon, S. S. Robey-Cafferty, J. Y. Ro, and Y. B. Lee.** 1990. The gastric juice urea and ammonia levels in patients with *Campylobacter pylori*. *Am. J. Clin. Pathol.* **94:**187–191.

49. **Koch, A. L.** 1998. How did bacteria come to be? *Adv. Microb. Physiol.* **40:**353–399.

50. **Kolenbrander, P. E.** 1989. Surface recognition among oral bacteria: multigeneric coaggregations and their mediators. *CRC Crit. Rev. Microbiol.* **17:**137–159.

51. **Kolenbrander, P. E., R. N. Andersen, and L. V. Moore.** 1989. Coaggregation of *Fusobacterium nucleatum, Selenomonas flueggei, Selenomonas infelix, Selenomonas noxia*, and *Selenomonas sputigena* with strains from 11 genera of oral bacteria. *Infect. Immun.* **57:**3194–3203.

52. **Kolenbrander, P. E., and J. London.** 1993. Adhere today, here tomorrow: oral bacterial adherence. *J. Bacteriol.* **175:**3247–3252.

53. **Kolenbrander, P. E., and J. London.** 1992. Ecological significance of coaggregation among oral bacteria. *Adv. Microb. Ecol.* **12:**183–217.

54. **Kolenbrander, P. E., K. D. Parrish, R. N. Andersen, and E. P. Greenberg.** 1995. Intergeneric coaggregation of oral *Treponema* spp. with *Fusobacterium* spp. and intrageneric coaggregation among *Fusobacterium* spp. *Infect. Immun.* **63:**4584–4588.

55. **Lancy, P., Jr., B. Appelbaum, S. C. Holt, and B. Rosan.** 1980. Quantitative in vitro assay for "corncob" formation. *Infect. Immun.* **29:**663–670.

56. **Lee, A., J. Fox, and S. Hazell.** 1993. Pathogenicity of *Helicobacter pylori*: a perspective. *Infect. Immun.* **61:**1601–1610.

57. **Lee, C. K., K. Soike, J. Hill, K. Georgakopoulos, T. Tibbitts, J. Ingrassia, H. Gray, J. Boden, H. Kleanthous, P. Giannasca, T. Ermak, R. Weltzin, J. Blanchard, and T. P. Monath.** 1999. Immunization with recombinant *Helicobacter pylori* urease decreases colonization levels following experimental infection of rhesus monkeys. *Vaccine* **17.**1493–1505.

58. **Lin, J., I. S. Lee, J. Frey, J. L. Slonczewski, and J. W. Foster.** 1995. Comparative analysis of extreme acid survival in *Salmonella typhimurium, Shigella flexneri*, and *Escherichia coli*. *J. Bacteriol.* **177:**4097–4104.

59. **Lin, J., M. P. Smith, K. C. Chapin, H. S. Baik, G. N. Bennett, and J. W. Foster.** 1996. Mechanisms of acid resistance in enterohemorrhagic *Escherichia coli*. *Appl. Environ. Microbiol.* **62:** 3094–3100.

60. **Loewen, P. C., and R. Hengge-Aronis.** 1994. The role of the sigma factor sigma S (KatF) in bacterial global regulation. *Annu. Rev. Microbiol.* **48:**53–80.

61. **Marshall, B. J., L. J. Barrett, C. Prakash, R. W. McCallum, and R. L. Guerrant.** 1990. Urea protects *Helicobacter (Campylobacter) pylori* from the bactericidal effect of acid. *Gastroenterology* **99:**697–702.

62. **Martin, D. D., R. A. Ciulla, and M. F. Roberts.** 1999. Osmoadaptation in archaea. *Appl. Environ. Microbiol.* **65:**1815–1825.

63. **McBride, B. C., and J. S. van der Hoeven.** 1981. Role of interbacterial adherence in colonization of the oral cavities of gnotobiotic rats infected with *Streptococcus mutans* and *Veillonella alcalescens*. *Infect. Immun.* **33:**467–472.

64. **Mera, S. L.** 1995. Peptic ulcers and gastric cancer. *Br. J. Biomed. Sci.* **52:**271–281.

65. **Mikx, F. H., and J. S. van der Hoeven.** 1975. Symbiosis of *Streptococcus mutans* and *Veillonella alcalescens* in mixed continuous cultures. *Arch. Oral Biol.* **20:**407–410.

66. **Mobley, H. L., M. J. Cortesia, L. E. Rosenthal, and B. D. Jones.** 1988. Characterization

of urease from *Campylobacter pylori*. *J. Clin. Microbiol.* **26**:831–836.

67. **Mobley, H. L. T., L.-T. Hu, and P. A. Foxall.** 1991. *Helicobacter pylori* urease: properties and role in pathogenesis. *Scand. J. Gastroenterol.* **S187:** 39–46.

68. **Moter, A., C. Hoenig, B. K. Choi, B. Riep, and U. B. Gobel.** 1998. Molecular epidemiology of oral treponemes associated with periodontal disease. *J. Clin. Microbiol.* **36**:1399–1403.

69. **Mouton, C., H. Reynolds, and R. J. Genco.** 1977. Combined micromanipulation, culture, and immunofluorescent techniques for isolation of the coccal organisms comprising the "corn cob" configuration of human dental plaque. *J. Biol. Buccale.* **5**:321–332.

70. **Mouton, C., H. S. Reynolds, E. A. Gasiecki, and R. J. Genco.** 1979. In vitro adhesion of tufted oral streptococci to *Bacterionema matruchotii*. *Curr. Microbiol.* **3**:181–186.

71. **Myers, G. A., T. H. Ermak, K. Georgakopoulos, T. Tibbitts, J. Ingrassia, H. Gray, H. Kleanthous, C. K. Lee, and T. P. Monath.** 1999. Oral immunization with recombinant *Helicobacter pylori* urease confers long-lasting immunity against *Helicobacter felis* infection. *Vaccine* **17:** 1394–1403.

72. **Nakamura, H., H. Yoshiyama, H. Takeuchi, T. Mizote, K. Okita, and T. Nakazawa.** 1998. Urease plays an important role in the chemotactic motility of *Helicobacter pylori* in a viscous environment. *Infect. Immun.* **66**:4832–4837.

73. **National Institutes of Health Consensus Conference.** 1994. *Helicobacter pylori* in peptic ulcer disease. *JAMA* **272**:65–69.

74. **Nilius, A. M., S. C. Spencer, and L. G. Simonson.** 1993. Stimulation of in vitro growth of *Treponema denticola* by extracellular growth factors produced by *Porphyromonas gingivalis*. *J. Dent. Res.* **72**:1027–1031.

75. **O'Driscoll, B., C. G. M. Gahan, and C. Hill.** 1996. Adaptive acid tolerance response in *Listeria monocytogenes*: isolation of an acid-tolerant mutant which demonstrates increased virulence. *Appl. Environ. Microbiol.* **62**:1693–1698.

76. **Oliveira, A. A. P., L. M. Farias, J. R. Nicoli, J. E. Costa, and M. A. R. Carvalho.** 1998. Bacteriocin production by *Fusobacterium* isolates recovered from the oral cavity of human subjects with and without periodontal disease and of marmosets. *Res. Microbiol.* **149**:585–594.

77. **Parsonnet, J., G. D. Friedman, D. P. Vandersteen, Y. Chang, J. H. Vogelman, N. Orentreich, and R. K. Sibley.** 1991. *Helicobacter pylori* infection and the risk of gastric carcinoma. *N. Engl. J. Med.* **325**:1127–1131.

78. **Paster, B. J., F. E. Dewhirst, B. C. Coleman, C. N. Lau, and R. L. Ericson.** 1998. Phylogenetic analysis of cultivable oral treponemes from the Smibert collection. *Int. J. Syst. Bacteriol.* **48**:713–722.

79. **Peddie, B. A., J. Wong-She, K. Randall, M. Lever, and S. T. Chambers.** 1998. Osmoprotective properties and accumulation of betaine analogues by *Staphylococcus aureus*. *FEMS Microbiol. Lett.* **160**:25–30.

80. **Perez-Perez, G. I., A. Z. Olivares, T. L. Cover, and M. J. Blaser.** 1992. Characteristics of *Helicobacter pylori* variants selected for urease deficiency. *Infect. Immun.* **60**:3658–3663.

81. **Phadnis, S. H., M. H. Parlow, M. Levy, D. Ilver, C. M. Caulkins, J. B. Connors, and B. E. Dunn.** 1996. Surface localization of *Helicobacter pylori* urease and a heat shock protein homolog requires bacterial autolysis. *Infect. Immun.* **64:** 905–912.

82. **Quigley, E. M., and L. A. Turnberg.** 1987. pH of the microclimate lining human gastric and duodenal mucosa in vivo. Studies in control subjects and in duodenal ulcer patients. *Gastroenterology* **92**:1876–1884.

83. **Record, M. T., Jr., E. S. Courtenay, D. S. Cayley, and H. J. Guttman.** 1998. Responses of *E. coli* to osmotic stress: large changes in amounts of cytoplasmic solutes and water. *Trends Biochem. Sci.* **23**:143–148.

84. **Record, M. T., Jr., E. S. Courtenay, S. Cayley, and H. J. Guttman.** 1998. Biophysical compensation mechanisms buffering *E. coli* protein-nucleic acid interactions against changing environments. *Trends Biochem. Sci.* **23**:190–194.

85. **Reid, G., A. W. Bruce, M. Llano, J. A. McGroarty, and M. Blake.** 1990. Bacterial aggregation in sepsis. *Curr. Microbiol.* **20**:185–190.

86. **Reid, G., J. A. McGroarty, R. Angotti, and R. L. Cook.** 1988. *Lactobacillus* inhibitor production against *Escherichia coli* and coaggregation ability with uropathogens. *Can. J. Microbiol.* **34:** 344–351.

87. **Riviere, G. R., T. A. DeRouen, S. L. Kay, S. P. Avera, V. K. Stouffer, and N. R. Hawkins.** 1997. Association of oral spirochetes from sites of periodontal health with development of periodontitis. *J. Periodontol.* **68**:1210–1214.

88. **Saiman, L., G. Cacalano, and A. Prince.** 1990. *Pseudomonas cepacia* adherence to respiratory epithelial cells is enhanced by *Pseudomonas aeruginosa*. *Infect. Immun.* **58**:2578–2584.

89. **Scott, D. R., D. Weeks, C. Hong, S. Postius, K. Melchers, and G. Sachs.** 1998. The role of internal urease in acid resistance of *Helicobacter pylori*. *Gastroenterology* **114**:58–70.

90. **Simonson, L. G., K. T. McMahon, D. W. Childers, and H. E. Morton.** 1992. Bacterial synergy of *Treponema denticola* and *Porphyromonas gingivalis* in a multinational population. *Oral Microbiol. Immunol.* **7:**111–112.

91. **Solnick, J. V., C. Josenhans, S. Suerbaum, L. S. Tompkins, and A. Labigne.** 1995. Construction and characterization of an isogenic urease-negative mutant of *Helicobacter mustelae*. *Infect. Immun.* **63:**3718–3721.

92. **Stock, J. B., B. Rauch, and S. Roseman.** 1977. Periplasmic space in *Salmonella typhimurium* and *Escherichia coli*. *J. Biol. Chem.* **252:**7850–7861.

93. **Strom, A. R., and I. Kaasen.** 1993. Trehalose metabolism in *Escherichia coli*: stress protection and stress regulation of gene expression. *Mol. Microbiol.* **8:**205–210.

94. **ter Steeg, P. F., and J. S. van der Hoeven.** 1990. Growth stimulation of *Treponema denticola* by periodontal microorganisms. *Antonie Leeuwenhoek* **57:**63–70.

95. **Tsuda, M., M. Karita, M. G. Morshed, K. Okita, and T. Nakazawa.** 1994. A urease-negative mutant of *Helicobacter pylori* constructed by allelic exchange mutagenesis lacks the ability to colonize the nude mouse stomach. *Infect. Immun.* **62:**3586–3589.

96. **Vandevoorde, L., H. Christiaens, and W. Verstraete.** 1992. Prevalence of coaggregation reactions among chicken lactobacilli. *J. Appl. Bacteriol.* **72:**214–219.

97. **Vanet, A., and A. Labigne.** 1998. Evidence for specific secretion rather than autolysis in the release of some *Helicobacter pylori* proteins. *Infect. Immun.* **66:**1023–1027.

98. **Weerkamp, A. H.** 1985. Coaggregation of *Streptococcus salivarius* with gram-negative oral bacteria: mechanism and ecological significance, p. 177–183. *In* S. E. Mergenhagen (ed.), *Molecular Basis of Oral Microbial Adhesion.* ASM Press, Washington, D.C.

99. **Whatmore, A. M., and R. H. Reed.** 1990. Determination of turgor pressure in *Bacillus subtilis*: a possible risk for K+ in turgor regulation. *J. Gen. Microbiol.* **136:**2521–2526.

100. **Whittaker, C. J., C. M. Klier, and P. E. Kolenbrander.** 1996. Mechanisms of adhesion by oral bacteria. *Annu. Rev. Microbiol.* **50:**513–553.

101. **Wood, J. M.** 1999. Osmosensing by bacteria: signals and membrane-based sensors. *Microbiol. Mol. Biol. Rev.* **63:**230–262.

102. **Wrong, O. M., C. J. Edmonds, and V. S. Chadwick.** 1981. *The Large Intestine: Its Role in Mammalian Nutrition and Homeostasis.* John Wiley and Sons, New York, N.Y.

103. **Yan, D., T. P. Ikeda, A. E. Shauger, and S. Kustu.** 1996. Glutamate is required to maintain the steady-state potassium pool in *Salmonella typhimurium*. *Proc. Natl. Acad. Sci. USA* **93:**6527–6531.

104. **Yancey, P. H.** 1994. Compatible and counteracting solutes, p. 81–109. *In* K. Strange (ed.), *Cellular and Molecular Physiology of Cell Volume Regulation.* CRC Press, Inc., Boca Raton, Fla.

105. **Yao, E. S., R. J. Lamont, S. P. Leu, and A. Weinberg.** 1996. Interbacterial binding among strains of pathogenic and commensal oral bacterial species. *Oral Microbiol. Immunol.* **11:**35–41.

106. **Yeomans, N. D., R. W. Brimblecombe, J. Elder, R. V. Heatley, J. J. Misiewicz, T. C. Northfield, and A. Pottage.** 1995. Effects of acid suppression on microbial flora of upper gut. *Dig. Dis. Sci.* **40:**81S–95S.

107. **Young, K. A., R. P. Allaker, J. M. Hardie, and R. A. Whiley.** 1996. Interactions between *Eikenella corrodens* and 'Streptococcus milleri-group' organisms: possible mechanisms of pathogenicity in mixed infections. *Antonie Leeuwenhoek Int. J. Gen. Mol. Microbiol.* **69:**371–373.

RELATIONSHIPS BETWEEN COMMUNITY BEHAVIOR AND PATHOGENESIS IN *PSEUDOMONAS AERUGINOSA*

Matthew R. Parsek and E. P. Greenberg

5

Pseudomonads are known for their broad distribution in the environment. Their ability to colonize a variety of environments is largely due to their nutritional versatility. Many can utilize a number of compounds as sole sources of carbon and energy without the need of vitamins or growth factors. Therefore, pseudomonads play a key role in the mineralization of organic compounds in nature (85). Many are also capable of growing at temperatures that range from 4 to 42°C, and they can persist for weeks in moist environments with little or no nutrients (85). These characteristics partially explain why *Pseudomonas aeruginosa* is an effective opportunistic pathogen. *P. aeruginosa* can be isolated from soil, water, and the skin of healthy human beings. However, *P. aeruginosa* has been implicated in a variety of nosocomial infections, including infections of burn tissue and colonization of indwelling medical devices such as catheters. *P. aeruginosa* is also the primary infectious agent in the lungs of individuals suffering from cystic fibrosis (CF) (43, 46). *P. aeruginosa* infections usually

involve someone who is immunocompromised. The ability of *P. aeruginosa* to adapt from an environment outside the body to a pathogenic situation within a human host has been the subject of extensive study. One of the many goals of this research has been to understand the key factors that allow *P. aeruginosa* to make this transition, in order to both combat and prevent infections. The sequence of the *P. aeruginosa* genome has been completed. This should lead to great advances in understanding this versatile opportunistic pathogen.

Quite a bit is known about the numerous virulence factors that contribute to the pathogenicity of *P. aeruginosa*. Much of this research has involved careful genetic and biochemical characterizations of the virulence factor in question. Many times these studies included the analysis of *P. aeruginosa* virulence in animal models using mutant strains unable to produce a particular virulence factor. This chapter will review some of these virulence factors and the roles they are thought to play in pathogenesis. However, until recently, these studies were viewed in the context of the individual cell. We now know that *P. aeruginosa* utilizes acyl-HSL quorum sensing to regulate virulence gene expression. Quorum sensing allows the bacteria to coordinate gene

Matthew R. Parsek Department of Civil Engineering, Northwestern University, 2145 Sheridan Rd., Evanston, IL 60208. *E. P. Greenberg* Department of Microbiology, University of Iowa, Iowa City, IA 52246.

Virulence Mechanisms of Bacterial Pathogens, 3rd ed., Edited by K. A. Brogden et al.
©2000 ASM Press, Washington, D.C.

expression as a group. We will review quorum sensing in this organism with an emphasis on quorum sensing-signal generation.

Another example of community behavior among bacteria is the formation of bacterial biofilms. Biofilms are populations of bacteria attached to a surface and enmeshed in a matrix of extracellular polymeric substances (EPS) (15). It is estimated that most bacteria in the environment exist as biofilms. These include biofilms of a clinical nature, which are implicated in many types of disease, including those caused by *P. aeruginosa* (16). Biofilm populations are subjected to a variety of gradients (e.g., oxygen, carbon source, pH, etc.), and cells within a biofilm exhibit heterogeneous physiologies. This heterogeneity may in part be reflected by the highly differentiated structures seen in biofilms (15, 23, 62). A *P. aeruginosa* biofilm consists of towers and mushrooms of cells and EPS permeated by water channels, which are thought to be involved in nutrient acquisition and export of waste products from the biofilm (15). In *P. aeruginosa*, quorum sensing has been shown to be involved in coordinating the community behavior required for the formation of these structures (22). Finally, we address the role of biofilm formation in pathogenesis and its relationship to quorum sensing. The apparent importance of quorum sensing and biofilm formation in the virulence of *P. aeruginosa* gives one pause to reevaluate the context of the individual cell. The individuals are members of a community trying to survive.

VIRULENCE FACTORS OF *PSEUDOMONAS AERUGINOSA*

The interaction of higher eukaryotic organisms and bacteria has been described as "détente" (69). This analogy is important to consider when discussing an opportunistic pathogen. The intestinal tract is a good example of a mutually beneficial coexistence of human beings and microorganisms. Slight perturbations in this environment can upset the balance between the host and the microbial flora, resulting in deleterious effects on both.

P. aeruginosa is commonly found on the skin of a healthy human being, where it is a rather poor pathogen (63). However, it produces a number of toxins and extracellular virulence factors that make it a formidable pathogen when the immune system is impaired. This section describes some of these factors and their roles in pathogenesis.

Colonization Factors

The skin/mucous membrane barrier of a healthy human being is a daunting obstacle to *P. aeruginosa*. When this barrier is breached by severe trauma, such as burn or injury, it provides an avenue for infection. The ability of *P. aeruginosa* to colonize an infection site depends upon an array of determinants. We will list a few of these and what is known of their function.

The single polar flagellum produced by *P. aeruginosa* has been implicated as an important factor in infection. Motility allows the bacterium to spread within the host and obtain nutrients using chemotaxis. A mutation in the *fliC* gene, encoding flagellin (the structural component of a flagellum), resulted in reduced virulence in a mouse pulmonary model of infection (36). Tissue infected by the *fliC* mutant showed discrete foci of infection, suggesting that the bacteria were impaired in their ability to move within the tissue. In a separate study, a mutant unable to produce flagella was less virulent than its isogenic parental strain in a mouse burn model of infection (113). The flagellum may also be important for adherence to eukaryotic cells, binding to specific glycolipids such as GM1 (14). The flagellar cap protein, FliD, binds to mucin, a component of human airways (5). Flagella have also been shown to be important for the formation of *P. aeruginosa* biofilms on abiotic surfaces (80). However, the flagellum is highly immunogenic, and isolates of *P. aeruginosa* from chronic infections have often lost their flagella (64, 66).

Type IV pili in *P. aeruginosa* are polar and have been shown to be required for adherence to eukaryotic cells as well as for surface motility, or twitching motility (47, 96). Twitching motility is poorly understood, but one proposed mechanism is thought to involve extension and retraction of the pilus (10). There are numerous genes at several loci on the chromosome required for the biogenesis and function of type IV pili (2, 3, 19, 20, 67). Nonpiliated mutants have been shown to decrease virulence in several animal models (13, 35, 101). Type IV pili mutants were also deficient in biofilm development on abiotic surfaces, suggesting that they may play an important role in biofilm infections (80).

Proteases, Hemolysins, and Toxins

P. aeruginosa produces a battery of extracellular proteases. Two of the best studied are the elastolytic proteases, LasA (staphylolysin or LasA protease) and LasB (elastase). They are both zinc metalloendopeptidases. LasB is synthesized as a 54-kDa preproenzyme (6, 58). It is transported into the periplasm with a 2.4-kDa signal sequence. In the periplasm it undergoes autocleavage to yield the 33-kDa holoenzyme (70, 72). The propetide remains complexed with the holoenzyme and acts as a chaperone for proper LasB folding (71). The type II secretion apparatus, encoded by the *xcp* genes, then secretes elastase from the cell (114). The maturation of the 20-kDa LasA is less understood. It is also believed to be exported by the general secretory apparatus. Both elastase and LasA protease have been shown to be exported with their propeptides (7, 59). The removal of the propeptide by other secreted proteases and generation of the active forms of these enzymes occur extracellularly.

Elastin is a constituent of both lung tissue and blood vessels and is a substrate of both proteases. LasA protease and elastase act synergistically to degrade elastin (39). The localized tissue damage and ensuing inflammatory response caused by these proteases presumably contribute to compromising the host and aid in the dissemination of the organism. Besides elastin, elastase has been shown to proteolyze a variety of host proteins, including components of the immune system, such as comple-

ment factors and immunoglobulins (52, 53). This suggests that in addition to causing localized tissue damage, secreted proteases also play a role in hindering the immune response during infection. LasA protease has limited substrate specificity compared to elastase. It nicks elastin, which increases its susceptibility to other proteases (39). A second staphylolytic protease encoded by the *lasD* gene has been reported (86). This protease apparently shares a number of the enzymatic properties with the LasA protease.

Two other less-studied extracellular *P. aeruginosa* proteases, a lysine-specific protease (Ps-1) and alkaline protease, may also play a role in pathogenesis (31, 77). The lysine-specific protease cleaves peptides specifically on the carboxyl side of lysine residues and is thought to play a role in the processing of other *P. aeruginosa* proteases (31). Alkaline protease, encoded by *aprA*, also appears to be involved in the extracellular processing of proteases, such as the LasA protease. Although it is 10-fold less active than elastase, it has a fairly broad substrate range and is thought to act synergistically with the other proteases (77).

There have been several studies that indicate the importance of these proteases in pathogenesis. Strains of *P. aeruginosa* that produce less elastase have reduced virulence in animal models (79, 112, 113). Also, clinical isolates have protease profiles consistent with elastase as an important virulence factor. Recent studies have suggested that the role of elastase as a virulence factor may vary depending upon the type of infection. Preston et al., using a murine corneal scratch infection model, showed that wild-type *P. aeruginosa* and an isogenic *lasB* knockout strain had similar virulence properties (94). The same study suggested that LasA protease might be vital for pathogenesis. Alkaline protease has also been implicated as an important virulence factor in corneal keratitis; however, apparently it does not cause damage to lung epithelium (54).

Exotoxin A is a secreted toxin that inhibits host cell protein synthesis (115). This 66-kDa ADP-ribosyltransferase is encoded by the *toxA* gene and is similar in overall mechanism to the pertussis, cholera, and diptheria toxins. It acts by catalyzing the transfer of the ADP-ribose moiety of NAD^+ to elongation factor 2 (115). This effectively shuts off host protein synthesis at the translational level by inhibiting polypeptide elongation, which results in eventual cell death. Purified exotoxin A is lethal to mice and has been identified as an important virulence factor in several pathogenicity models (98, 117). Exotoxin A toxicity presumably results in significant localized tissue damage and aids in spreading the bacteria. Another ADP-ribosyltransferase produced by *P. aeruginosa* is exoenzyme S (38). The targets for ADP-ribosylation by exoenzyme S are different from those for exotoxin A. Exoenzyme S preferentially modifies vimentin, a structural component of the host cell, and GTP-binding proteins such as Ras (56). Exoenzyme S has been shown to cause tissue damage in situ and may contribute to pathogenesis in a similar manner as exotoxin A, although its actual role in infection remains unclear (37).

Pyocyanin is one of the secreted pigments that give *P. aeruginosa* laboratory cultures their characteristic blue-green color. Pyocyanin is a redox-cycling compound consisting of a modified phenazine ring that is derived from anthrinilate. The biosynthetic pathway intermediates and genes required for its synthesis have not all been identified. Pyocyanin toxicity to bacterial and eukaryotic cells is due to the reactive oxygen intermediates it generates, such as superoxide radical and hydrogen peroxide (65). Studies have shown that mutants unable to make pyocyanin have reduced virulence in an animal model (65). Pyocyanin has also been implicated in impairing host defense mechanisms in chronic infection of the CF lung; however, the actual role pyocyanin plays in pathogenesis remains uncertain (9, 25, 107).

Phospholipase C, a hemolysin encoded by the *plcS* gene, is another virulence factor in *P. aeruginosa* infections. Expression of *plcS* is activated by *phoB* in response to low phosphate concentrations (97). One of the functions proposed for phospholipase C in respiratory infections is the degradation of phosphotidyl choline, a major surfactant of the lung (84). It also causes increased adherence to lung epithelial cells and localized tissue damage due to its cytotoxicity. Phospholipase C has also been demonstrated to be required for virulence in both plant and animal model systems (98). Rhamnolipids are another hemolysin produced by *P. aeruginosa*. Rhamnolipids are mainly acylated rhamnose moieties synthesized by four enzymes using substrates derived from fatty acid biosynthesis and thymidine-diphospho-L-rhamnose (11). Rhamnolipids can act to reduce water surface tension and to solubilize nonpolar compounds that *P. aeruginosa* can use as growth substrates, such as hexadecane (4, 105).

Nutrient Acquisition

Many of the exoproteases and hemolysins listed above play a role in the generation of nutrients for *P. aeruginosa* at the site of infection. The various peptides and biomolecules released by the action of these enzymes generate an array of substrates that can be used as carbon, energy, and nitrogen sources. Therefore, the local environment in an infection is rich in many of the compounds necessary for growth. The same is not true for iron. *P. aeruginosa* requires iron for growth and faces stiff competition for iron from host proteins, such as ferritin and transferrin (118). *P. aeruginosa* adapts to low iron environments by producing secreted siderophores, such as pyoverdine and pyochelin. These siderophores can chelate iron and then bind receptor proteins in the membrane, which transport the iron into the cell.

Pyoverdine, a greenish-yellow compound, is the best studied of the siderophores. It has a hydroxyquinolone chromophore to which an amino acid tail is attached (111). The tail can vary in length. The synthesis of pyoverdine is directed by at least three separate loci on the chromosome and requires a special sigma factor, PvdS (17, 74). This sigma factor is in turn regulated by the Fur repressor, a regulatory protein sensitive to environmental iron concentrations (51). The importance of pyoverdine as a virulence factor has been demonstrated in a burned mouse animal model (73). Strains that are unable to make pyoverdine cause significantly less mortality in an animal model. Full virulence is restored when purified pyoverdine is added to the initial inoculum of the pyoverdine-deficient strains.

QUORUM SENSING IN *P. AERUGINOSA*

Quorum sensing is a term now used to describe systems that regulate gene expression in a cell density–dependent manner. These systems utilize a variety of mechanisms to sense population numbers. For example, various gram-positive bacteria use secreted peptides (27). The system we will discuss here is used by several gram-negative bacteria and involves extracellular signaling molecules called acyl-homoserine lactones (acyl-HSLs). Many *P. aeruginosa* genes have been described as quorum sensing–regulated. A common thread to many of the virulence factors listed in the previous section is that quorum sensing either directly or indirectly regulates them. Virulence factors reported to be regulated by quorum sensing include LasA protease, elastase, exotoxin A, alkaline protease, exoenzyme S, type IV pili, pyoverdine, pyocyanin, and rhamnolipids (41, 45, 61, 83, 89, 99, 104, 110). The importance of quorum sensing to the virulence of *P. aeruginosa* has also been demonstrated in animal models. Why would the cell regulate the production of many of these factors in correlation with cell density? One possible explanation is that until a community of bacteria reaches a sufficient size, it would be energetically unfavorable for a cell to produce some of its secreted virulence factors. For example, the amounts of exoprotease produced by a single cell probably wouldn't alter its environment. However, the cumulative efforts of a cluster of a few hundred cells might be effective. Another possible quorum-sensing strategy is to delay the production of extra-

cellular virulence factors until there is a critical mass of bacterial cells to produce a significant quantity of the factors to overwhelm the host defenses. In some infections this cluster of cells may exist as a surface-associated community called a biofilm, which is discussed later in this chapter. Cells within a biofilm have been shown to be much more resistant to host defense mechanisms.

The identification of the *lasR* gene and its role in regulating *lasB* expression led to the discovery of quorum sensing in *P. aeruginosa* (40). This was based upon the sequence similarity LasR exhibited with LuxR, a transcriptional regulator of the quorum-sensing system of *Vibrio fischeri*. It was in *V. fischeri*, a marine luminescent bacterium, that acyl HSL based quorum sensing was first described and *V. fischeri* remains the paradigm system. This bacterium can be found in seawater at very low cell densities (a few cells per ml); however, it is also capable of colonizing the light organs of a variety of marine organisms, such as the squid, *Euprymna scolopes* (100). In the light organ, *V. fischeri* exists in pure culture at very high cell densities (on the order of 10^{11} cells per ml). *V. fischeri* produces light by way of the enzyme luciferase. Light production is energy intensive and, not surprisingly, highly regulated. In 1970, Nealson et al. determined that *V. fischeri* produced a type of extracellular factor that they called an autoinducer, which induced luciferase synthesis (78). In 1981 Eberhard et al. determined that the structure of the *V. fischeri* autoinducer (VAI) was 3-oxo-hexanoyl homoserine lactone (28). Subsequent studies led to the cloning of a DNA segment from *V. fischeri* that directed *Escherichia coli* to make light (32). Genetic analysis of this DNA led to the identification of genes encoding the enzymes involved in light production, *luxA*, *luxB*, *luxC*, *luxD*, and *luxE*, and two regulatory genes, *luxI* and *luxR* (33). Most of these genes are in an operon, *luxICDABE*, with *luxR* being divergently transcribed from this operon. LuxI is an acyl-homoserine lactone synthase, which directs the synthesis of VAI (28, 102). LuxR is a transcriptional regulator, capable of interacting with VAI and regulating expression of the *luxICDABE* operon and autoregulating its own expression (26, 32, 48). LuxR does this by binding to DNA upstream of the promoter for the *luxICDABE* operon at a 20-bp regulatory sequence called the "*lux* box" (30, 108, 109). Therefore, the working model for quorum sensing is that at low cell densities, the cell produces a basal level of autoinducer, which diffuses freely both into and out of the cell and eventually diffuses away down the concentration gradient into the local environment. At higher cell densities, the autoinducer concentration builds to a critical level at which it interacts with the LuxR protein, activating transcription of quorum sensing–regulated genes. This results in activation of the bioluminescence genes, as well as *luxI*, which induces a higher level of signal production by the cell. Thus, at low cell densities such as those found in seawater, *V. fischeri* does not spend energy on light production, whereas the light organ is a nutritional environment that can support higher cell densities and bioluminescence.

Since the observation that LasR was homologous to LuxR, *P. aeruginosa* has been shown to possess two quorum-sensing systems, *las* and *rhl* (Fig. 1). Each quorum-sensing system has its own transcriptional regulator (LasR and RhlR), acyl-HSL synthase (LasI and RhlI), and acyl-HSL (3-oxo-dodecanoyl-HSL for the *las* system and butyryl-HSL for the *rhl* system) (Fig. 1) (30, 40, 61, 82, 83, 89-91, 108, 109, 116). These systems regulate a variety of genes. Therefore, quorum sensing in *P. aeruginosa* constitutes a global regulatory network (93, 103).

The *las* quorum-sensing system was the first described in *P. aeruginosa*. The *las* system has been reported to regulate a number of genes, including *lasB, toxA, lasA, lasI, rhlR,* and two genes of the general secretory pathway, *xcpP* and *xcpR* (12, 41, 103). Similar to the *V. fischeri* system, *lasI*, which is a *luxI* homolog, has been shown to direct the synthesis of an acyl-

HSL, 3-oxo-dodecanoyl HSL (89, 90). Unlike the *V. fischeri* signal, this acyl–HSL does not diffuse freely into and out of the cell. This may be in part due to its longer acyl side chain, which may cause it to partition into the lipid bilayer. 3-Oxo-dodecanoyl-HSL leaves the cell in part through an active process driven by the *mexAB-oprM* multidrug efflux pump (34, 92). At high concentrations, 3-oxododecanoyl-HSL presumably interacts with the transcriptional regulator, LasR. However, the concentration of 3-oxo-dodecanoyl-HSL required for activation of *lasB* expression is approximately 10 times higher than that required for *lasI* (103). These data, and the fact that the *las* system is required for expression of the *rhl* system, suggest that quorum sensing is a regulatory hierarchy. The *las* quorum-sensing system has been shown to be regulated by three other elements, GacA, Vfr, and RsaL (1, 24, 99). GacA shows homology to a two-component transcriptional regulator, while Vfr is a CRP-like protein. Both Vfr and GacA are thought to positively regulate *lasR* expression. Recently, the *rsaL* locus was shown to negatively regulate expression of *lasI*. Although the exact roles of these regulatory components are not clear, the expression of quorum-sensing genes can be modulated in response to other environmental signals. Therefore, the number of cells that constitute a quorum depends upon the environment.

Shortly after the discovery of the *las* system, a second quorum-sensing system in *P. aeruginosa* involved in the regulation of rhamnolipid biosynthesis, the *rhl* system, was identified (this system has also been called the *vsmR/I* system for virulence and secondary metabolites) (8, 82, 83, 91, 116). The RhlI protein directs the synthesis of butyryl-HSL. This particular acyl-HSL diffuses into and out of the cell, and at a critical concentration interacts with RhlR to activate *rhl*-controlled genes (92). These include *lasB, rhlAB,* and genes involved in pyocyanin and pyoverdin synthesis. The *rhl* system is regulated by elements other than the *las* sys-

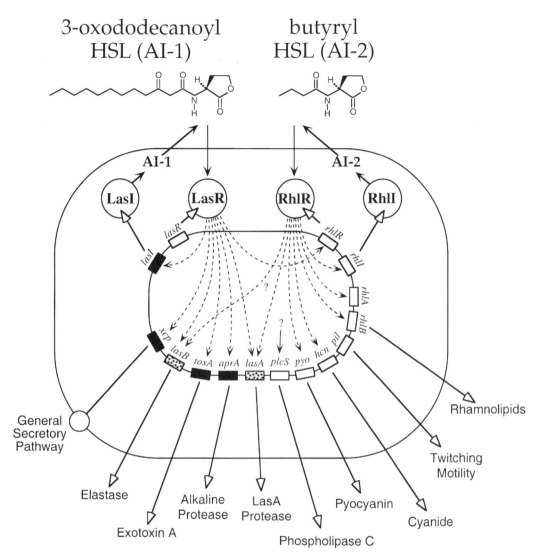

FIGURE 1 A model of the quorum-sensing regulatory network in *P. aeruginosa*. This diagram illustrates the components of the *las* and *rhl* quorum-sensing systems. Each system has its own autoinducer synthase (LasI and RhlI), transcriptional regulator (LasR and RhlR), and acyl-HSL (3-oxo-dodecanoyl-HSL and butyryl-HSL). The *las* system regulates the *rhl* system, which in turn regulates *rpoS*, a sigma factor of stationary phase genes. The clear, black, and stippled boxes indicate genes thought regulated by the *rhl*, *las*, and both systems.

tem; for example, this system has also been shown to be regulated by *gacA/S* (99). The structural differences between the cognate autoinducers lend specificity to the *las* and *rhl* systems, although there is evidence that at high levels, 3-oxo-dodecanoyl-HSL could block the binding of butyryl-HSL to RhlR

(93). This would provide an additional level of control of the *las* system over the *rhl* system.

QUORUM-SENSING SIGNAL GENERATION

The role quorum sensing plays in regulating the expression of many virulence factors produced by *P. aeruginosa* makes it an attractive target for antimicrobial therapy. The use of inhibitors that would block the binding of acyl-HSLs to the transcriptional regulators could be a way to inhibit quorum sensing. Another potential avenue would be to inhibit quorum-sensing signal generation. Such an approach would require an understanding of what substrates autoinducer synthases use for acyl-HSL synthesis and the enzymatic mechanism for synthesis, as well as inhibitors of acyl-HSL synthesis.

Initial studies conducted by Engebrecht et al. determined that the *V. fischeri luxI* gene was the only gene required for *E. coli* to direct the synthesis of VAI (32). Later studies with *traI* of *Agrobacterium tumefaciens* and *lasI* of *P. aeruginosa* demonstrated that they also directed *E. coli* to make their cognate acyl-HSLs, suggesting that the I genes not only directed the synthesis of acyl-HSLs, but provided the specificity for the acyl-HSL product (119). However, these studies did not exclude the possibility that the I proteins were not directly involved in catalysis but may instead direct *E. coli* enzymes to synthesize acyl-HSLs. Studies conducted with crude extracts of *V. fischeri* showed that the synthesis of VAI was stimulated by the addition of *S*-adenosylmethionine (SAM) and 3-oxo-hexanoyl-CoA, implying that these common cellular compounds may be substrates for synthesis of VAI (29). Two separate studies conducted with amino acid auxotrophs of *E. coli* suggested that different substrates were the source of the homoserine lactone ring moiety of VAI. One group indicated that homoserine lactone was a substrate, whereas another suggested SAM (49, 55). A breakthrough occurred at about the same time when two groups were able to purify acyl-HSL synthases and demonstrate acyl-HSL synthesis in vitro. Fusion proteins of TraI and LuxI were purified and shown in vitro to synthesize acyl-HSLs from SAM and acylated-acyl carrier protein (ACP) (76, 102). Previously, a model for the enzymatic reaction mechanism was proposed that involved an acylated-enzyme intermediate, characteristic

of a ping-pong reaction mechanism (106). Such a reaction mechanism would first involve an initial transfer of the acyl group to an active site residue on the acyl-HSL synthase. This acyl-enzyme intermediate would then interact with SAM, resulting in formation of an amide bond between SAM and the acyl group. This may occur before, after, or simultaneously with lactonization of SAM to yield the homoserine lactone ring. The finding that acyl-ACP is a substrate of acyl-HSL synthesis supported the proposed ping-pong reaction mechanism, since enzymes involved in fatty acid biosynthesis that also use acyl-ACP as a substrate utilize a similar mechanism (18, 42). However, attempts to isolate an acyl-enzyme intermediate in vitro were unsuccessful and mutagenesis of acyl-HSL synthases failed to identify conserved amino acids that could act as an acceptor site for acyl group transfer (50, 87).

A report by Jiang et al. suggested that purified, albeit mostly insoluble, RhlI could catalyze the synthesis of acyl-HSLs using either SAM or homoserine lactone as a source of the homoserine lactone ring, and either acyl-ACP or acyl-CoA as a source of the acyl side chain (57). It was proposed that RhlI belongs to a unique class of acyl-HSL synthases that have unique substrate ranges. These in vitro studies, as well as the studies done with the LuxI and TraI fusion proteins, were hampered because these enzymes had very low specific activities (57, 76, 102). This was presumably due either to the insolubility of the proteins or the fact that some were fused at the N terminus to heterologous proteins. Another limitation of these studies is that detection and quantitation of acyl-HSL product was dependent upon biological reporter strains.

To identify inhibitors of acyl-HSL synthesis and the enzymatic reaction mechanism, we purified soluble, active RhlI. Enzymatic assays were radiometric, avoiding the need for biological reporter strains. Subsequent experiments demonstrated that only SAM and various acyl-ACPs and acyl-CoAs could be used as substrates for acyl-HSL synthesis (88).

Kinetic data suggested that the probable substrates in vivo for the synthesis butyryl-HSL are SAM and butyryl-ACP. The same study identified the reaction products, 5'-methylthioadenosine and holo-ACP, as inhibitors of the reaction. Various analogs of SAM, such as S-adenosylhomocysteine and sinefungin, were also found to be potent inhibitors of acyl-HSL synthesis. A kinetic analysis using the substrates and inhibitors suggested that RhlI uses a sequential ordered reaction mechanism to synthesize butyryl-HSL (88). This reaction proceeds initially through the binding of SAM to RhlI (Fig. 2). The binding of butyryl-ACP follows this, then amide bond formation occurs with the subsequent release of holo ACP. Lactonization of the homoserine lactone ring occurs and then butyryl-HSL release; finally, 5'-methylthioadenosine is released.

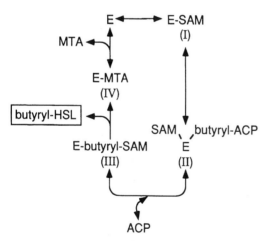

FIGURE 2 Acyl-HSL biosynthesis by RhlI. The substrates for butyryl-HSL synthesis are SAM and butyryl-ACP. The first step in acyl-HSL synthesis is the binding of SAM to RhlI, which is followed by binding of butyryl-ACP. Amide bond formation occurs between SAM and the acyl group of butyryl-ACP, accompanied by the release of ACP. The next step is lactonization of the homoserine lactone ring and subsequent release of butyryl-HSL and finally 5'-methylthioadenosine. A number of SAM analogs have been identified as inhibitors of this reaction. Reprinted from reference 88 with permission. Copyright 1999 National Academy of Sciences, U.S.A.

The potential use of SAM analogs as quorum-sensing inhibitors shows some promise. Although SAM is a necessary component of both prokaryotic and eukaryotic cell function, the reaction chemistry of acyl-HSL synthases with SAM is unique. No other reaction uses the same portion of the SAM molecule as acyl-HSL synthases. This may allow the use of a SAM analog that specifically inhibits quorum sensing, while sparing vital eukaryotic enzymes that use SAM as a substrate.

QUORUM SENSING AND BIOFILM FORMATION

Bacteria are known for their ability to adhere to and colonize surfaces, forming exopolysaccharide-enmeshed communities called biofilms. Biofilms have been implicated in many infections caused by *P. aeruginosa*, including CF lung infections and corneal keratitis (16). In most cases biofilms form on inert surfaces or dead tissue, and once established, they are notoriously difficult to eliminate. Biofilm bacteria, as evidenced by studies with *P. aeruginosa*, are generally much more resistant to antibiotics and host defense mechanisms than nonattached bacteria of the same species (15, 44). Although the mechanism of resistance is not completely clear, it is probably due to the contribution of many factors, including the nature of the exopolysaccharide matrix and the unique physiology of biofilm bacteria. During the course of chronic infection in CF, *P. aeruginosa* undergoes a switch to a mucoid phenotype, which results from the production of the exopolysaccharide alginate (46). This switch may facilitate the biofilm mode of existence within the CF lung (60, 68). In many cases, people suffering from CF exhibit inflammation at the site of infection that damages host tissue while failing to eliminate biofilm bacteria.

The biofilms formed by *P. aeruginosa* are extremely heterogeneous, both structurally and in the physiologies of the bacterial cells within them (23, 62). This fact was not fully appreciated until developing technologies, such as scanning confocal laser microscopy

(SCLM) and microsensor probes, were applied to biofilms. SCLM allowed the visualization of living biofilms without the artifacts associated with other types of microscopy, and microsensor probes allowed researchers to demonstrate that members of a biofilm population are subjected to a variety of chemical gradients. Mature biofilms of *P. aeruginosa* consist of "towers" or "mushrooms" of cells enmeshed in exopolysaccharide and separated by water channels. These structures arise from a few attached bacteria, which multiply on the attachment surface. Some researchers have hypothesized that this process may entail a developmental program of gene expression (22, 80, 81). Previous studies have shown that gene expression can be induced upon the attachment of a single cell to a surface and can vary

within members of a biofilm population (21, 75). The initial step in biofilm formation is the attachment of free-swimming bacteria to a surface. O'Toole and Kolter have shown that biofilm formation on an abiotic surface by both *P. aeruginosa* and *Pseudomonas fluorescens* requires specific sets of genes (81). A recent study by Prigent-Combaret et al. demonstrated that following attachment a specific set of genes is induced in *E. coli* (95). These genes are involved in responses to high osmolarity, oxygen limitation, and increased cell density. Microcolony formation occurs subsequent to attachment. Microcolonies are groups or mounds of clustered cells (15). The process of microcolony formation is poorly defined, but it is thought to involve twitching motility mediated by type IV pili (80). Microcolonies

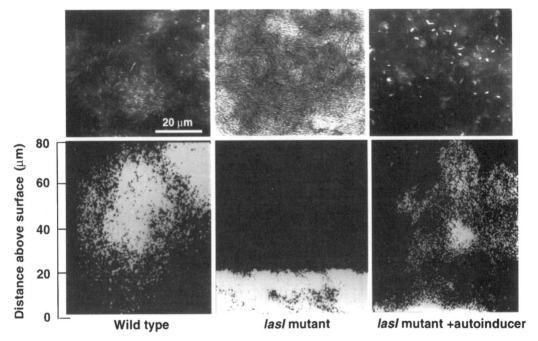

FIGURE 3. Epifluorescence and scanning confocal photomicrographs of wild-type *P. aeruginosa* and a *lasI* quorum-sensing mutant. The top three panels represent epifluorescence micrographs of the wild-type PAO1, a *lasI* mutant, and a *lasI* mutant complemented with exogenously added autoinducer (3-oxododecanoyl-HSL) as indicated. The bottom three panels represent saggital or side views of the biofilms generated with SCLM. The attachment surface is the *x* axis. The bacteria are tagged with the green fluorescent protein. The mutant biofilm is thinner and more densely packed than the wild-type biofilm. Addition of autoinducer to the growth medium complements the mutant phenotype and restores a wild-type biofilm phenotype. Reprinted with permission from reference 22. Copyright 1998 American Association for the Advancement of Science.

eventually develop into the towers and mushrooms of a mature biofilm. Detachment of cells is the last step in the biofilm developmental cycle. Detachment is thought to result from both active processes on the part of the bacteria and passive events resulting from mechanical forces such as shear.

Recently, it was reported that normal biofilm development in *P. aeruginosa* required a functional *las* quorum-sensing system (22). Biofilms of isogenic *lasI* mutant strains were flat, undifferentiated, and densely packed (Fig. 3). The coordination of gene expression by the bacterial community during the development of a biofilm is not completely unexpected. Examination of a mature biofilm poses some questions. Why do the bacteria form such elaborate structures instead of a homogeneous layer of cells and EPS? One can imagine that maintenance of clear water channels would allow nutrient penetration and prevent waste product build up. Quorum sensing would be a means of controlling cell density and preventing water channels from filling in with bacteria. What the *las* system is regulating, either directly or indirectly, that plays a role in biofilm development remains to be determined. Of considerable importance is the finding that the *lasI* mutant biofilms were susceptible to sodium dodecyl sulfate (SDS) (22). Wild-type biofilms were resistant to SDS treatment for periods up to 19 h. However, the mutant biofilms were washed away within minutes of SDS treatment. This finding suggests that some property of the mutant biofilm, such as the undifferentiated architecture, renders it amenable to antimicrobial treatment. Therefore, disrupting quorum sensing could potentially provide an effective strategy for combating *P. aeruginosa* biofilm infections. There are many questions that remain to be answered concerning quorum sensing in a biofilm. What constitutes a quorum in an attached population of bacteria, especially one that is subjected to high flow rates? Can acyl-HSL production even be measured in a biofilm community? Ultimately, by understanding the program of gene expression during biofilm development and the role quorum sensing plays in this process, we may be able to steer *P. aeruginosa* away from forming biofilm infections.

CONCLUSIONS

P. aeruginosa is a formidable opportunistic human pathogen that possesses an arsenal of virulence factors. These factors are important for different aspects of pathogenesis, such as colonization and cytotoxicity. Perhaps its most significant weapon is its ability to act as a community. The expression of many of its virulence factors depends upon quorum sensing, which allows the community to modulate gene expression in accordance with cell density. Quorum sensing also plays a role in the formation of biofilms. The challenge is to understand the biofilm lifestyle of *P. aeruginosa* and how quorum sensing is important for this mode of existence.

REFERENCES

1. **Albus, A. M., E. C. Pesci, L. J. Runyen-Janecky, S. E. West, and B. H. Iglewski.** 1997. Vfr controls quorum sensing in *Pseudomonas aeruginosa. J. Bacteriol.* **179:**3928–3935.
2. **Alm, R., A. Bodero, P. Free, and J. Mattick.** 1996. Identification of a novel gene, *pilZ,* essential for type 4 fimbrial biogenesis in *Pseudomonas aeruginosa. J. Bacteriol.* **178:**46–53.
3. **Alm, R., and J. Mattick.** 1997. Genes involved in the biogenesis and function of type-4 fimbriae in *Pseudomonas aeruginosa. Gene* **192:**89–98.
4. **Arino, S., R. Marchal, and J. P. Vandecasteele.** 1998. Involvement of a rhamnolipid-producing strain of *Pseudomonas aeruginosa* in the degradation of polycyclic aromatic hydrocarbons by a bacterial community. *J. Appl. Microbiol.* **84:**769–776.
5. **Arora, S. K., B. W. Ritchings, E. C. Almira, S. Lory, and R. Ramphal.** 1998. The *Pseudomonas aeruginosa* flagellar cap protein, FliD, is responsible for mucin adhesion. *Infect. Immun.* **66:**1000–1007.
6. **Bever, R. A., and B. H. Iglewski.** 1988. Molecular characterization and nucleotide sequence of the *Pseudomonas aeruginosa* elastase structural gene. *J. Bacteriol.* **170:**4309–4314.
7. **Braun, P., A. de Groot, W. Bitter, and J. Tommassen.** 1998. Secretion of elastinolytic

enzymes and their propeptides by *Pseudomonas aeruginosa*. *J. Bacteriol.* **180**:3467–3469.

8. **Brint, J. M., and D. E. Ohman.** 1995. Synthesis of multiple exoproducts in *Pseudomonas aeruginosa* is under control of RhlR-RhlI, another set of regulators in strain PAO1 with homology to the autoinducer-responsive LuxR-LuxI family. *J. Bacteriol.* **177**:7155–7163.

9. **Britigan, B. E., M. A. Railsback, and C. D. Cox.** 1999. The *Pseudomonas aeruginosa* secretory product pyocyanin inactivates alpha1 protease inhibitor: implications for the pathogenesis of cystic fibrosis lung disease. *Infect. Immun.* **67**:1207–1212.

10. **Burchard, R. P.** 1981. Gliding motility in prokaryotes: ultrastructure, physiology, and genetics. *Annu. Rev. Microbiol.* **35**:497–529.

11. **Campos-Garcia, J., A. D. Caro, R. Najera, R. M. Miller-Maier, R. A. Al-Tahhan, and G. Soberon-Chavez.** 1998. The *Pseudomonas aeruginosa rhlG* gene encodes an NADPH-dependent beta-ketoacyl reductase which is specifically involved in rhamnolipid synthesis. *J. Bacteriol.* **180**:4442–4451.

12. **Chapon-Herve, V., M. Akrim, A. Latifi, P. Williams, A. Lazdunski, and M. Bally.** 1997. Regulation of the *xcp* secretion pathway by multiple quorum-sensing modulons in *Pseudomonas aeruginosa*. *Mol. Microbiol.* **24**:1169–1178.

13. **Comolli, J. C., A. R. Hauser, L. Waite, C. B. Whitchurch, J. S. Mattick, and J. N. Engel.** 1999. *Pseudomonas aeruginosa* gene products PilT and PilU are required for cytotoxicity in vitro and virulence in a mouse model of acute pneumonia. *Infect. Immun.* **67**:3625–3630.

14. **Comolli, J. C., L. L. Waite, K. E. Mostov, and J. N. Engel.** 1999. Pili binding to asialo-GM1 on epithelial cells can mediate cytotoxicity or bacterial internalization by *Pseudomonas aeruginosa*. *Infect. Immun.* **67**:3207–3214.

15. **Costerton, J. W., Z. Lewandowski, D. E. Caldwell, D. R. Korber, and H. M. Lappin-Scott.** 1995. Microbial biofilms. *Annu. Rev. Microbiol.* **49**:711–745.

16. **Costerton, J. W., P. Stewart, and E. P. Greenberg.** 1999. Bacterial biofilms: a common cause of persistent infections. *Science* **284**:1318–1322.

17. **Cunliffe, H. E., T. R. Merriman, and I. L. Lamont.** 1995. Cloning and characterization of *pvdS*, a gene required for pyoverdine synthesis in *Pseudomonas aeruginosa*: PvdS is probably an alternative sigma factor. *J. Bacteriol.* **177**:2744–2750.

18. **D'Agnolo, G., I. S. Rosenfeld, and P. R. Vagelos.** 1973. β-Ketoacyl-acyl carrier protein synthase. Characterization of the acyl-enzyme intermediate. *Biochim. Biophys. Acta* **326**:155–166.

19. **Darzins, A.** 1994. Characterization of a *Pseudomonas aeruginosa* gene cluster involved in pilus biosynthesis and twitching motility: sequence similarity to the chemotaxis proteins of enterics and the gliding bacterium *Myxococcus xanthus*. *Mol. Microbiol.* **11**:137–153.

20. **Darzins, A.** 1993. The *pilG* gene product, required for *Pseudomonas aeruginosa* pilus production and twitching motility, is homologous to the enteric, single-domain response regulator CheY. *Mol. Microbiol.* **175**:5934–5944.

21. **Davies, D. G., and G. G. Geesey.** 1995. Regulation of the alginate biosynthesis gene *algC* in *Pseudomonas aeruginosa* during biofilm development in continuous culture. *Appl. Environ. Microbiol.* **61**:860–867.

22. **Davies, D. G., M. R. Parsek, J. P. Pearson, B. H. Iglewski, J. W. Costerton, and E. P. Greenberg.** 1998. The involvement of cell-to-cell signals in the development of a bacterial biofilm. *Science* **280**:295–298.

23. **DeBeer, D., P. Stoodley, and Z. Lewandowski.** 1994. Liquid flow in heterogenous biofilms. *Biotech. Bioeng.* **44**:636–641.

24. **de Kievit, T., P. C. Seed, J. Nezezon, L. Passador, and B. H. Iglewski.** 1999. RsaL, a novel repressor of virulence gene expression in *Pseudomonas aeruginosa*. *J. Bacteriol.* **181**:2175–2184.

25. **Denning, G. M., L. A. Wollenweber, M. A. Railsback, C. D. Cox, L. L. Stoll, and B. E. Britigan.** 1998. Pseudomonas pyocyanin increases interleukin-8 expression by human airway epithelial cells. *Infect. Immun.* **66**:5777–5784.

26. **Dunlap, P. V., and E. P. Greenberg.** 1988. Control of *Vibrio fischeri lux* gene transcription by a cyclic AMP receptor protein-LuxR protein regulatory circuit. *J. Bacteriol.* **170**:4040–4046.

27. **Dunny, G. M., and B. A. Leonard.** 1997. Cell-cell communication in gram-positive bacteria. *Annu. Rev. Microbiol.* **51**:527–564.

28. **Eberhard, A., A. L. Burlingame, C. Eberhard, G. L. Kenyon, K. H. Nealson, and N. J. Oppenheimer.** 1981. Structural identification of autoinducer of *Photobacterium fischeri* luciferase. *Biochemistry* **20**:2444–2449.

29. **Eberhard, A., T. Longin, C. A. Widrig, and S. J. Stanick.** 1991. Synthesis of the *lux* gene autoinducer in *Vibrio fischeri* is positively autoregulated. *Arch. Microbiol.* **155**:294–297.

30. **Egland, K. A., and E. P. Greenberg.** 1999. Quorum sensing in *Vibrio fischeri*: elements of the *luxI* promoter. *Mol. Microbiol.* **31**:1197–1204.

31. **Elliot, B. W., and C. Cohen.** 1986. Isolation and characterization of a lysine-specific protease from *Pseudomonas aeruginosa*. *J. Biol. Chem.* **261**:11259–11265.

32. **Engebrecht, J., K. H. Nealson, and M. Silverman.** 1983. Bacterial bioluminescence: isolation and genetic analysis of the functions from *Vibrio fischeri*. *Cell* **32**:773–781.

33. **Engebrecht, J., and M. Silverman.** 1984. Identification of genes and gene products necessary for bacterial bioluminescence. *Proc. Natl. Acad. Sci. USA* **81**:4154–4158.

34. **Evans, K., L. Passador, R. Srikumar, E. Tsang, J. Nezezou, and K. Poole.** 1998. Influence of the MexAB-OprM multidrug efflux system on quorum sensing in *Pseudomonas aeruginosa*. *J. Bacteriol.* **180**:5443–5447.

35. **Farinha, M. A., B. D. Conway, L. M. Glasier, N.W. Ellert, R.T. Irvin, R. Sherburne, and W. Paranchych.** 1994. Alteration of the pilin adhesin of *Pseudomonas aeruginosa* PAO results in normal pilus biogenesis but a loss of adherence to human pneumocyte cells and decreased virulence in mice. *Infect. Immun.* **62**: 4118–4123.

36. **Feldman, M., R. Bryan, S. Rajan, L. Scheffler, S. Brunnert, H. Tang, and A. Prince.** 1998. Role of flagella in pathogenesis of *Pseudomonas aeruginosa* pulmonary infection. *Infect. Immun.* **66**:43–51.

37. **Fleiszig, S. M. J., J. P. Wiener-Kronish, H. Miyazaki, V. Vallas, K. E. Mostov, D. Kanada, T. Sawa, T. S. Yen, and D. W. Frank.** 1997. *Pseudomonas aeruginosa*–mediated cytotoxicity and invasion correlate with distinct genotypes at the loci encoding exoenzyme S. *Infect. Immun.* **65**:579–586.

38. **Frank, D. W.** 1997. The exoenzyme S regulon of *Pseudomonas aeruginosa*. *Mol. Microbiol.* **26**:621–629.

39. **Galloway, D. R.** 1991. *Pseudomonas aeruginosa* elastase and elastolysis revisited: recent developments. *Mol. Microbiol.* **5**:2315–2321.

40. **Gambello, M. J., and B. H. Iglewski.** 1991. Cloning and characterization of the *Pseudomonas aeruginosa lasR* gene, a transcriptional activator of elastase expression. *J. Bacteriol.* **173**:3000–3009.

41. **Gambello, M. J., S. Kaye, and B. H. Iglewski.** 1993. LasR of *Pseudomonas aeruginosa* is a transcriptional activator of the alkaline protease gene (*apr*) and an enhancer of exotoxin A expression. *Infect. Immun.* **61**:1180–1184.

42. **Garwin, J. L., A. L. Klages, and J. J. E. Cronan.** 1980. Beta-ketoacyl-acyl carrier protein synthase II of *Escherichia coli*. Evidence for function in the thermal regulation of fatty acid synthesis. *J. Biol. Chem.* **255**:3263–3265.

43. **Gilligan, P. H.** 1991. Microbiology of airway disease in patients with cystic fibrosis. *Microbiol. Rev.* **4**:35–51.

44. **Gilbert, P., J. Das, and I. Foley.** 1997. Biofilm susceptibility to antimicrobials. *Adv. Dent. Res.* **11**:160–167.

45. **Glessner, A., R. S. Smith, B. H. Iglewski, and J. B. Robinson.** 1999. Roles of *Pseudomonas aeruginosa las* and *rhl* quorum-sensing systems in control of twitching motility. *J. Bacteriol.* **181**:1623–1629.

46. **Govan, J. R. W., and V. Deretic.** 1996. Microbial pathogenesis in cystic fibrosis: mucoid *Pseudomonas aeruginosa* and *Burkholderia cepacia*. *Microbiol. Rev.* **60**:539–574.

47. **Hahn, H. P.** 1997. The type-4 pilus is the major virulence-associated adhesin of *Pseudomonas aeruginosa*—a review. *Gene* **192**:99–108.

48. **Hanzelka, B. L., and E. P. Greenberg.** 1995. Evidence that the N-terminal region of the *Vibrio fischeri* LuxR protein constitutes an autoinducer-binding domain. *J. Bacteriol.* **177**:815–817.

49. **Hanzelka, B. L., and E. P. Greenberg.** 1996. Quorum sensing in *Vibrio fischeri*: evidence that S-adenosylmethionine is the amino acid substrate for autoinducer synthesis. *J. Bacteriol.* **178**:5291–5294.

50. **Hanzelka, B. L., A. M. Stevens, M. R. Parsek, and E. P. Greenberg.** 1997. Mutational analysis of the *Vibrio fischeri* LuxI polypeptide: critical regions of an autoinducer synthase. *J. Bacteriol.* **179**:4882–4887.

51. **Hassett, D. J., P. A. Sokol, M. L. Howell, J. F. Ma, H. T. Schweizer, U. Ochsner, and M. L. Vasil.** 1996. Ferric uptake regulator (Fur) mutants of *Pseudomonas aeruginosa* demonstrate defective siderophore-mediated iron uptake, altered aerobic growth, and decreased superoxide dismutase and catalase activities. *J. Bacteriol.* **178**: 3996–4003.

52. **Heck, L. W., P. G. Alarcon, R. M. Kulhavy, K. Morihara, M. W. Russell, and J. F. Mestecky.** 1990. Degradation of IgA proteins by *Pseudomonas aeruginosa* elastase. *J. Immunol.* **144**: 2253–2257.

53. **Hong, Y. Q., and B. Ghebrehiwet.** 1992. Effect of *Pseudomonas aeruginosa* elastase and alkaline protease on serum complement and isolated components C1q and C3. *Clin. Immunol. Immunopathol.* **62**:133–138.

54. **Howe, T. R., and B. H. Iglewski.** 1984. Isolation and characterization of alkaline protease-deficient mutants of *Pseudomonas aeruginosa* in vitro and in a mouse eye model. *Infect. Immun.* **43**:1058–1063.

55. **Huisman, G. W., and R. Kolter.** 1994. Sensing starvation: a homoserine lactone dependent signalling pathway in *Escherichia coli*. *Science* **265**: 537–539.

56. **Iglewski, B. H., J. Sadoff, M. J. Bjorn, and E. S. Maxwell.** 1978. *Pseudomonas aeruginosa* exoenzyme S: an adenosine diphosphate ribosyltransferase distinct from toxin A. *Proc. Natl. Acad. Sci. USA* **75:**3211–3215.

57. **Jiang, Y., M. Camara, S. R. Chhabra, K. R. Hardie, B. W. Bycroft, A. Lazdunski, G. P. C. Salmond, G. S. A. B. Stewart, and P. Williams.** 1998. *In vitro* biosynthesis of the *Pseudomonas aeruginosa* quorum-sensing signal molecule N-butanoyl-L-homoserine lactone. *Mol. Microbiol.* **28:**193–203.

58. **Kessler, E., and M. Safrin.** 1988. Partial purification and characterization of an inactive precursor of *Pseudomonas aeruginosa* elastase. *J. Bacteriol.* **170:**1215–1219.

59. **Kessler, E., M. Safrin, J. K. Gustin, and D. E. Ohman.** 1998. Elastase and the LasA protease of *Pseudomonas aeruginosa* are secreted with their propeptides. *J. Biol. Chem.* **273:**30225–30231.

60. **Lam, J., R. Chan, K. Lam, and J. W. Costerton.** 1980. Production of mucoid microcolonies by *Pseudomonas aeruginosa* within infected lungs in cystic fibrosis. *Infect. Immun.* **28:**546–556.

61. **Latifi, A., K. M. Winson, M. Foglino, B. W. Bycroft, G. S. A. B. Stewart, A. Lazdunski, and P. Williams.** 1995. Multiple homologues of LuxR and LuxI control expression of virulence determinants and secondary metabolites through quorum sensing in *Pseudomonas aeruginosa* PAO1. *Mol. Microbiol.* **17:**333–344.

62. **Lawrence, J. R., D. R. Korber, B. D. Hoyle, J. W. Costerton, and D. E. Caldwell.** 1991. Optical sectioning of microbial biofilms. *J. Bacteriol.* **173:**6558–6567.

63. **Lowbury, E.** 1972. Prevention and treatment of sepsis in burns. *Proc. R. Soc. Med.* **65:**25–27.

64. **Luzar, M., M. Thomassen, and T. Montie.** 1985. Flagella and motility alterations in *Pseudomonas aeruginosa* strains from patients with cystic fibrosis: relationship to patient clinical condition. *Infect. Immun.* **50:**577–582.

65. **Mahajan-Miklos, S., M. W. Tan, L. G. Rahme, and F. M. Ausubel.** 1999. Molecular mechanisms of bacterial virulence elucidated using a *Pseudomonas aeruginosa–Caenorhabditis elegans* pathogenesis model. *Cell* **96:**47–56.

66. **Mahenthiralingam, E., M. Campbell, and D. P. Speert.** 1994. Nonmotility and phagocytic resistance of *Pseudomonas aeruginosa* isolates from chronically colonized patients with cystic fibrosis. *Infect. Immun.* **62:**596–605.

67. **Martin, P., A. A. Watson, T. F. McCaul, and J. S. Mattick.** 1995. Characterization of a five-gene cluster required for the biogenesis of type 4 fimbriae in *Pseudomonas aeruginosa*. *Mol. Microbiol.* **16:**497–508.

68. **Mathee, K., O. Ciofu, C. Sternberg, P. W. Lindum, J. I. Campbell, P. Jensen, A. H. Johnsen, M. Givskov, D. E. Ohman, S. Molin, N. Hoiby, and A. Kharazmi.** 1999. Mucoid conversion of *Pseudomonas aeruginosa* by hydrogen peroxide: a mechanism for virulence activation in the cystic fibrosis lung. *Microbiology* **145:**1349–1357.

69. **McFall-Ngai, M.** 1998. The development of cooperative associations between animals and bacteria: establishing detente between domains. *Am. Zool.* **38:**593–608.

70. **McIver, K. S., E. Kessler, and D. E. Ohman.** 1991. Substitution of active-site His-223 in *Pseudomonas aeruginosa* elastase and expression of the mutated *lasB* alleles in *Escherichia coli* show evidence for autoproteolytic processing of proelastase. *J. Bacteriol.* **173:**7781–7789.

71. **McIver, K. S., E. Kessler, J. C. Olson, and D. E. Ohman.** 1995. The elastase propeptide functions as an intramolecular chaperone required for elastase activity and secretion in *Pseudomonas aeruginosa*. *Mol. Microbiol.* **18:**877–889.

72. **McIver, K. S., J. C. Olson, and D. E. Ohman.** 1993. *Pseudomonas aeruginosa lasB1* mutants produce an elastase, substituted at active-site His-223, that is defective in activity, processing, and secretion. *J. Bacteriol.* **175:**4008–4015.

73. **Meyer, J. M., A. Neely, A. Stintzi, C. Georges, and I. A. Holder.** 1996. Pyoverdin is essential for virulence of *Pseudomonas aeruginosa*. *Infect. Immun.* **64:**518–523.

74. **Miyazaki, H., H. Kato, T. Nakazawa, and M. Tsuda.** 1995. A positive regulatory gene, *pvdS*, for expression of pyoverdin biosynthetic genes in *Pseudomonas aeruginosa* PAO. *Mol. Gen. Genet.* **248:**17–24.

75. **Møller, S., C. Sternberg, J. B. Andersen, B. B. Christensen, J. L. Ramos, M. Givskov, and S. Mølin.** 1998. In situ gene expression in mixed-culture biofilms: evidence of metabolic interactions between community members. *Appl. Environ. Microbiol.* **64:**721–732.

76. **Moré, M. I., D. Finger, J. L. Stryker, C. Fuqua, A. Eberhard, and S. C. Winans.** 1996. Enzymatic synthesis of a quorum-sensing autoinducer through the use of defined substrates. *Science* **272:**1655–1658.

77. **Morihara, K., and J. Y. Homma.** 1985. Pseudomonas proteases, p. 41–79. *In* I. A. Holder (ed.), *Bacterial Enzymes and Virulence.* CRC Press, Boca Raton, Fla.

78. **Nealson, K. H., T. Platt, and J. W. Hastings.** 1970. Cellular control of the synthesis and activ-

ity of the bacterial luminescence system. *J. Bacteriol.* **104**:313–322.

79. **Nicas, T. I., and B. H. Iglewski.** 1985. The contribution of exoproducts to virulence of *Pseudomonas aeruginosa. Can. J. Microbiol.* **31**:387–392.

80. **O'Toole, G. A., and R. Kolter.** 1998. Flagellar and twitching motility are necessary for *Pseudomonas aeruginosa* biofilm development. *Mol. Microbiol.* **30**:295–304.

81. **O'Toole, G. A., and R. Kolter.** 1998. Initiation of biofilm formation in *Pseudomonas fluorescens* WCS365 proceeds via multiple, convergent signalling pathways: a genetic analysis. *Mol. Microbiol.* **28**:449–461.

82. **Ochsner, U. A., A. K. Koch, A. Fiechter, and J. Reiser.** 1994. Isolation and characterization of a regulatory gene affecting rhamnolipid biosurfactant synthesis in *Pseudomonas aeruginosa. J. Bacteriol.* **176**:2044–2054.

83. **Ochsner, U. A., and J. Reiser.** 1995. Autoinducer-mediated regulation of rhamnolipid biosurfactant synthesis in *Pseudomonas aeruginosa. Proc. Natl. Acad. Sci. USA* **92**:6424–6428.

84. **Ostroff, R. M., and M. L. Vasil.** 1987. Identification of a new phospholipase C activity by analysis of an insertional mutation in the hemolytic phospholipase C structural gene of *Pseudomonas aeruginosa. J. Bacteriol.* **169**:4597–4601.

85. **Palleroni, N.** 1984. Family I. Pseudomonadaceae. *In* N. R. Krieg and J. G. Holt (ed.), *Bergey's Manual of Systematic Bacteriology*, vol. 1. The Williams & Wilkins Co., Baltimore, Md.

86. **Park, S., and D. R. Galloway.** 1995. Purification and characterization of LasD: a second staphylolytic proteinase produced by *Pseudomonas aeruginosa. Mol. Microbiol.* **16**:263–270.

87. **Parsek, M. R., A. L. Schaefer, and E. P. Greenberg.** 1997. Analysis of random and site-directed mutations in *rhlI*, a *Pseudomonas aeruginosa* gene encoding an acylhomoserine lactone synthase. *Mol. Microbiol.* **26**:301–310.

88. **Parsek, M. R., D. L. Val, B. L. Hanzelka, J. E. Cronan, Jr., and E. P. Greenberg.** 1999. Acyl homoserine-lactone quorum-sensing signal generation. *Proc. Natl. Acad. Sci. USA* **96**:4360–4365.

89. **Passador, L., J. M. Cook, M. J. Gambello, L. Rust, and B. H. Iglewski.** 1993. Expression of *Pseudomonas aeruginosa* virulence genes requires cell-to-cell communication. *Science* **260**:1127–1130.

90. **Pearson, J. P., K. M. Gray, L. Passador, K. D. Tucker, A. Eberhard, B. H. Iglewski, and E. P. Greenberg.** 1994. Structure of the autoinducer required for expression of *Pseudomonas aeruginosa* virulence genes. *Proc. Natl. Acad. Sci. USA* **91**:197–201.

91. **Pearson, J. P., L. Passador, B. H. Iglewski, and E. P. Greenberg.** 1995. A second *N*-acylhomoserine lactone signal produced by *Pseudomonas aeruginosa. Proc. Natl. Acad. Sci. USA* **92**:1490–1494.

92. **Pearson, J. P., C. Van Delden, and B. H. Iglewski.** 1999. Active efflux and diffusion are involved in transport of *Pseudomonas aeruginosa* cell-to-cell signals. *J. Bacteriol.* **181**:1203–1210.

93. **Pesci, E. C., J. P. Pearson, P. C. Seed, and B. H. Iglewski.** 1997. Regulation of *las* and *rhl* quorum sensing systems in *Pseudomonas aeruginosa. J. Bacteriol.* **179**:3127–3132.

94. **Preston, M. J., P. C. Seed, D. S. Toder, B. H. Iglewski, D. E. Ohman, J. K. Gustin, J. B. Goldberg, and G. B. Pier.** 1997. Contribution of proteases and LasR to the virulence of *Pseudomonas aeruginosa* during corneal infections. *Infect. Immun.* **65**:3086–3090.

95. **Prigent-Combaret, C., O. Vidal, C. Dorel, and P. Lejeune.** 1999. Abiotic surface sensing and biofilm-dependent regulation of gene expression in *Escherichia coli. J. Bacteriol.* **181**:5993–6002.

96. **Prince, A.** 1992. Adhesins and receptors of *Pseudomonas aeruginosa* associated with infection of the respiratory tract. *Microb. Pathog.* **13**:251–260.

97. **Pritchard, A. E., and M. L. Vasil.** 1986. Nucleotide sequence and expression of a phosphate-regulated gene encoding a secreted hemolysin of *Pseudomonas aeruginosa. J. Bacteriol.* **167**:291–298.

98. **Rahme, L. G., E. J. Stevens, S. F. Wolfort, J. Shao, R. G. Tompkins, and F. M. Ausubel.** 1995. Common virulence factors for bacterial pathogenicity in plants and animals. *Science* **268**:1899–1902.

99. **Reimmann, C., M. Beyeler, A. Latifi, H. Winteler, M. Foglino, A. Lazdunski, and D. Haas.** 1997. The global activator GacA of *Pseudomonas aeruginosa* PAO positively controls the production of the autoinducer N-butyryl-homoserine lactone and the formation of the virulence factors pyocyanin, cyanide, and lipase. *Mol. Microbiol.* **24**:309–319.

100. **Ruby, E. G.** 1996. Lessons from a cooperative bacterial-animal association: the *Vibrio fischeri–Euprymna scolopes* light organ symbioses. *Annu. Rev. Microbiol.* **50**:591–624.

101. **Sata, H., K. Okinda, and H. Saiton.** 1988. Role of pilin in the pathogenesis of *Pseudomonas aeruginosa* burn infection. *Microbiol. Immunol.* **32**:131–139.

102. **Schaefer, A. L., D. L. Val, B. L. Hanzelka, J. E. Cronan, Jr., and E. P. Greenberg.** 1996. Generation of cell-to-cell signals in quo-

rum sensing: acyl homoserine lactone synthase activity of a purified *Vibrio fischeri* LuxI protein. *Proc. Natl. Acad. Sci. USA* **93**:9505–9509.

103. **Seed, P. C., L. Passador, and B. H. Iglewski.** 1995. Activation of the *Pseudomonas aeruginosa lasI* gene by LasR and the *Pseudomonas* autoinducer PAI: an autoinduction regulatory hierarchy. *J. Bacteriol.* **177**:654–659.

104. **Shadel, G. S., R. Young, and T. Baldwin.** 1990. Use of regulated cell lysis in a lethal genetic selection in *Escherichia coli*: identification of the autoinducer-binding region of the LuxR protein from *Vibrio fischeri* ATCC 7744. *J. Bacteriol.* **172**:3980–3987.

105. **Shreve, G. S., S. Inguva, and S. Gunnam.** 1995. Rhamnolipid biosurfactant enhancement of hexadecane biodegradation by *Pseudomonas aeruginosa. Mol. Mar. Biol. Biotechnol.* **4**:331–337.

106. **Sitnikov, D. M., J. B. Schineller, and T. O. Baldwin.** 1995. Transcriptional regulation of bioluminescence genes from *Vibrio fischeri. Mol. Microbiol.* **17**:801–812.

107. **Sorensen, R. U., and F. J. Joseph.** 1993. Phenazine pugments in *Pseudomonas aeruginosa* infection, p. 43–57. *In* M. Campa, M. Bendenelli, and H. Friedman (ed.), *Pseudomonas aeruginosa as an Opportunistic Pathogen.* Plenum Press, New York, N.Y.

108. **Stevens, A. M., K. M. Dolan, and E. P. Greenberg.** 1994. Synergistic binding of the *Vibrio fischeri* LuxR transcriptional activator domain and RNA polymerase to the *lux* promoter region. *Proc. Natl. Acad. Sci. USA* **91**:12619–12623.

109. **Stevens, A. M., and E. P. Greenberg.** 1997. Quorum sensing in *Vibrio fischeri*: essential elements for activation of the luminescence genes. *J. Bacteriol.* **179**:557–562.

110. **Stintzi, A., K. Evans, J. M. Meyer, and K. Poole.** 1998. Quorum-sensing and siderophore biosynthesis in *Pseudomonas aeruginosa: lasR/lasI* mutants exhibit reduced pyoverdine biosynthesis. *FEMS Microbiol. Lett.* **166**:341–345.

111. **Stintzi, A., Z. Johnson, M. Stonehouse, U. Ochsner, J. M. Meyer, M. L. Vasil, and K.** Poole. 1999. The pvc gene cluster of *Pseudomonas aeruginosa*: role in synthesis of the pyoverdine chromophore and regulation by PtxR and PvdS. *J. Bacteriol.* **181**:4118–4124.

112. **Tamura, Y., S. Suzuki, and T. Sawada.** 1992. Role of elastase as a virulence factor in experimental *Pseudomonas aeruginosa* infection in mice. *Microb. Pathog.* **12**:237–244.

113. **Tang, H. B., E. DiMango, R. Bryan, M. Gambello, B. H. Iglewski, J. B. Goldberg, and A. Prince.** 1996. Contribution of specific *Pseudomonas aeruginosa* virulence factors to pathogenesis of pneumonia in a neonatal mouse model of infection. *Infect. Immun.* **64**:37–43.

114. **Tommassen, J., A. Filloux, M. Bally, M. Murgier, and A. Lazdunski.** 1992. Protein secretion in *Pseudomonas aeruginosa. FEMS Microbiol. Lett.* **9**:73–90.

115. **Wick, M. J., A. N. Hamood, and B. H. Iglewski.** 1990. Analysis of the structure-function relationship of *Pseudomonas aeruginosa* exotoxin A. *Mol. Microbiol.* **4**:527–535.

116. **Winson, M. K., M. Camara, A. Latifi, M. Foglino, S. R. Chhabra, M. Daykin, M. Bally, V. Chapon, G. P. C. Salmond, B. W. Bycroft, A. Lazdunski, G. S. A. B. Stewart, and P. Williams.** 1995. Multiple *N*-acyl-L-homoserine lactone signal molecules regulate production of virulence determinants and secondary metabolites in *Pseudomonas aeruginosa. Proc. Natl. Acad. Sci. USA* **92**:9427–9431.

117. **Woods, D. E., S. J. Cryz, R. L. Friedman, and B. H. Iglewski.** 1982. Contribution of toxin A and elastase to virulence of *Pseudomonas aeruginosa* in chronic lung infection of rats. *Infect. Immun.* **36**:1223–1228.

118. **Xiao, R., and W. S. Kisaalita.** 1997. Iron acquisition from transferrin and lactoferrin by *Pseudomonas aeruginosa* pyoverdin. *Microbiology* **143**:2509–2515.

119. **Zhang, L., P. J. Murphy, A. Kerr, and M. E. Tate.** 1993. *Agrobacterium* conjugation and gene regulation by *N*-acyl-L-homoserine lactones. *Nature* (London) **362**:446–448.

ANTIBIOTIC RESISTANCE AND SURVIVAL IN THE HOST

Emma L. A. Macfarlane and Robert E. W. Hancock

6

At the start of the antibiotic era it was predicted that antibiotics would effectively eradicate bacterial diseases. The failure of these drugs in this respect is directly attributable to the exceptional adaptive abilities of bacteria. The alarming increase in the number of multidrug-resistant microorganisms isolated in both clinical and nonclinical settings over the past decade parallels the rise in antibiotic use and exemplifies these adaptive abilities. What is rapidly becoming clear, therefore, is that the emergence of antibiotic resistance in bacteria is a logical consequence of their ability to overcome host defenses and successfully colonize a variety of species, including humans.

ANTIBIOTICS AND THEIR UPTAKE

Although a large number of antibiotic drugs are currently in clinical use, these drugs have limited structural diversity and fall into one of only several chemical classes (Table 1). Moreover, while many new drugs have appeared in the clinic over the past decade, they are all variations of these classical structures. No new chemical class of antimicrobials has been introduced since the quinolones more than 30 years ago. The new chemical classes of drugs currently undergoing clinical trials, including the oxazolidinones and the streptogramins, are bacteriostatic and of rather narrow specificity (versus gram-positive organisms) and thus seem unlikely to make a major impact on infectious diseases medicine.

Current antibiotic drugs target a variety of cellular processes and result in cell stasis or death through inhibition of protein, RNA or DNA synthesis, disruption of permeability barriers, or inhibition of cell wall peptidoglycan biosynthesis (Table 1). The effects of bacteriostatic drugs can often be enhanced by combination with a second drug; such "combination therapy" can yield synergistic bacteriocidal effects.

Antimicrobial agents gain access into the bacterial cell by one of several methods. Uptake across the outer membrane of gram-negative bacteria often relies on passive diffusion through the nonspecific protein channels of porins. Certain antibiotics can pass through gated or specific porins (e.g., imipenem uptake through porin OprD in *Pseudomonas aeruginosa*). Hydrophobic antibiotics are thought to directly cross the outer membrane bilayer, whereas large polycationic antibiotics

Emma L. A. Macfarlane and Robert E. W. Hancock Department of Microbiology, 300-6174 University Boulevard, University of British Columbia, Vancouver, British Columbia, Canada V6T 1Z3.

Virulence Mechanisms of Bacterial Pathogens, 3rd ed., Edited by K. A. Brogden et al.
©2000 ASM Press, Washington, D.C.

TABLE 1 Major classes of antibiotics in current medical use

Antibiotic class	Mechanism of action	Mechanism(s) of resistance
Aminoglycosides	Inhibition of protein synthesis	Enzymatic modification of drug, efflux
	Inhibition of initiation of DNA synthesis	Alterations in energy of uptake, mutation of target
Tetracyclines	Inhibition of protein synthesis	Increased efflux
Chloramphenicol	Inhibition of protein synthesis	Enzymatic modification of drug
Macrolides	Inhibition of protein synthesis	Target modification
Fusidanes	Inhibition of polypeptide chain elongation	Efflux
β-Lactams	Inhibition of cell wall biosynthesis and assembly. Stimulation of autolysins.	Enzymatic modification of drug
Glycopeptides	Inhibition of cell-wall biosynthesis	Structural modification of peptidoglycan
Quinolones and fluoroquinolones	Inhibition of DNA gyrase	Target modification, decreased uptake due to porin modifications
Novobiocin	Inhibition of DNA gyrase	Efflux
Rifamycins	Inhibition of nucleic acid synthesis	Target modification
Polymyxins and colistin	Disruption of outer membrane	LPS modifications
Sulfonamides	Inhibition of folic acid synthesis	Target modification, efflux
Trimethoprim	Inhibition of folic acid synthesis	Target modification, efflux

such as polymyxin B, as well as aminoglycosides and antimicrobial peptides, utilize "self-promoted uptake" (28) to traverse this barrier. During self-promoted uptake, the divalent cations that normally stabilize the outer membrane by cross-bridging binding sites on adjacent lipopolysaccharide (LPS) molecules are competitively displaced by the polycations. The large size of these antibiotic molecules leads to a disruption of the outer membrane, resulting in increased membrane permeability and entry of the polycation into the cell.

Many drugs passively diffuse across the cytoplasmic membranes of both gram-positive and -negative bacteria. For zwitterionic drugs, such as tetracyclines or fluoroquinolones, it is the occasional uncharged drug molecule that successfully crosses the membrane (45). While active uptake of drugs is rare in bacteria, some little-used antibiotics such as cycloserine and phosphonomycin mimic natural substrates and can enter the cell through specific active transport systems (27).

BACTERIAL INTRINSIC RESISTANCE

Bacteria possess certain intrinsic properties that provide natural resistance to some classes of antibiotics. Gram-negative bacteria, for example, are intrinsically more resistant than gram-positive organisms to a variety of antibiotics. This is due in part to the relative impermeability of their outer membrane, which provides a barrier to the initial uptake of drugs. Modifying or hydrolytic enzymes provide bacteria with a method of neutralizing certain drugs that have gained access to the cell. For example, β-lactamases that hydrolyze a range of β-lactam-based antibiotics have been identified in a number of species. At least four major classes of these enzymes with differing specificities have now been identified (12), and while some are the product of a gene induced by contact of the cell with low levels of β-lactams, others are constitutively expressed from plasmid-borne genes. Enzymes that modify (and therefore render ineffective) certain classes of aminoglycosides and chloramphenicol are also a part of the natural arsenal of bacteria (56, 67). The majority of genes encoding these enzymes are located on mobile genetic elements (plasmids or transposons), but they are also occasionally found in the chromosome (24).

An important contribution to the intrinsic resistance of bacteria is made by efflux pump systems. Such systems efficiently transport unwanted antimicrobial agents out of the cell by an active process that derives its energy from the proton motive force. Many efflux pumps have a broad specificity that allows them to export an array of antimicrobial compounds including detergents and organic solvents. The ability of these substances to achieve intracellular concentrations equivalent to toxic levels is thus limited. Efflux pumps found in gram-positive bacteria tend to fall into the major facilitator (MF) family and have a relatively restricted substrate profile (leading to a relatively restricted resistance pattern). Resistance to tetracyclines, quinolones, and macrolides can be mediated by such systems. Tetracycline efflux pumps of the MF family are often plasmid-encoded and are a common mechanism of tetracycline resistance in gram-negative bacteria. Small efflux pumps of the staphylococcal multidrug resistance (SMR) family also occur in both gram-positive and gram-negative bacteria. However, these pumps transport mainly quaternary ammonium salts, basic dyes, and lipophilic cations and do not contribute significantly to clinical resistance (48). Most pumps that give rise to clinically relevant antibiotic resistance fall into the resistance-nodulation-division (RND) efflux family (47). Thus *Escherichia coli* K-12 carrying a mutation in the *acr* operon is deficient in the AcrA-AcrB-TolC RND efflux system (17, 36) and displays hypersusceptibility to a broad range of dyes, detergents, and antibiotics (42). Likewise, the intrinsic resistance of *P. aeruginosa*, an opportunistic pathogen that frequently causes problematic infections in patients with cystic fibrosis, shows a strong dependence on the MexA-MexB-OprM RND system (50). The recent identification of multidrug-resistant clinical isolates of *P. aeruginosa* as overexpression mutants of the MexA-MexB-OprM system (68) underlines the importance of this type of resistance mechanism. Similar broad-range efflux pumps have been identified in a number of other gram-negative bacteria (for a review, see reference 44).

ANTIBIOTIC RESISTANCE AND SURVIVAL IN THE HOST

While research into the acquisition, regulation, and mechanisms of antibiotic resistance has been intense, the interrelationship of antibiotic resistance and virulence is a largely neglected area. A close relationship between these two factors is intuitive. The nature of modern medicine dictates that, at some point, all pathogenic bacteria will encounter antimicrobials; thus, resistance will inevitably arise (possibly based on host-defense evasion mechanisms). The emergence of nonintrinsic antibiotic resistance in a bacterial population is generally due to any of three basic resistance mechanisms: alteration of membrane-permeability, including modifications to efflux pumps and porins; modification, or even replacement, of the drug target; or increased expression of enzymes capable of modifying or hydrolyzing the drug. A fourth mechanism, overexpression of the drug target that results in the antibiotic being "swamped," is relevant only to bacteria such as the mycobacteria, for which interspecies genetic exchange is rare (13).

Cost of Resistance

Of the resistance mechanisms mentioned, both membrane alteration and target modification can call for a high expenditure of energy and resources. Therefore, maintenance of these resistance phenotypes may impose a "cost" on bacteria that results in a decreased ability to survive once the antibiotic pressure is removed. Numerous studies lend support to the "cost of resistance" theory, demonstrating that resistant bacteria grow relatively slowly and are less virulent than their antibiotic-susceptible counterparts (8, 61). It is worth noting that in certain circumstances, decreased virulence may be only a minor disadvantage. The occurrence of *Staphylococcus aureus* bacteremic infections, for example, is rare in healthy humans but is frequently problematic in immunocompromised individuals. Infections due to methicillin-resistant *S. aureus* (MRSA) arising in these individuals can be life-threatening. Mizobuchi et al. (41) used a mouse model to demonstrate that while MRSA shows decreased virulence in normal hosts, in neutropenic hosts its virulence is comparable to the methicillin-sensitive strain. The authors postulated that this decreased virulence may actually be advantageous by allowing MRSA to more easily colonize healthy "carriers," thereby facilitating its transport to new sites of infection.

However, the cost of resistance is often high. It follows that the highly adaptive nature of bacteria may foster the acquisition of secondary mutations that restore fitness and virulence while allowing resistance to be maintained. Indeed, this phenomenon is becoming increasingly well documented. Antibiotic resistance frequently results from the acquisition of mobile genetic elements carrying one or more resistance genes. Horizontal transfer of these elements, usually conjugative plasmids or transposons, readily occurs between bacteria of different species and genera and is a major contributor to the emergence of antibiotic-resistant strains. Resistance genes on plasmids or transposons often occur in clusters that are generated by integrons (26), and the acquisition of a single genetic element can thus confer resistance to several antibiotics (15). Maintenance of this element may impose a cost on the host bacterium (34), particularly if it is large and encodes many proteins. However, several studies have shown that, rather than losing these elements in the absence of antibiotic selective pressure, many bacteria succeed in stably maintaining them (for a review, see reference 54). This may be due in part to the ability of these plasmids to replicate in multiple copies (up to 50 per cell). In addition, genetic elements carrying clusters of resistance genes regularly carry other genes that confer a selective advantage, such as heavy metal resistance or certain virulence factors (1, 62). Thus, bacteria harboring these elements maintain a survival advantage over antibiotic-susceptible strains even in the absence of antibiotics.

When resistance is the result of a chromosomal mutation and imposes a large cost on the bacterial cell, the most logical mechanism to restore fitness once the antibiotic is removed is reversion to wild type. However, studies on streptomycin-resistant *rpsL* mutants of *E. coli* demonstrated that within 180 generations of growth in the absence of antibiotic, secondary site mutations had occurred that reduced the cost of resistance by more than 50% without any reduction in antibiotic resistance (S. J. Shrag and V. Perrot, Letter, *Nature* **381**:120–121, 1996). A more recent study using *rpsL, rpoB,* and *gyrA* mutants of *Salmonella enterica* serovar Typhimurium resistant to streptomycin, rifampin, and nalidixic acid, respectively, indicated that secondary-site compensatory mutants occurred rapidly (within 18 to 36 generations). These mutations restored wild-type virulence while retaining high levels of resistance and were significantly more frequent than reversions to wild type (9).

Persistence and Tolerance—Nonclassical Resistance Mechanisms

Failure of antibiotic therapy in the treatment of bacterial infections despite achieving in vivo drug concentrations above the in vitro MICs can often be traced to transient phenotypic changes. On removal from the host bacteria revert to full susceptibility, making this type of resistance (termed persistence) (11) difficult to study.

Significant advances in our understanding of bacterial persistence have come from studies of *S. aureus* small colony variants (SCVs) (for a review, see reference 51). *S. aureus* SCVs are characterized by their small size, atypical colony morphology, weak coagulase activity, lack of pigmentation, and lack of hemolytic activity. They are frequently associated with very persistent infections and show increased resistance to most antibiotics. Proctor et al. have recently proposed that all these phenotypic characteristics can be accounted for by a deficient electron transport system in these variants (52). The effects of reduced electron transport on the cell are substantial due to a decreased synthesis of ATP, which is essential for many cellular reactions. Reduction of the electrochemical gradient across the membrane would also decrease the ability of the cell to take up positively charged molecules, including aminoglycosides. SCVs also demonstrate reduced virulence, possibly due to their slow growth rate, and decreased α-toxin production, which enable them to persist in mammalian cells (4). Decreased exotoxin production may also contribute to the ability of these variants to evade host defenses and to their increased resistance to antibiotics in vivo.

The cost of this type of resistance is evidently high given the rapid reversion rate in the absence of antibiotics, but it confers significant advantages by allowing bacteria to persist in the host despite rigorous antibiotic therapy. Interestingly, both antibiotic therapy and the internal milieu of mammalian cells can select for SCVs (65), and thus S. aureus SCVs with increased antibiotic resistance can emerge even in the absence of antibiotic pressure.

Bryan et al. first used the term persistence in 1985 to describe the unstable forms of resistance they observed in mutants of P. aeruginosa defective in energy generation (11). These mutants demonstrated slow growth, aminoglycoside resistance, reduced virulence in iron-dextran treated mice, and very high reversion rates. Variants with similar phenotypic traits, together with decreased pyocyanin production, were isolated during unsuccessful antibiotic therapy of experimental endocarditis in a rabbit model (5). The strains isolated in both studies were most probably SCVs of P. aeruginosa. Small colony variants of several other bacterial species have also been reported (51), indicating that persistence is likely to be a widespread resistance mechanism in bacteria.

Persistent infections are also closely associated with bacterial biofilm production by the infecting organism. Biofilms are produced by a wide range of both gram-negative and gram-positive bacteria and are defined as structured communities of cells enclosed in a self-produced exopolysaccharide matrix that ad-

here to inert or living surfaces (for a review, see reference 14). Biofilms are inherently resistant to both antibiotics and host defenses. Exopolysaccharide, which encases the adherent bacteria in biofilms, provides a permeability barrier that may be difficult for antibiotics and other compounds to penetrate. In addition, many bacterial cells toward the center of the biofilm will be in a slow-growing or nutrient-starved state that greatly reduces the effectiveness of any compound targeting metabolic processes.

Production of mature biofilms involves a complex regulatory pathway. In Pseudomonas species, the initial attachment of planktonic cells to an appropriate surface involves the sad genes, flagella, and other motility factors (46). Subsequent formation of microcolonies is dependent on twitching motility and type IV pili, while development of the mature biofilm appears to be initiated by the LasR-LasI quorum-sensing system. Quorum sensing is a method of detecting cell density that is widespread in bacteria and involves the release of chemical signal molecules, the acylhomoserine lactones (also called autoinducers). In the P. aeruginosa LasR-LasI system, LasI directs the synthesis of 3-oxododecanoylhomoserine lactone, while LasR functions as a response regulator. LasR regulates a number of important virulence genes in addition to a second quorum-sensing system, RhlR-RhlI. When adherent bacteria reach a critical cell density, the accumulation of autoinducer within an individual cell triggers the transcription of genes required for the transition of undifferentiated microcolonies into mature biofilms.

Biofilms are of particular relevance to infections in cystic fibrosis patients, in whom conversion of P. aeruginosa to a mucoid phenotype is associated with the development of biofilms and leads to chronic infection (for a review, see reference 18). Conversion to mucoidy is usually due to a regulatory mutation and results in the overproduction of alginate exopolysaccharide. Alginate has been associated with the suppression of lymphocyte function and stimulation of proinflammatory cytokines, IL-1, and tumor necrosis factor in host cells (18). Regulation of alginate production involves the response regulator AlgR (18), which is involved in twitching motility, and the alternative sigma factor AlgU (55) that is part of the stress response pathway in *P. aeruginosa*. This intricate interplay between quorum-sensing, biofilm development, stress response, and production of virulence factors is typical of bacterial responses to survival pressure within host cells and demonstrates how antibiotic resistance can arise as a corollary of such adaptive mechanisms.

Like persistence, antibiotic tolerance can also be viewed as a nonclassical mechanism of antibiotic resistance. Whereas the SCVs that give rise to persistent infections continue to grow at a reduced rate, tolerant strains cease to divide in the presence of antibiotics (63). However, these strains remain viable and resume growth once the antibiotic pressure is removed. In a recent study, a vancomycin-tolerant mutant of *Streptococcus pneumoniae* was isolated by screening a library of loss-of-function mutants for loss of penicillin-induced autolysis (R. Novak, B. Henriques, E. Charpentier, S. Normark, and E. Tuomanen, Letter, *Nature* **399**:590–593, 1999). The mutation responsible for the tolerant phenotype was mapped to the histidine-kinase of a two-component regulatory system, VncR–VncS. The *vncS* mutant also displayed tolerance to other classes of antibiotics and was more readily transformed to high-level resistance than its parent. In a rabbit meningitis model, the *vncS* mutant survived vancomycin treatment that effectively eradicated the parent strain and resumed growth 12 h after treatment ceased. The *vncS* mutant still produced active LytA autolysin, indicating that the tolerant phenotype was due to an interruption in the autolysis pathway rather than repression of *lytA* expression. The authors also demonstrated the presence of vancomycin-tolerant *S. pneumoniae* among clinical isolates. These strains all contained a mutation in *vncS* and were cross-resistant to penicillin.

Two-Component Signal Transduction—a Link between Antibiotic Resistance and Virulence?

Two-component regulatory systems, such as VncR–VncS, are ubiquitous in bacteria. Similar systems are noticeably absent from mammalian cells. These systems allow the global regulation of a large number of genes in response to an environmental stimulus and thus play an essential role in the rapid adaptation of bacteria to new environments. Classic two-component regulators comprise a sensor histidine-kinase protein, which is usually located in the cytoplasmic membrane, and a response regulator protein which, when activated, is capable of recognizing and binding certain DNA sequences. Signal transduction is achieved by a series of phosphorylation/dephosphorylation reactions. The sensor-kinase protein responds to an environmental signal by interacting directly with a signal ligand, or with a receptor that binds the ligand. This step induces autophosphorylation of a conserved histidine residue in the kinase at the expense of ATP. Transfer of this phosphate to the response regulator produces the active phosphorylated form of this protein that can bind DNA and activate or repress its target genes. More complex, expanded versions of two-component regulatory systems also exist, such as the phosphorelay system that controls sporulation in *Bacillus subtilis* (31). These involve intermediate proteins to transduce the signal from the histidine-kinase to the final response regulator.

One of the most extensively studied two-component regulatory systems is the PhoP-PhoQ system in *S. enterica* serovar Typhimurium, which provides an archetype for the role of these regulatory systems in virulence and antibiotic resistance. *S. enterica* serovar Typhimurium PhoP-PhoQ has been shown to be involved in the regulation of virulence (38), magnesium uptake (60), invasion of mammalian cells (7), LPS structural alterations (30), and resistance to defensins (16) and polymyxin B (21). *S. enterica* serovar Typhimurium PhoP-null mutants are avirulent (16), unable to survive in macrophages (38), and susceptible to defensins (38) and polymyxin (22). A mutant with a single base change in *phoQ*, which results in a threonine to valine change in the *N* terminus of the PhoQ protein, constitutively phosphorylates PhoP and is both attenuated for virulence and deficient in invasion of mammalian cells (7, 39). However, in contrast to the PhoP-null strain, this mutant shows increased resistance to polymyxin B (22).

The environmental stimulus sensed by PhoQ has been shown to be a decrease in extracellular magnesium ion concentrations (64). Recent studies have also indicated that PhoQ will respond to mild acid stress (6). Both low-magnesium concentrations and low pH are conditions encountered by the bacterium upon entry into host macrophages and thus will serve to activate the PhoP-PhoQ regulon and initiate adaptive changes required for survival in this hostile environment.

Some of the phenotypic characteristics controlled by *S. enterica* serovar Typhimurium PhoP-PhoQ are not regulated directly by PhoP, but are mediated by a second two-component regulatory system, PmrA-PmrB (22, 58). Alterations to the outer membrane LPS and the polymyxin B resistance that is associated with these changes are under the control of PmrA-PmrB (23, 30). While this system itself responds to low pH (through PmrA-activation) (58), it responds most strongly to PhoP-activation under low-magnesium conditions (58). Thus, the increased polymyxin resistance observed in the mutant that constitutively produces phospho-PhoP was due to increased activation of *pmrA-pmrB* (22). Phospho-PhoP is also directly responsible for transcription of the *phoP-phoQ* operon under inducing conditions (59).

The reduced invasion abilities of the PhoP constitutive mutant were shown to result from down-regulation of a PhoP-repressed gene, *hilA* (2). The *hilA* gene is part of a cluster of genes important for invasion of host epithelial cells found at centisome 63 of the *S. enterica* serovar Typhimurium genome and known as *Salmonella* pathogenicity island 1 (SPI-1) (40). HilA is itself a response regulator that coordinates expression of the genes on this pathogenicity island and has been shown to be under the control of both PhoP and a second response-regulator, SirA, which is believed to be part of another two-component regulatory system (32). The genes that encode both PhoP and SirA are located outside the pathogenicity island. The functions and interactions of the *S. enterica* serovar Typhimurium PhoP-PhoQ system are summarized in Figure 1.

PhoP-PhoQ homologs have been identified in a number of other bacterial species including *E. coli* (20, 33), *Shigella flexneri* (19), and *Salmonella enterica* serovar Typhi (3). In our laboratory, we have recently identified a PhoP-PhoQ homolog in *P. aeruginosa* (37). Through construction of PhoP- and PhoQ-null mutants, we demonstrated that this system is involved in polymyxin B resistance and virulence in *P. aeruginosa*. However, the system shows subtle differences from that in *S. enterica* serovar Typhimurium. In *P. aeruginosa*, the *phoP-phoQ* genes are located immediately downstream of *oprH*, which encodes a small outer membrane protein overexpressed under magnesium starvation conditions. We were able to show that *oprH-phoP-phoQ* form an operon that is activated by PhoP under low-magnesium conditions.

When grown under magnesium starvation conditions, *P. aeruginosa* is resistant to polymyxin B (10, 43). However, the PhoP-null strain showed polymyxin resistance equal to that of the wild-type strain, suggesting that this resistance is independent of the PhoP protein. In addition, the virulence of this mutant was equivalent to that of wild type in a neutropenic mouse model. In contrast, the PhoQ-null strain showed high levels of polymyxin B resistance under both high- and low-magnesium growth conditions and was 100-fold less virulent than wild type. This

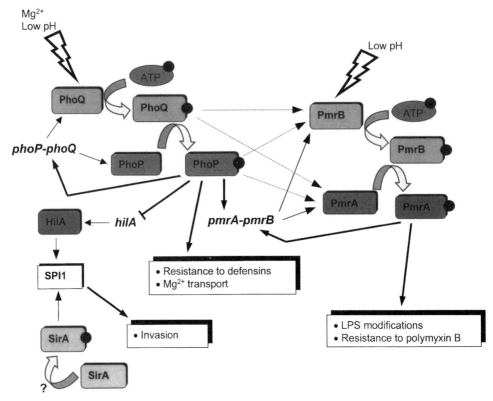

FIGURE 1 Schematic diagram summarizing the functions and interactions of the PhoP-PhoQ two-component regulatory system in *S. enterica* serovar Typhimurium.

strain was also shown to constitutively over-express both OprH and PhoP, which implies an indirect role for PhoP in polymyxin resistance and virulence and suggests that PhoQ may act primarily as a phosphatase that re-presses PhoP activation under high-magnesium conditions. A similar nontraditional role has been proposed for the $VanS_B$ histidine-kinase of the *Enterococcus faecalis* $VanS_B$-$VanR_B$ two-component regulatory system (57), as well as the recently identified *S. pneumoniae* VncS (R. Novak et al., Letter, *Nature* **399:**590–593, 1999).

Recent work has demonstrated the importance of antimicrobial cationic peptides in the nonspecific defense of host organisms against bacterial infections (29). As with polymyxin B resistance, we found that resistance of *P. aeruginosa* to the model cationic peptide CEMA

(49) was Mg^{2+}-regulated but not dependent upon a functional PhoP protein, with both the PhoP-null and an OprH-null mutant showing resistance to this peptide only in low-magnesium medium (E. L. A. Macfarlane and R. E. W. Hancock, unpublished data). The PhoQ-null mutant retained resistance and also showed significant resistance under high-magnesium conditions. Resistance to polymyxin B and to CEMA, therefore, is likely to follow similar pathways. The lack of phenotypic changes exhibited by the PhoP-null mutant with respect to both polymyxin B resistance and virulence suggests that regulation of these factors in *P. aeruginosa* may involve more than one regulatory system.

An interesting feature of the *P. aeruginosa* PhoQ-null mutant was its slow growth rate and consequent small colony size. Taken together with the increased antibiotic resistance of this mutant, this raises the interesting possibility that PhoQ may be involved in the development of *P. aeruginosa* SCVs.

Another illustration of the importance of two-component regulatory systems in the complex global responses of bacteria is provided by recent work on the *Pseudomonas* GacA-GacS system. In *P. aeruginosa,* GacA positively controls the production of the autoinducer *N*-butyrylhomoserine lactone produced by the RhlR-RhlI quorum-sensing system, together with the production of the virulence factors pyocyanin, cyanide, and lipase (53). In addition, GacA positively regulates production of both LasR and RhlR, although whether this is by direct interaction with the gene promoters has not been determined (53). The response regulator RhlR is known to influence expression of *rpoS*, the stationary phase sigma factor, in *P. aeruginosa* (35), and both GacA and GacS have been shown to influence RpoS accumulation in *Pseudomonas fluorescens* (66). Thus, in a complex regulatory cascade, GacA-GacS influences virulence, biofilm development, and antibiotic resistance (through LasR and RhlR) together with the RpoS-mediated stress response pathway. GacA homologs have been identified in a number of other species. Of particular note is the GacA homolog in *S. enterica* serovar Typhimurium, SirA, which is one of the regulators responsible for the transcription of virulence genes in SPI-1 (32).

Two-component regulatory systems provide bacteria with a way to integrate regulation of expression of virulence factors and antibiotic resistance into their general stress response pathways. The ability of bacteria to invade, colonize, and survive within a host is dependent on their adaptive capabilities when confronted with a variety of host defenses. Since these factors include both various forms of environmental stress (e.g., low osmolarity [25], mild acid, low cation concentrations) and antimicrobial compounds (such as cationic peptide defensins), the coordination of these bacterial responses to these separate challenges

provides a method of extremely rapid adaptation.

CONCLUSIONS

The ability of bacteria to acquire antibiotic resistance by horizontal transfer of resistance plasmids or transposons or by specific mutations is a well-documented reality of our antibiotic era. Resistance to antibiotics is also an inevitable consequence of the activation or mutation of the general stress response and adaptation pathways of bacteria. As most pathogenic bacteria will encounter antimicrobials during the course of an infection, the survival advantage conferred by this resistance in the face of increasing antibiotic selective pressure is significant. A more judicious use of current antimicrobial agents, together with the development of new drugs targeting other cellular processes (including two-component regulatory systems), may reduce the emergence of new multidrug-resistant bacterial strains. However, the frequency with which bacteria acquire secondary-site mutations to accommodate resistance factors, and the tendency for resistance genes to cluster and recruit other genes to confer selective advantages, suggests that bacteria could retain resistance even if antibiotic selective pressure is removed.

REFERENCES

1. **Amabile-Cuevas, C. F., and M. Cardenas-Garcia.** 1996. Antibiotic resistance: merely the tip of the iceberg of plasmid-driven bacterial evolution, p. 35–56. *In* C. F. Amabile-Cuevas (ed.), *Antibiotic Resistance: From Molecular Basics to Therapeutic Options.* R. G. Londes Co., Austin, Tex.
2. **Bajaj, V., R. L. Lucas, C. Hwang, and C. A. Lee.** 1996. Co-ordinate regulation of *Salmonella typhimurium* invasion genes by environmental and regulatory factors is mediated by control of *hilA* expression. *Mol. Microbiol.* **22:** 703–714.
3. **Baker, S. J., J. S. Gunn, and R. Morona.** 1999. The *Salmonella typhi* melittin resistance gene *pqaB* affects intracellular growth in PMA-differentiated U937 cells, polymyxin B resistance and lipopolysaccharide. *Microbiology* **145:**367–378.
4. **Balwit, J. M., P. van Langervelde, J. M. Vann, and R. A. Proctor.** 1994. Gentamicin-

resistant menadione and hemin auxotrophic *Staphylococcus aureus* persist within cultured epithelial cells. *J. Infect. Dis.* **170:**1033–1037.

5. **Bayer, A. S., D. C. Norman, and K. S. Kim.** 1987. Characterization of impermeability variants of *Pseudomonas aeruginosa* isolated during unsuccessful therapy of experimental endocarditis. *Antimicrob. Agents Chemother.* **31:**70–75.

6. **Bearson, B., L. Wilson, and J. W. Foster.** 1998. A low pH–inducible, PhoPQ-dependent acid tolerance response protects *Salmonella typhimurium* against inorganic stress. *J. Bacteriol.* **180:** 2409–2417.

7. **Behlau, I., and S. I. Miller.** 1993. A PhoP-repressed gene promotes *Salmonella typhimurium* invasion of epithelial cells. *J. Bacteriol.* **175:**4475–4484.

8. **Bennett, P. M., and A. H. Linton.** 1986. Do plasmids influence the survival of bacteria? *J. Antimicrob. Chemother.* **18**(Suppl C):123–126.

9. **Björkman, J., D. Hughes, and D. I. Andersson.** 1998. Virulence of antibiotic-resistant *Salmonella typhimurium. Proc. Natl. Acad. Sci. USA* **95:**3949–3953.

10. **Brown, M. R. W., and J. Melling.** 1969. Role of divalent cations in the action of polymyxin B and EDTA on *Pseudomonas aeruginosa. J. Gen. Microbiol.* **59:**263–271.

11. **Bryan, L. E., A. J. Godfrey, and T. Schollardt.** 1985. Virulence of *Pseudomonas aeruginosa* strains with mechanisms of microbial persistence for β-lactam and aminoglycoside antibiotics in a mouse infection model. *Can. J. Microbiol.* **31:** 377–380.

12. **Bush, K., G. A. Jacoby, and A. A. Medeiros.** 1995. A functional classification scheme for beta-lactamases and its correlation with molecular structure. *Antimicrob. Agents Chemother.* **39:**1211–1233.

13. **Chopra, I.** 1998. Over-expression of target genes as a mechanism of antibiotic resistance in bacteria. *J. Antimicrob. Chemother.* **41:**584–588.

14. **Costerton, J. W., P. S. Stewart, and E. P. Greenberg.** 1999. Bacterial biofilms: a common cause of persistent infections. *Science* **284:**1318–1322.

15. **Davies, J. E.** 1997. Origins, acquisition and dissemination of antibiotic resistance determinants. *Ciba Found. Symp.* **207:**15–27.

16. **Fields, P. I., E. A. Groisman, and F. Heffron.** 1989. A *Salmonella* locus that controls resistance to microbicidal proteins from phagocytic cells. *Science* **243:**1059–1062.

17. **Fralick, J. A.** 1996. Evidence that TolC is required for the functioning of the Mar / AcrAB efflux pump of *Escherichia coli. J. Bacteriol.* **178:** 5803–5805.

18. **Govan, J. R. W., and V. Deretic.** 1996. Microbial pathogenesis in cystic fibrosis: mucoid *Pseudomonas aeruginosa* and *Burkholderia cepacia. Microbiol. Rev.* **60:**539–574.

19. **Groisman, E. A., E. Chiao, C. J. Lipps, and F. Heffron.** 1989. *Salmonella typhimurium phoP* virulence gene is a transcriptional regulator. *Proc. Natl. Acad. Sci. USA* **86:**7077–7081.

20. **Groisman, E. A., F. Heffron, and F. Soloman.** 1992. Molecular genetic analysis of the *Escherichia coli phoP* locus. *J. Bacteriol.* **174:**486–491.

21. **Groisman, E. A., J. Kayser, and F. C. Soncini.** 1997. Regulation of polymyxin resistance and adaptation to low-Mg^{2+} environments. *J. Bacteriol.* **179:**7040–7045.

22. **Gunn, J. S., and S. I. Miller.** 1996. PhoP-PhoQ activates transcription of *pmrAB*, encoding a two-component regulatory system involved in *Salmonella typhimurium* antimicrobial peptide resistance. *J. Bacteriol.* **178:**6857–6864.

23. **Gunn, J. S., K. B. Lim, J. Krueger, K. Kim, L. Guo, M. Hackett, and S. I. Miller.** 1998. PmrA-PmrB-regulated genes necessary for 4-aminoarabinose modification and polymyxin B resistance. *Mol. Microbiol.* **27:**1171–1182.

24. **Hachler, H., P. Santarnam, and F. H. Kayser.** 1996. Sequence and characterization of a novel chromosomal aminoglycoside phosphotransferase gene *aph(3′)-IIb* in *Pseudomonas aeruginosa. Antimicrob. Agents Chemother.* **40:**1254–1256.

25. **Hall, M. N., and T. J. Sihavy.** 1981. Genetic analysis of the ompB locus in *Escherichia coli* K-12. *J. Mol. Biol.* **151:**1–15.

26. **Hall, R. M., and C. M. Collis.** 1995. Mobile gene cassettes and integrons: capture and spread of genes by site-specific recombination. *Mol. Microbiol.* **15:**593–600.

27. **Hancock, R. E. W.** 1994. Bacterial transport as an import mechanism and target for antimicrobials, p. 289–306. *In* N. H. Georgopapadakou (ed.), *Drug Transport in Antimicrobial and Anticancer Chemotherapy.* Marcel Dekker Inc., New York, N.Y.

28. **Hancock, R. E. W., and A. Bell.** 1988. Antibiotic uptake in Gram negative bacteria. *Eur. J. Microbiol. Infect. Dis.* **7:**713–720.

29. **Hancock, R. E. W., and R. Lehrer.** 1998. Cationic peptides: a new source of antibiotics. *TIBTECH* **16:**82–88.

30. **Hëlander, I. M., I. Kilpeläinen, and M. Vaara.** 1994. Increased substitution of phosphate groups in lipopolysaccharides and lipid A of the polymyxin-resistant *pmrA* mutants of *Salmonella typhimurium*: a ^{31}P-NMR study. *Mol. Microbiol.* **11:**481–487.

31. **Hoch, J. A.** 1993. The phosphorelay signal transduction pathway in the initiation of *Bacillus subtilis* sporulation. *J. Cell Biochem.* **51**:55–61.

32. **Johnston, C., D. A. Pegues, C. J. Hueck, C. A. Lee, and S. I. Miller.** 1996. Transcriptional activation of *Salmonella typhimurium* invasion genes by a member of the phosphorylated response-regulator superfamily. *Mol. Microbiol.* **22**:715–727.

33. **Kasahara, M., A. Nakata, and H. Shinagawa.** 1992. Molecular analysis of the *Escherichia coli phoP-phoQ* operon. *J. Bacteriol.* **174**:492–498.

34. **Klemperer, R. M. M., N. T. A. Ismail, and M. R. W. Brown.** 1979. Effect of R plasmid RP1 on the nutritional requirements of *Escherichia coli* in batch culture. *J. Gen. Microbiol.* **115**:325–331.

35. **Latifi, A., M. Foglino, K. Tanaka, P. Williams, and A. Lazdunski.** 1996. A hierarchical quorum-sensing cascade in *Pseudomonas aeruginosa* links the transcriptional activators LasR and RhlR (VsmR) to the expression of the stationary phase sigma factor RpoS. *Mol. Microbiol.* **21**:1137–1146.

36. **Ma, D., D. N. Cook, M. Alberti, N. G. Pon, H. Nikaido, and J. E. Hearst.** 1993. Molecular cloning and characterization of *acrA* and *acrE* genes of *Escherichia coli*. *J. Bacteriol.* **175**:6299–6313.

37. **Macfarlane, E. L. A., A. Kwasnicka, M. M. Ochs, and R. E. W. Hancock.** 1999. PhoP-PhoQ homologues in *Pseudomonas aeruginosa* regulate expression of the outer-membrane protein OprH and polymyxin B resistance. *Mol. Microbiol.* **34**:305–316.

38. **Miller, S. I., A. M. Kukral, and J. J. Mekalanos.** 1989. A two-component regulatory system (*phoP phoQ*) controls *Salmonella typhimurium* virulence. *Proc. Natl. Acad. Sci. USA* **86**:5054–5058.

39. **Miller, S. I., and J. J. Mekalanos.** 1990. Constitutive expression of the *phoP* regulon attenuates *Salmonella* virulence and survival within macrophages. *J. Bacteriol.* **172**:2485–2490.

40. **Mills, D. M., V. Bajaj, and C. A. Lee.** 1995. A 40kB chromosomal fragment encoding *Salmonella typhimurium* invasion genes is absent from the corresponding region of the *Escherichia coli* K-12 chromosome. *Mol. Microbiol.* **15**:749–759.

41. **Mizobuchi, S., J. Minami, F. Jin, O. Matsushita, and A. Okabe.** 1994. Comparison of the virulence of methicillin-resistant and methicillin-sensitive *Staphylococcus aureus*. *Microbiol. Immunol.* **38**:599–605.

42. **Nakamura, H.** 1968. Genetic determination of resistance to acriflavine, phenethyl alcohol, and sodium dodecyl sulfate in *Escherichia coli*. *J. Bacteriol.* **96**:987–996.

43. **Nicas, T. I., and R. E. W. Hancock.** 1980. Outer-membrane protein H1 of *Pseudomonas aeruginosa*: involvement in adaptive and mutational resistance to ethylenediaminetetraacetic acid, polymyxin B, and gentamicin. *J. Bacteriol.* **143**:872–878.

44. **Nikaido, H.** 1998. Antibiotic resistance caused by Gram negative multidrug efflux pumps. *Clin. Infect. Dis.* **27**:S32–S41.

45. **Nikaido, H., and D. G. Thanassi.** 1993. Penetration of lipophilic agents with multiple protonation sites into bacterial cells: tetracyclines and fluoroquinolones as examples. *Antimicrob. Agents Chemother.* **37**:1393–1399.

46. **O'Toole, G. A., and R. Kolter.** 1998. Initiation of biofilm development in *Pseudomonas fluorescens* WCS365 proceeds via multiple convergent signaling pathways: a genetic analysis. *Mol. Microbiol.* **30**:295–299.

47. **Paulsen, I. T., M. W. Brown, and R. A. Skurray.** 1996. Proton-dependent multidrug efflux systems. *Microbiol. Rev.* **60**:575–608.

48. **Paulsen, I. T., R. A. Skurray, R. Tam, M. H. Saier, Jr., R. J. Turner, J. H. Weiner, E. B. Goldberg, and L. L. Grinius.** 1996. The SMR family: a novel family of multidrug efflux proteins involved with the efflux of lipophilic drugs. *Mol. Microbiol.* **19**:1167–1175.

49. **Piers, K. L., M. H. Brown, and R. E. W. Hancock.** 1994. Improvement of outer membrane-permeabilizing and lipopolysaccharide-binding activities of an antimicrobial cationic peptide by C-terminal modification. *Antimicrob. Agents Chemother.* **38**:2311–2316.

50. **Poole, K., K. Krebes, C. McNally, and S. Neshat.** 1993. Multiple antibiotic resistance in *Pseudomonas aeruginosa*: evidence for involvement of an efflux operon. *J. Bacteriol.* **175**:7363–7372.

51. **Proctor, R. A., B. Kahl, C. von Eiff, P. E. Vandaux, D. P. Lew, and G. Peters.** 1998. Staphylococcal small colony variants have novel mechanisms for antibiotic resistance. *Clin. Infect. Dis.* **27**(Suppl 1):S68–S74.

52. **Proctor, R. A., J. M. Balwit, and O. Vesga.** 1994. Variant populations of *Staphylococcus aureus* can cause persistent and recurrent infections. *Infect. Agents Dis.* **3**:302–312.

53. **Reimmann, C., M. Beyeler, A. Latifi, H. Winteler, M. Foglino, A. Lazdunski, and D. Haas.** 1997. The global activator GacA of *Pseudomonas aeruginosa* PAO positively controls the production of the autoinducer *N*-butyrylhomoserine lactone and the formation of virulence factors pyocyanin, cyanide, and lipase. *Mol. Microbiol.* **24**:309–319.

54. **Salyers, A. A., and C. F. Amabile-Cuevas.** 1997. Why are antibiotic resistance genes so re-

sistant to elimination? *Antimicrob. Agents Chemother.* **41:**2321–2325.

55. **Schurr, M. J., H. Yu, J. C. Boucher, N. S. Hibler, and V. Deretic.** 1995. Multiple promoters and induction by heat shock of the gene encoding the alternative sigma factor AlgU (σ^E) which controls mucoidy in cystic fibrosis isolates of *Pseudomonas aeruginosa. J. Bacteriol.* **177:**5670–5679.

56. **Shaw, K. J., P. N. Rather, R. S. Hare, and G. H. Miller.** 1993. Molecular genetics of aminoglycoside resistance genes and familial relationships of the aminoglycoside-modifying enzymes. *Microbiol. Rev.* **57:**138–163.

57. **Silva, J. C., A. Haldimann, M. K. Prahalad, C. T. Walsh, and B. L. Wanner.** 1998. In vivo characterization of the type A and B vancomycin-resistant enterococci (VRE) VanRS two-component systems in *Escherichia coli*: a nonpathogenic model for studying the VRE signal transduction pathways. *Proc. Natl. Acad. Sci. USA* **95:**11951–11956.

58. **Soncini, F. C., and E. A. Groisman.** 1996. Two-component regulatory systems can interact to process multiple environmental signals. *J. Bacteriol.* **178:**6796–6801.

59. **Soncini, F. C., E. G. Véscovi, and E. A. Groisman.** 1995. Transcriptional autoregulation of the *Salmonella typhimurium phoPQ* operon. *J. Bacteriol.* **177:**4364–4371.

60. **Soncini, F. C., E. G. Véscovi, F. Soloman, and E. A. Groisman.** 1996. Molecular basis of the magnesium deprivation response in *Salmonella typhimurium*: identification of PhoP-regulated genes. *J. Bacteriol.* **178:**5092–5099.

61. **Spratt, B. G.** 1996. Antibiotic resistance: counting the cost. *Curr. Biol.* **6:**1219–1221.

62. **Summers, A. O., J. Wireman, M. J. Vimy, F. L. Lorsheider, B. Marshall, S. B. Levy, S. Bennet, and L. Billard.** 1993. Mercury released from dental "silver" fillings provokes an increase in mercury- and antibiotic-resistant bacteria in oral and intestinal floras of primates. *Antimicrob. Agents Chemother.* **37:**825–834.

63. **Tomasz, A., A. Albino, and E. Zanati.** 1970. Multiple antibiotic resistance in a bacterium with suppressed autolytic systems. *Nature* **227:**138–140.

64. **Véscovi, E. G., F. C. Soncini, and E. A. Groisman.** 1996. Mg^{2+} as an extracellular signal: environmental regulation of Salmonella virulence. *Cell* **84:**165–174.

65. **Vesga, O., M. C. Groeschel, M. F. Otten, D. W. Brar, J. M. Vann, and R. A. Proctor.** 1996. *Staphylococcus aureus* small colony variants are induced by the endothelial cell intracellular milieu. *J. Infect. Dis.* **173:**739–742.

66. **Whistler, C. A., N. A. Corbell, A. Sarniguet, W. Ream, and J. E. Loper.** 1998. The two-component regulators GacS and GacA influence accumulation of the stationary phase sigma factor σ^S and the stress response in *Pseudomonas fluorescens* Pf-5. *J. Bacteriol.* **180:**6635–6641.

67. **Wright, G. D., and P. R. Thompson.** 1999. Aminoglycoside phosphotransferases: proteins, structure, and mechanism. *Frontiers Biosci.* **9:**d9–d21.

68. **Ziha-Zarifi, I., C. Llanes, T. Köhler, J.-C. Pechere, and P. Plesiat.** 1999. In vivo emergence of multidrug resistant mutants of *Pseudomonas aeruginosa* overexpressing the active efflux system MexA-MexB-OprM. *Antimicrob. Agents Chemother.* **43:**287–291.

BACTERIAL EVASION
OF HOST DEFENSE
MECHANISMS

WHAT IS THE VERY MODEL OF A MODERN MACROPHAGE PATHOGEN?

David G. Russell

7

Our appreciation of how bacterial pathogens avoid killing by phagocytes has expanded greatly over the last few years, but it still falls woefully short of providing a rationale for tipping the balance in favor of the macrophage. There are multiple reasons for this that reflect the diversity of strategies that bacteria have evolved to maximize their chances of survival upon encountering a phagocyte. The fact that most pathogenic bacilli are capable of surviving at least limited exposure to the phagocyte bears witness to the powerful evolutionary pressure that the macrophage, which may in some instances have been preceded by free-living amoebae, has placed on these microbes.

Understanding the mechanisms of resistance to killing requires knowledge of the different routes of killing employed by the macrophage. Some are innate, while others are regulated by the host's immune response and illustrate the intimate association between the innate and acquired immune responses. This chapter describes the major mechanisms of killing used by the macrophage, how these killing mechanisms can be modulated by the

host immune response, and how different bacteria have learned to deal with this potential nemesis.

INNATE KILLING MECHANISMS

Macrophages are extremely degradative cells. One of their prime tasks while they migrate through body tissues is to recognize, internalize, and digest "unwanted" material. This material may be cellular debris or apoptotic cells that are recognized by phagocytes (31, 59) or may be microbes in the process of invading the tissues (56). Much like popular brands of lavatory cleaner, macrophages are capable of killing 99.9% of all known germs, most of which fall victim to the active lysosomal system present in these cells. Macrophages are equipped with a lysosomal network abundant with hydrolases, such as the endopeptidases cathepsins B, C, D, L, and S; exopeptidases such as cathepsins A and H; a range of lysosomal glycosidases and sulfatases; and various lipid-processing enzymes such as sphingomyelinase, ceramidase, and phospholipases A1, A2, and C (41, 44, 83). The vast majority of these enzymes show a relatively narrow pH optimum with maximal activity at an acid pH (pH 4.5 to 5.5), which is in keeping with the acidic nature of the host's lysosome. This impressive battery of lytic enzymes will rapidly attack any

David G. Russell Department of Molecular Microbiology, Washington University School of Medicine, 660 S. Euclid Avenue, St. Louis, MO 63110.

Virulence Mechanisms of Bacterial Pathogens, 3rd ed., Edited by K. A. Brogden et al.
©2000 ASM Press, Washington, D.C.

accessible targets on the surface of bacteria that are phagocytosed and delivered to the macrophage's lysosome.

In addition to the lysosomal system, resident, unstimulated macrophages are also capable of producing reactive oxygen intermediates (ROIs) such as superoxide and hydrogen peroxide through the translocation and activation of an NADPH oxidase and cytochrome C to the cell surface (12, 18, 81). Release of these ROIs from resting macrophages is dependent on the identity of receptors ligated by the bound bacilli. Receptors for the Fc portion of immunoglobulin (FcRs) and the mannose receptors trigger such release, whereas complement receptors (CR1, CR3, and CR4) will not unless the macrophage has been exposed to macrophage-activating cytokines such as gamma interferon (3).

One additional bacteriostatic or bacteriocidal property of the macrophage is the pH of the lysosome. Many bacteria show limited tolerance to low pH and intriguingly, one of the host genetic loci implicated in innate resistance to the intramacrophage pathogens *Leishmania donovana*, *Salmonella enterica* serovar Typhimurium, and *Mycobacterium bovis* BCG encodes the protein Nramp1, which has been shown to exert a profound effect on phagosome acidification in culture (28). The robust nature of this phenotype does raise the question as to why such an impressive effect shows little influence over the related pathogens *Leishmania major* and *Mycobacterium tuberculosis*.

IMMUNE-REGULATED KILLING MECHANISMS

Activation of macrophages by cytokines such as gamma interferon and tumor necrosis factor alpha will up-regulate the killing capacity of the phagocyte (48, 58, 61). These cytokines have pleotropic effects, affecting the killing pathways of the macrophages themselves and enhancing the responsiveness of the host's cellular immune system. Gamma interferon and tumor necrosis factor alpha up-regulate the superoxide burst, both in magnitude and in the range of phagocytic receptors capable of triggering the response. Moreover, treatment of macrophages with these cytokines will, after a delay of around 24 h, culminate in the expression of the inducible nitric oxide synthase (iNOS), which has been shown to play a critical role in the regulation and clearance of most intramacrophage pathogens (47, 63). iNOS metabolizes L-arginine to produce nitric oxide and citrulline. The nitric oxide produces nitrous acid and can combine with ROIs to produce peroxynitrite (85) or lead to formation of hypervalent iron—all rather nasty molecules!

This ability of macrophages to respond to cytokine stimulation and up-regulate their microbicidal battery renders them potent mediators of protection. However, note that the final effector molecules in these cascades are toxic and not restricted in site of action to the microbes themselves (53). In consequence, the host has negative regulators that will suppress these activities. This helps to protect the host from extraneous tissue damage, but also provides a potential route of modulation that the pathogen could exploit to inactivate a potentially protective immune response.

HOW DO THE PATHOGENS DO IT?

It is clear that the ability of any pathogen to survive and persist in macrophages as their primary host cell requires a fine balancing act. An "ideal" pathogen might induce an infection that caused little overt damage to the host cell, yet impaired its ability to converse with or respond to the host's immune system, thus ensuring the pathogen's enduring presence. Although some pathogens do follow this route, it is by no means the norm. Certain bacteria employ a shock approach and avoid clearance by the macrophage through the induction of an extensive inflammatory response and massive host cell death. Although these particular pathogens do not usually cause persistent infection, they are nonetheless extremely successful in penetrating the host population. The next section explores the range of strategies evolved by different bacterial species to ensure their success. Although adherence and uptake are definitely part of the process, these will not be dealt with in depth here. Most pathogens enter via complement receptors following deposition of C3b and iC3b on their surface. These receptors are relatively benign, showing minimal stimulation of microbicidal responses. Moreover, in all systems studied in depth, the identity of the receptor(s) ligated during the phagocytic process does not appear to play a dominant role in determining the outcome of the infection process (21, 62). The three predominant issues that affect intracellular survival involve (i) the choice of intracellular location, (ii) avoidance or suppression of the induction of a productive immune response, and (iii) suppression or resistance to the host cell's microbicidal responses.

THE INTRACELLULAR LOCATION
Bacterial pathogens exploit a range of different niches in their host cell. Some bacteria persist in lysosomes (*Coxiella burnetii*), some reside in modified endosomal compartments (*Mycobacterium* spp., *Ehrlichia*, and *Legionella*), while others manage to modify their entry vacuole to such a degree that the compartment exhibits minimal interaction with the host cell's endosomal system (*Chlamydia* spp., reviewed in reference 56). In addition to these bacilli that remain intravacuolar, other bacterial pathogens (*Listeria* and *Shigella*) rapidly lyse their entry vacuole and establish an infection within the host cell cytosol.

The parasitophorous vacuole in which *Coxiella* resides is acidic, freely accessible to lysosomal tracers and is therefore likely to be actively hydrolytic (29, 34). Despite this, the bacilli thrive and multiply in this compartment. How the bacteria resist degradation is unclear; however, a body of evidence is accumulating suggesting that another intralysosomal pathogen, the protist *Leishmania*, survives through restricting the exposure of potential substrates on its surface, which is covered in a truncated lipidoglycan that appears to be resistant to degradation in the mac-

rophage (84). This scenario is aided by the parasite's flagellar pocket, which allows uptake of host macromolecules but restricts access of hydrolases to the parasite (80).

Another bacterium that enters via a phagosome that exhibits communication with lysosomal compartments is serovar Typhimurium. However, the data on the nature of the vacuoles in which these bacilli reside are contradictory and indicate that the vacuole population is extremely heterogeneous (1, 2, 9, 40, 51). Despite reports of restricted fusion of the bacterial phagosome with lysosomes, treatment of infected cells with inhibitors of the H-ATPase (which acidifies intracellular compartments) actually reduces bacterial viability, suggesting that the bacterium requires acidification to stimulate expression of genes important for intracellular survival (51). However, recent studies on the substrates for the SPI-2 type III secretion complex have indicated that one of these proteins, SpiC, is capable of suppressing vesicle-vesicle fusion in vitro (72), and mutation at this locus attenuates the virulence of the bacterium. The data are robust, suggesting that there is still much to do to elucidate the physical characteristics of the *Salmonella*-containing vacuole.

Of the bacteria that enter via phagocytosis yet modulate their vacuole following uptake, the best studied are *Mycobacterium*, *Legionella*, and *Chlamydia* spp., all of which exhibit differing degrees of sequestration of their vacuoles within the host cell (64). The vacuoles in which *M. tuberculosis*, *Mycobacterium avium*, and *M. bovis* BCG reside show several common features (16, 58). These vacuoles exhibit limited acidification, show minimal fusion with lysosomes, yet remain highly dynamic and readily accessible to certain elements of the host cell plasmalemma (13, 14, 49, 57, 67, 68, 86). Closer examination demonstrated that the vacuoles remain accessible to the recycling endosomal pathway of the host macrophage, lying within the continuum defined by the routing of the transferrin–transferrin receptor complex. This retention within a peripheral compartment is reflected in the persistence of the GTPase rab 5 on the vacuole membrane (75). rab 5 is a small GTP-binding protein that modulates the membrane fusion machinery on early endocytic compartments. Lysosomal proteins that do enter mycobacterial vacuoles, such as cathepsin D, appear to have originated from the host cell's synthetic pathway following exit from the trans-Golgi network (67). These data indicate that *Mycobacterium* spp. avoid the ramifications of remaining within the endocytic network through arresting the normal progression of their vacuoles to lysosomes. These vacuoles also possess a unique cytoplasmic "coat" protein called TACO (24), which is a coronin, or actin-binding protein. It is unclear if the presence of the protein, which is another potential regulator of membrane fusion, is symptomatic of or causal to the properties of these vesicles. How this is achieved is unknown, although speculation has implicated cell-wall lipids (65) or the activity of the bacterium's urease or glutamine synthase (32) as potential modulators of acidification. It is interesting that compounds that prevent acidification, such as bafilomycin A or weak bases, also reduce the maturation of phagosomal or endosomal compartments. This suggests that, in this instance, suppression of the symptom (acidification) may reduce the cause, i.e., phagosomal maturation and recruitment of H-ATPase complexes (11, 74).

Legionellae modulate their vacuole to a greater degree than mycobacteria. Vacuoles containing *Legionella* bacilli exhibit limited and transient acquisition of endosomal proteins and rapidly sequester their vacuole out with the endosomal system (38, 78). The vacuoles become intimately associated with the host cell's rough endoplasmic reticulum and avoid any subsequent interaction with lysosomes. The successful sequestration of the bacterial vacuoles requires expression of a set of genes defined by the *dot* or *icm* locus that encodes a type III–like secretion system, although the activities of the proteins exported into the host cell cytosol have not been defined (70, 77). Unlike *Mycobacterium* spp., which appear to have to actively maintain the characteristics of their vacuole (killing the bacilli with antibi-

otics or activating the macrophage will lead to acidification of the phagosome), the vacuoles containing *Legionella* do not "re-enter" the endosomal continuum if the bacteria are killed by antibiotics after establishment of the infection.

The consummate master of intracellular remodeling has to be *Chlamydia* spp., including *Chlamydia trachomatis*, which may win the prize for the most common human bacterial infection. It is believed to cause a usually asymptomatic genital infection with 3 to 10 million new cases per year in the United States alone. Following internalization by macrophages or epithelial cells, chlamydiae induce formation of a large inclusion body, the limiting membrane of which appears to be completely lacking in any host-derived proteins (5, 25, 29, 30, 34). Moreover, with the exception of the acquisition of sphingomyelin from the Golgi, the bacterial inclusion body appears to derive virtually all its components from the infecting microbe.

The final group of bacterial pathogens have opted out entirely of dealing with the problems that beset survival within a vacuole. These pathogens, namely *Listeria, Shigella*, and *Rickettsia*, lyse their phagosome shortly after uptake and escape into the cytosol of the macrophage. *Listeria* expresses a porin, listeriolysin O, and two phospholipases that exhibit acidic pH optima, facilitate rapid dissolution of the phagosome membrane, and allow the bacterium to hijack the host's cytoskeleton for motility and metastasis (15, 19, 27, 43, 50). *Shigella* expresses IpaC and IpaD, which are required for exit from the vacuole; however, the mode of action of these proteins has yet to be determined (6, 46). Obviously, the translocation to the cytosol bypasses any potential problems that the bacteria might have with the host cell's lysosomal system. Bacterial mutants defective in vacuolar escape are avirulent and are rapidly degraded within the lysosome, thus demonstrating the importance of this alternate route to these species of bacteria.

Most of the data discussed above were garnered from studies of infections in macrophages, or macrophagelike cell lines, maintained in culture. The next section re-examines these interplays in the context of the host's immune system.

MODULATION OF THE HOST'S IMMUNE RESPONSE

The cellular immune response is the product of the presentation of foreign antigens in the context of class I and class II major histocompatibility (MHC) antigens, or CD1 molecules. All of these molecules are expressed by macrophages, rendering them highly competent antigen-presenting cells. Although not exclusively, class I MHC molecules sample peptide antigens from the cell cytosol, class II MHC molecules display peptide antigens acquired from within the endosomal/lysosomal continuum, and CD1 molecules present nontraditional antigens such as lipidoglycan moieties. Once stimulated, T cells are capable of lysing the target cell (cytolytic T cells that are mainly CD8+), or releasing cytokines that lead to macrophage activation (these are TH1 helper cells that release gamma interferon and tumor necrosis factor alpha), or cytokines that stimulate B cell maturation and antibody production (these are TH2 helper cells that release IL4, 5, and 6). Not surprisingly, the two arms of the T helper cell response produce antagonistic cytokines that suppress the opposing response. IL10 is released by TH2 cells. It down-regulates macrophage responsiveness and depresses expansion of TH1 cells. In contrast, IL12 is released by phagocytes in response to gamma interferon and suppresses production of IL4 and TH2 expansion while promoting responsiveness to gamma interferon.

Pathogens that want to modulate this system have two obvious routes: they may exploit the host's own cytokine network to bias the immune response away from a protective reaction, or they can directly interrupt the production or response to cytokines that induce TH1 expansion or macrophage activation.

Exploitation of the host's cytokine pathways is best illustrated by examination of pathogens that induce extremely acute symptomatic infections, such as bacillary dysentery caused by *Shigella*. *Shigella* appears to have a vested interest in inducing a massive inflammatory response within the host's intestine to trigger epithelial disruption and the rapid release of viable bacteria, thus spreading the infection. *Shigella* resides in the host cell cytosol and releases proteins such as IpaB, which activate caspase 1 (or ICE) and trigger apoptosis in the host cell (8, 35, 36, 60, 87). The apoptotic response is accompanied by a massive proinflammatory reaction and release of IL1 and IL8, which lead to recruitment of polymorphonuclear monocytes to the site and to increased tissue damage. In addition to the Ipa proteins, lipopolysaccharide from *Shigella* is a potent amplifier of the proinflammatory reaction responsible for bacillary dysentery. In the presence of lipopolysaccharide-binding protein in the serum, bacterial lipopolysaccharide is transcytosed by epithelial cells and induces release of IL8 from cells of the subtending tissue. Although this reaction will undoubtedly lead to activation of bystander cells and death of some bacteria, it would appear that the benefits from the increased shedding associated with the induced dysentery far outweigh the negative repercussions.

Obviously, this type of shock tactic is of little use to pathogens that cause chronic or enduring infections. These bacteria must remain within their host cells and induce minimal inflammatory responses to avoid the attention of the immune system. The paradigm for this approach is *M. tuberculosis*, which persists in approximately one-third of the human population, but only induces overt disease in a minority. The anomaly about *Mycobacterium* spp. is that the cell wall constituents are capable of potentiating immune responses; they are, after all, the active ingredient of Freund's complete adjuvant. Nonetheless, clearance or control of mycobacterial infections is expressed at the level of the infected macrophage through activation and up-regulation of microbicidal responses (61, 76). The answer may lie, in part, in the granulomatous nature of the infection foci. While it has been reported that *Mycobacterium* and its cell wall components can induce proinflammatory cytokines, some of these constituents are also thought to suppress T-cell proliferation (4, 26, 37, 69, 73, 79). Moreover, the proinflammatory cytokine cascade induced by *Mycobacterium* spp. can itself cause anergy through extremely high localized concentrations of cytokines such as IL6 (73). In addition to blocking T-cell responsiveness, high localized concentrations of IL6 may also bias T-cell development or expansion to favor production of an impotent TH2 helper cell response (54). Histology of early granulomas prior to their caseation reveals a highly ordered structure of infected macrophage, giant cells, and normal macrophages with interacting rafts of T cells (52). Central caseation appears in later granulomas that have limited lymphocyte infiltration. In these granulomas, the T cells are restricted to a mantle surrounding the macrophages and giant cells. Such a structure will limit productive access of the TH1 cells capable of activating macrophages at the center of the infection site. I view the granuloma as an equilibrium that is the product of coevolution of host and pathogen whereby the pathogen is restricted and unable to metastasize, yet the host is unable to totally resolve the infection. This equilibrium persists until disturbed by factors that compromise the immune response, such as human immunodeficiency virus infection, old age, or malnutrition. Unfortunately, the complexity of cytokine networks active in granuloma formation and maintenance renders the model resistant to resolution through experimentation. Once again, bacterial cell wall lipidoglycans may be involved in mediating these effects. Recent analysis of mycobacterial cell wall lipids has also shown that they can exploit host cell trafficking pathways to spread from cell to cell (7). Mycobacterial lipids are released in infected macrophages and form multivesicular structures that can be exocytosed and internalized by bystander, uninfected macrophages.

In addition to the regulation of the cytokine networks themselves, pathogenic bacteria are known to produce molecules or toxins that actively alter the behavior of the host cell. Although certain examples such as the antiphagocytic effector proteins inoculated by *Yersinia* into the host cell have been studied extensively, the less obvious mechanisms whereby certain intracellular pathogens render their host macrophages anergic to gamma interferon and tumor necrosis factor alpha are not well understood. There are numerous reports regarding the nonresponsiveness of macrophages infected with *M. tuberculosis*, and some data indicating that cell wall constituents may play a role (4, 45). Recent experiments indicate that the blockage comes downstream of receptor activation, at the level of the intranuclear interaction between STAT 1 and the transcriptional coactivators CREB-binding protein and p300 (71). The benefits of blocking macrophage activation are clear for pathogens like *Mycobacterium* that maintain a prolonged intracellular infection.

DIRECT RESISTANCE TO KILLING

All intramacrophage pathogens have evolved mechanisms that confer a measure of direct resistance to the major routes of killing mobilized by the phagocyte. The killing mechanisms that appear most active are those that rely on the generation of reactive oxygen or nitrogen intermediates. The most compelling data illustrating the central role of nitric oxide (NO) come from studies on iNOS knockout mice, or mice or macrophages treated with nonhydrolysable analogs of arginine that block NO production. In both instances, the ability of the mice or macrophages to regulate bacterial infection has been severely compromised. However, these data indicate that NO production is required, but may not in itself be sufficient for an effective microbicidal response. The ability of the macrophage to upregulate its oxidative burst and to acidify its endosomes and lysosomes more effectively will potentiate the activity of the ROIs. However, much work remains to be done before the direct routes of killing can be demonstrated.

The mechanisms of resistance to NO–mediated killing are dealt with in depth in chapter 9. However, I will briefly discuss some main routes of resistance to free radicals. Resistance can be mediated through either passive, scavenging mechanisms or active, detoxifying or repairing mechanisms. The cell wall components of some microbes are effective scavengers of ROIs. With respect to detoxification, catalase activity and superoxide dismutases are the more ubiquitous means of deactivation of radicals (22, 23, 33, 42, 55, 66). It is interesting that the catalase activity of KatG of *M. tuberculosis* is known to confer resistance to radical-mediated killing both in culture and in infected macrophages. However, the same catalase activity converts isoniazid from a prodrug to its active form; therefore, many of the drug-resistant strains of *M. tuberculosis* show a loss of catalase activity and an impaired OxyR response in general (17, 82). Other proteins such as AphC, the c subunit of the alkyl hydroperoxide reductase, can reduce organic peroxides and provide an alternate means of protection (10, 42). Mn and Cu/Zn superoxide dismutases are found in most intracellular bacteria and in *Salmonella* and *Mycobacterium* spp. These are found either in the periplasmic space or outside the bacterium, suggesting that they act outside the bacilli in the bacterial phagosomes (22, 33).

The recurring theme in all antioxidative and nitrative-resistance responses is redundancy. Genetic screens for bacterial mutants with impaired ability to withstand exposure to different radicals keep implicating new loci in the process, yet the mode of action of many of these gene products, such as the noxR1 gene (20) or the hmp (flavohemoglobin) gene of *M. tuberculosis* (39), remain to be elucidated.

CONCLUSIONS

This chapter provided an overview of the field of macrophage/microbe interplay, but it is limited in the depth with which it treats each example. Despite this, it is obvious that there are many recurring themes resulting from conserved genes, convergent evolution, or horizontal gene transfer. The strength of the evolutionary pressure operating on these microbes is such that most of the pathogens have acquired and retained multiple, functionally discrete methods of subverting the macrophage's microbicidal tendencies.

ACKNOWLEDGMENTS
The author and his lab are supported by US PHS grants AI33348 and HL55936.

REFERENCES

1. **Alpuche-Aranda, C. M., E. P. Berthiaume, B. Mock, J. A. Swanson, and S. I. Miller.** 1995. Spacious phagosome formation within mouse macrophages correlates with *Salmonella* serotype pathogenicity and host susceptibility. *Infect. Immun.* **63:**4456–4462.

2. **Alpuche-Aranda, C. M., E. L. Racoosin, J. A. Swanson, and S. I. Miller.** 1994. *Salmonella* stimulate macrophage macropinocytosis and persist within spacious phagosomes. *J. Exp. Med.* **179:**601–608.

3. **Astarie-Dequeker, C., E. N. N'Diaye, V. Le Cabec, M. G. Rittig, J. Prandi, and I. Maridonneau-Parini.** 1999. The mannose receptor mediates uptake of pathogenic and nonpathogenic mycobacteria and bypasses bactericidal responses in human macrophages. *Infect. Immun.* **67:**469–477.

4. **Baba, T., Y. Natsuhara, K. Kaneda, and I. Yano.** 1997. Granuloma formation activity and mycolic acid composition of mycobacterial cord factor. *Cell. Mol. Life Sci.* **53:**227–232.

5. **Bannantine, J. P., D. D. Rockey, and T. Hackstadt.** 1998. Tandem genes of *Chlamydia psittaci* that encode proteins localized to the inclusion membrane. *Mol. Microbiol.* **28:**1017–1026.

6. **Barzu, S., Z. Benjelloun-Touimi, A. Phalipon, P. Sansonetti, and C. Parsot.** 1997. Functional analysis of the *Shigella flexneri* IpaC invasin by insertional mutagenesis. *Infect. Immun.* **65:**1599–1605.

7. **Beatty, W. L., E. R. Rhoades, H.-J. Ullrich, D. Chatterjee, J. E. Heuser, and D. G. Russell.** Trafficking and release of mycobacterial lipids from infected macrophages. *Traffic*, in press.

8. **Beatty, W. L., and P. J. Sansonetti.** 1997. Role of lipopolysaccharide in signaling to subepithelial polymorphonuclear leukocytes. *Infect. Immun.* **65:**4395–4404.

9. **Buchmeier, N. A., and F. Heffron.** 1991. Inhibition of macrophage phagosome-lysosome fusion by *Salmonella typhimurium. Infect. Immun.* **59:**2232–2238.

10. **Chen, L., Q. W. Xie, and C. Nathan.** 1998. Alkyl hydroperoxide reductase subunit C (AhpC) protects bacterial and human cells against reactive nitrogen intermediates. *Mol. Cell* **1:**795–805.

11. **Clague, M. J., S. Urbe, F. Aniento, and J. Gruenberg.** 1994. Vacuolar ATPase activity is required for endosomal carrier vesicle formation. *J. Biol. Chem.* **269**:21–24.

12. **Clark, R. A.** 1999. Activation of the neutrophil respiratory burst oxidase. *J. Infect. Dis.* **179**(Suppl 2):S309–S317.

13. **Clemens, D. L.** 1996. Characterization of the *Mycobacterium tuberculosis* phagosome. *Trends Microbiol.* **4**:113–118.

14. **Clemens, D. L., and M. A. Horwitz.** 1995. Characterization of the *Mycobacterium tuberculosis* phagosome and evidence that phagosomal maturation is inhibited. *J. Exp. Med.* **181**:257–270.

15. **Cossart, P.** 1998. Interactions of the bacterial pathogen *Listeria monocytogenes* with mammalian cells: bacterial factors, cellular ligands, and signaling. *Folia Microbiol.* **43**:291–303.

16. **de Chastellier, C., and L. Thilo.** 1998. Modulation of phagosome processing as a key strategy for *Mycobacterium avium* survival within macrophages. *Res. Immunol.* **149**:699–702.

17. **Deretic, V., J. Song, and E. Pagan-Ramos.** 1997. Loss of oxyR in *Mycobacterium tuberculosis*. *Trends Microbiol.* **5**:367–372.

18. **Dinauer, M. C.** 1993. The respiratory burst oxidase and the molecular genetics of chronic granulomatous disease. *Crit. Rev. Clin. Lab. Sci.* **30**:329–369.

19. **Dramsi, S., and P. Cossart.** 1998. Intracellular pathogens and the actin cytoskeleton. *Annu. Rev. Cell. Dev. Biol.* **14**:137–166.

20. **Ehrt, S., M. U. Shiloh, J. Ruan, M. Choi, S. Gunzburg, C. Nathan, Q. Xie, and L. W. Riley.** 1997. A novel antioxidant gene from *Mycobacterium tuberculosis J. Exp. Med.* **186**:1885–1896.

21. **Ernst, J. D.** 1998. Macrophage receptors for Mycobacterium tuberculosis. *Infect. Immun.* **66**:1277–1281.

22. **Fang, F. C., M. A. DeGroote, J. W. Foster, A. J. Baumler, U. Ochsner, T. Testerman, S. Bearson, J. C. Giard, Y. Xu, G. Campbell, and T. Laessig.** 1999. Virulent *Salmonella typhimurium* has two periplasmic Cu, Zn-superoxide dismutases. *Proc. Natl. Acad. Sci. USA* **96**:7502–7507.

23. **Farrant, J. L., A. Sansone, J. R. Canvin, M. J. Pallen, P. R. Langford, T. S. Wallis, G. Dougan, and J. S. Kroll.** 1997. Bacterial copper- and zinc-cofactored superoxide dismutase contributes to the pathogenesis of systemic salmonellosis. *Mol. Microbiol.* **25**:785–796.

24. **Ferrari, G., H. Langen, M. Naito, and J. Pieters.** 1999. A coat protein on phagosomes involved in the intracellular survival of mycobacteria. *Cell* **97**:435–447.

25. **Gaydos, C. A., J. T. Summersgill, N. N. Sahney, J. A. Ramirez, and T. C. Quinn.** 1996. Replication of *Chlamydia pneumoniae* in vitro in human macrophages, endothelial cells, and aortic artery smooth muscle cells. *Infect. Immun.* **64**:1614–1620.

26. **Gilbertson, B., J. Zhong, and C. Cheers.** 1999. Anergy, IFN-gamma production, and apoptosis in terminal infection of mice with *Mycobacterium avium J. Immunol.* **163**:2073–2080.

27. **Goldfine, H., T. Bannam, N. C. Johnston, and W. R. Zuckert.** 1998. Bacterial phospholipases and intracellular growth: the two distinct phospholipases C of *Listeria monocytogenes. Soc. Appl. Bacteriol. Symp. Ser.* **27**:7S–14S.

28. **Hackam, D. J., O. D. Rotstein, W. Zhang, S. Gruenheid, P. Gros, and S. Grinstein.** 1998. Host resistance to intracellular infection: mutation of natural resistance-associated macrophage protein 1 (Nramp1) impairs phagosomal acidification. *J. Exp. Med.* **188**:351–364.

29. **Hackstadt, T.** 1998. The diverse habitats of obligate intracellular parasites. *Curr. Opin. Microbiol.* **1**:82–87.

30. **Hackstadt, T., E. R. Fischer, M. A. Scidmore, D. D. Rockey, and R. A. Heinzen.** 1997. Origins and functions of the chlamydial. *Trends Microbiol.* **5**:288–293.

31. **Hart, S. P., C. Haslett, and I. Dransfield.** 1996. Recognition of apoptotic cells by phagocytes. *Experientia* **52**:950–956.

32. **Harth, G., D. L. Clemens, and M. A. Horwitz.** 1994. Glutamine synthetase of *Mycobacterium tuberculosis*: extracellular release and characterization of its enzymatic activity. *Proc. Natl. Acad. Sci. USA* **91**:9342–9346.

33. **Harth, G., and M. A. Horwitz.** 1999. Export of recombinant *Mycobacterium tuberculosis* superoxide dismutase is dependent upon both information in the protein and mycobacterial export machinery. A model for studying export of leaderless proteins by pathogenic mycobacteria. *J. Biol. Chem.* **274**:4281–4292.

34. **Heinzen, R. A., M. A. Scidmore, D. D. Rockey, and T. Hackstadt.** 1996. Differential interaction with endocytic and exocytic pathways distinguish parasitophorous vacuoles of *Coxiella burnetii* and *Chlamydia trachomatis. Infect. Immun.* **64**:796–809.

35. **Hilbi, H., J. E. Moss, D. Hersh, Y. Chen, J. Arondel, S. Banerjee, R. A. Flavell, J. Yuan, P. J. Sansonetti, and A. Zychlinsky.** 1998. *Shigella*-induced apoptosis is dependent on caspase-1 which binds to IpaB. *J. Biol. Chem.* **273**:32895–32900.

36. **Hilbi, H., A. Zychlinsky, and P. J. Sanso-netti.** 1997. Macrophage apoptosis in microbial infections. *Parasitology* **115**(Suppl.):S79–S87.

37. **Hmama, Z., R. Gabathuler, W. A. Jefferies, G. de Jong, and N. E. Reiner.** 1998. Attenuation of HLA-DR expression by mononuclear phagocytes infected with *Mycobacterium tuberculosis* is related to intracellular sequestration of immature class II heterodimers. *J. Immunol.* **161**:4882–4893.

38. **Horwitz, M. A., and F. R. Maxfield.** 1984. *Legionella pneumophila* inhibits acidification of its phagosome in human monocytes. *J. Cell Biol.* **99**:1936–1943.

39. **Hu, Y., P. D. Butcher, J. A. Mangan, M. A. Rajandream, and A. R. Coates.** 1999. Regulation of *hmp* gene transcription in *Mycobacterium tuberculosis*: effects of oxygen limitation and nitrosative and oxidative stress. *J. Bacteriol.* **181**:3486–3493.

40. **Ishibashi, Y., and T. Arai.** 1995. *Salmonella typhi* does not inhibit phagosome-lysosome fusion in human monocyte-derived macrophages. *FEMS Immunol. Med. Microbiol.* **12**:55–61.

41. **Johnson, W. J., G. J. Warner, P. G. Yancey, and G. H. Rothblat.** 1996. Lysosomal metabolism of lipids. *Subcell. Biochem.* **27**:239–294.

42. **Manca, C., S. Paul, C. E. Barry 3rd, V. H. Freedman, and G. Kaplan.** 1999. *Mycobacterium tuberculosis* catalase and peroxidase activities and resistance to oxidative killing in human monocytes in vitro. *Infect. Immun.* **67**:74–79.

43. **Marquis, H., H. Goldfine, and D. A. Portnoy.** 1997. Proteolytic pathways of activation and degradation of a bacterial phospholipase C during intracellular infection by *Listeria monocytogenes*. *J. Cell Biol.* **137**:1381–1392.

44. **Mason, R. W.** 1996. Lysosomal metabolism of proteins. *Subcell. Biochem.* **27**:159–190.

45. **McGee, Z. A., and C. M. Clemens.** 1994. Effect of bacterial products on tumor necrosis factor production: quantitation in biological fluids or tissues. *Methods Enzymol.* **236**:23–31.

46. **Menard, R., P. J. Sansonetti, and C. Parsot.** 1993. Nonpolar mutagenesis of the ipa genes defines IpaB, IpaC, and IpaD as effectors of *Shigella flexneri* entry into epithelial cells. *J. Bacteriol.* **175**:5899–5906.

47. **Nathan, C.** 1997. Inducible nitric oxide synthase: what difference does it make? *J. Clin. Invest.* **100**:2417–2423.

48. **Nicod, L. P.** 1999. Pulmonary defence mechanisms. *Respiration* **66**:2–11.

49. **Oh, Y. K., and R. M. Straubinger.** 1996. Intracellular fate of *Mycobacterium avium*: use of dual-label spectrofluorometry to investigate the influence of bacterial viability and opsonization on phagosomal pH and phagosome-lysosome interaction. *Infect. Immun.* **64**:319–325.

50. **Portnoy, D. A., and S. Jones.** 1994. The cell biology of *Listeria monocytogenes* infection (escape from a vacuole). *Ann. N. Y. Acad. Sci.* **730**:15–25.

51. **Rathman, M., M. D. Sjaastad, and S. Falkow.** 1996. Acidification of phagosomes containing *Salmonella typhimurium* in murine macrophages. *Infect. Immun.* **64**:2765–2773.

52. **Rhoades, E. R., A. A. Frank, and I. M. Orme.** 1997. Progression of chronic pulmonary tuberculosis in mice aerogenically infected with virulent *Mycobacterium tuberculosis*. *Tuber. Lung Dis.* **78**:57–66.

53. **Ricevuti, G.** 1997. Host tissue damage by phagocytes. *Ann. N. Y. Acad. Sci.* **832**:426–448.

54. **Rincon, M., J. Anguita, T. Nakamura, E. Fikrig, and R. A. Flavell.** 1997. Interleukin (IL)-6 directs the differentiation of IL-4-producing CD4+ T cells. *J. Exp. Med.* **185**:461–469.

55. **Roggenkamp, A., T. Bittner, L. Leitritz, A. Sing, and J. Heesemann.** 1997. Contribution of the Mn-cofactored superoxide dismutase (SodA) to the virulence of *Yersinia enterocolitica* serotype O8. *Infect. Immun.* **65**:4705–4710.

56. **Russell, D. G.** 1999. Where to stay inside the cell: a homesteader's guide to intracellular parasitism, p. 131–152. *In* P. Cossart, S. Normark, R. Rappuoli, and P. Boquet (ed.), *Cellular Microbiology*. AMS Press, Washington, D.C.

57. **Russell, D. G., J. Dant, and S. Sturgill-Koszycki.** 1996. *Mycobacterium avium*- and *Mycobacterium tuberculosis*-containing vacuoles are dynamic, fusion-competent vesicles that are accessible to glycosphingolipids from the host cell plasmalemma. *J. Immunol.* **156**:4764–4773.

58. **Russell, D. G., S. Sturgill-Koszycki, T. Vanheyningen, H. Collins, and U. E. Schaible.** 1997. Why intracellular parasitism need not be a degrading experience for Mycobacterium. *Philos. Trans. R. Soc. London B Biol. Sci.* **352**:1303–1310.

59. **Sambrano, G. R., and D. Steinberg.** 1995. Recognition of oxidatively damaged and apoptotic cells by an oxidized low density lipoprotein receptor on mouse peritoneal macrophages: role of membrane phosphatidylserine. *Proc. Natl. Acad. Sci. USA* **92**:1396–1400.

60. **Sansonetti, P. J., J. Arondel, M. Huerre, A. Harada, and K. Matsushima.** 1999. Interleukin-8 controls bacterial transepithelial translocation at the cost of epithelial destruction in experimental shigellosis. *Infect. Immun.* **67**:1471–1480.

61. Schaible, U. E., S. Sturgill-Koszycki, P. H. Schlesinger, and D. G. Russell. 1998. Cytokine activation leads to acidification and increases maturation of *Mycobacterium avium*-containing phagosomes in murine macrophages. *J. Immunol.* **160:**1290–1296.

62. Scidmore, M. A., D. D. Rockey, E. R. Fischer, R. A. Heinzen, and T. Hackstadt. 1996. Vesicular interactions of the *Chlamydia trachomatis* inclusion are determined by chlamydial early protein synthesis rather than route of entry. *Infect. Immun.* **64:**5366–5372.

63. Shiloh, M. U., J. D. MacMicking, S. Nicholson, J. E. Brause, S. Potter, M. Marino, F. Fang, M. Dinauer, and C. Nathan. 1999. Phenotype of mice and macrophages deficient in both phagocyte oxidase and inducible nitric oxide synthase. *Immunity* **10:**29–38.

64. Sinai, A. P., and K. A. Joiner. 1997. Safe haven: the cell biology of nonfusogenic pathogen vacuoles. *Annu. Rev. Microbiol.* **51:**415–462.

65. Spargo, B. J., L. M. Crowe, T. Ioneda, B. L. Beaman, and J. H. Crowe. 1991. Cord factor (alpha,alpha-trehalose 6,6′-dimycolate) inhibits fusion between phospholipid vesicles. *Proc. Natl. Acad. Sci. USA* **88:**737–740.

66. St. John, G., and H. M. Steinman. 1996. Periplasmic copper-zinc superoxide dismutase of *Legionella pneumophila*: role in stationary-phase survival. *J. Bacteriol.* **178:**1578–1584.

67. Sturgill-Koszycki, S., U. E. Schaible, and D. G. Russell. 1996. *Mycobacterium*-containing phagosomes are accessible to early endosomes and reflect a transitional state in normal phagosome biogenesis. *EMBO J.* **15:**6960–6968.

68. Sturgill-Koszycki, S., P. H. Schlesinger, P. Chakraborty, P. L. Haddix, H. L. Collins, A. K. Fok, R. D. Allen, S. L. Gluck, J. Heuser, and D. G. Russell. 1994. Lack of acidification in *Mycobacterium* phagosomes produced by exclusion of the vesicular proton-ATPase. *Science* **263:**678–681.

69. Sussman, G., and A. A. Wadee. 1992. Supernatants derived from CD8+ lymphocytes activated by mycobacterial fractions inhibit cytokine production. The role of interleukin-6. *Biotherapy* **4:**87–95.

70. Swanson, M. S., and R. R. Isberg. 1996. Identification of *Legionella pneumophila* mutants that have aberrant intracellular fates. *Infect. Immun.* **64:**2585–2594.

71. Ting, L.-M., A. C. Kim, A. Cattamanchi, and J. D. Ernst. 1999. *Mycobacterium tuberculosis* inhibits IFN-g transcriptional responses without inhibiting activation of STAT1. *J. Immunol.* **163:**3898–3906.

72. Uchiya, K.-I., M. A. Barbieri, K. Funato, A. Shah, P. D. Stahl, and E. A. Groisman. 1999. A *Salmonella* virulence protein that inhibits cellular trafficking. *EMBO J.* **18:**101–110.

73. VanHeyningen, T. K., H. L. Collins, and D. G. Russell. 1997. IL-6 produced by macrophages infected with *Mycobacterium* species suppresses T cell responses. *J. Immunol.* **158:**330–337.

74. van Weert, A. W., K. W. Dunn, H. J. Gueze, F. R. Maxfield, and W. Stoorvogel. 1995. Transport from late endosomes to lysosomes, but not sorting of integral membrane proteins in endosomes, depends on the vacuolar proton pump. *J. Cell Biol.* **130:**821–834.

75. Via, L. E., D. Deretic, R. J. Ulmer, N. S. Hibler, L. A. Huber, and V. Deretic. 1997. Arrest of mycobacterial phagosome maturation is caused by a block in vesicle fusion between stages controlled by rab5 and rab7. *J. Biol. Chem.* **272:**13326–13331.

76. Via, L. E., R. A. Fratti, M. McFalone, E. Pagan-Ramos, D. Deretic, and V. Deretic. 1998. Effects of cytokines on mycobacterial phagosome maturation. *J. Cell Sci.* **111:**897–905.

77. Vogel, J. P., H. L. Andrews, S. K. Wong, and R. R. Isberg. 1998. Conjugative transfer by the virulence system of *Legionella pneumophila*. *Science* **279:**873–876.

78. Vogel, J. P., and R. R. Isberg. 1999. Cell biology of *Legionella pneumophila*. *Curr. Opin. Microbiol.* **2:**30–34.

79. Wadee, A. A., G. Sussman, R. H. Kuschke, and S. G. Reddy. 1993. Suppression of cytokine production by supernatants from CD8+ lymphocytes activated by mycobacterial fractions: the role of interleukins 4 and 6. *Biotherapy* **7:**125–136.

80. Webster, P., and D. G. Russell. 1993. Biology of the Trypanosomatid flagellar pocket. *Parasitol. Today* **9:**201–206.

81. Wientjes, F. B., and A. W. Segal. 1995. NADPH oxidase and the respiratory burst. *Semin. Cell Biol.* **6:**357–365.

82. Wilson, T., B. J. Wards, S. J. White, B. Skou, G. W. de Lisle, and D. M. Collins. 1997. Production of avirulent *Mycobacterium bovis* strains by illegitimate recombination with deoxyribonucleic acid fragments containing an interrupted ahpC gene. *Tuber. Lung Dis.* **78:**229–235.

83. Winchester, B. G. 1996. Lysosomal metabolism of glycoconjugates. *Subcell. Biochem.* **27:**191–238.

84. Winter, G., M. Fuchs, M. J. McConville, Y. D. Stierhof, and P. Overath. 1994. Surface antigens of *Leishmania mexicana* amastigotes: char-

acterization of glycoinositol phospholipids and a macrophage-derived glycosphingolipid. *J. Cell Sci.* **107:**2471–2482.

85. **Xia, Y., and J. L. Zweier.** 1997. Superoxide and peroxynitrite generation from inducible nitric oxide synthase in macrophages. *Proc. Natl. Acad. Sci. USA* **94:**6954–6958.

86. **Xu, S., A. Cooper, S. Sturgill-Koszycki, T. van Heyningen, D. Chatterjee, I. Orme, P. Allen, and D. G. Russell.** 1994. Intracellular trafficking in *Mycobacterium tuberculosis* and *Mycobacterium avium*-infected macrophages. *J. Immunol.* **153:**2568–2578.

87. **Yamasaki, C., Y. Natori, X. T. Zeng, M. Ohmura, S. Yamasaki, and Y. Takeda.** 1999. Induction of cytokines in a human colon epithelial cell line by Shiga toxin 1 (Stx1) and Stx2 but not by non-toxic mutant Stx1 which lacks N-glycosidase activity. *FEBS Lett.* **442:** 231–234.

BACTERIAL RESISTANCE TO ANTIBODY-DEPENDENT DEFENSES

Lynette B. Corbeil

8

Bacterial disease results from a dynamic interaction between pathogen and host. The complexity of this interaction is awe inspiring. Not only is there a vast array of microbial virulence factors and corresponding host defenses, but also the microbes have a multitude of ways to evade host responses. Surely, God loves variety. Steps in pathogenesis usually involve some combination of adherence, colonization, multiplication, invasion, and host damage. Host defenses include innate and acquired immunity. This chapter will not deal with the innate immune system, which consists of such defenses as the skin as a barrier, the pH of the stomach, the mucociliary elevator of the respiratory tract, phagocytic defense prior to a specific immune response, and other factors. Acquired or specific immunity is usually divided into cell-mediated and humoral immunity. Although the T-cell-mediated arm of the immune response is closely linked to the induction of humoral responses, this discussion will be limited to defensive functions of antibody responses and the pathogen's mechanisms of evading these defenses. The overview will cover examples of the great variety of types of evasive mechanisms. More data will be given in a few cases where conclusions can be drawn concerning in vivo relevance.

Antibody protects against bacterial infection in numerous ways. Agglutination (or clumping) and immobilization of motile organisms are among the conceptually simplest defense mechanisms. Clumps and immotile organisms may be unable to penetrate mucus and therefore may be cleared from mucous surfaces. Similarly, antibody neutralization of enzymes that digest mucus may prevent penetration to the epithelial cell layer of mucosal membranes, where most bacteria first encounter the host. Adherence to these epithelial cells is usually the first step in colonization, so inhibition of adherence by antibody is important to prevent establishment of infection. Virulence factors involved in invasion vary from organism to organism, but neutralization of each factor by antibody is another critical step. If organisms escape these defenses, then neutralization of toxins becomes important. Aside from combating these bacterial steps in pathogenesis, the humoral immune system also takes a direct approach by killing organisms outright or by inhibiting growth. Metabolic inhibition is perhaps best illustrated by antibody interaction with mycoplasmas (30).

Lynette B. Corbeil Department of Pathology, University of California San Diego School of Medicine, San Diego, CA 92103-8416.

Virulence Mechanisms of Bacterial Pathogens, 3rd ed., Edited by K. A. Brogden et al.
©2000 ASM Press, Washington, D.C.

Killing of bacteria can be accomplished by antibody-mediated phagocytosis (opsonization) and intracellular destruction or by antibody-mediated killing by complement C. The latter is relevant to some gram-negative bacteria and mycoplasmas. With this variety of humoral defense mechanisms, not to mention the innate and cell-mediated immune systems, it is surprising that the host ever succumbs to bacterial infection. Obviously, the balance must shift in favor of the pathogen when disease occurs. Evasion of host defense to shift this balance is accomplished by a variety of strategies, as discussed below.

AGGRESSIVE ATTACK
The most direct way to evade specific humoral immune response is to kill the host before the antibody response develops. Anthrax is a good example of this. Natural infection causes death in a few days, although inoculation of lethal toxin into laboratory animals causes death in minutes. The rapidity of death in a nonimmunized population circumvents a primary immune response because there is not time to develop such a response. This is one reason that *Bacillus anthracis* is useful for bioterrorism or biological warfare. It is thought that *B. anthracis* lethal toxin mounts this aggressive attack by interrupting a key cell-signaling pathway, the mitogen-activated protein kinase (MAPK) signal transduction pathway (29). Rapid disruption of such a critical cellular pathway not only may kill cells directly, but may be linked to cytokine pathways (68). The mechanisms causing rapid death are not entirely elucidated, but it is clear that death before stimulation of a humoral response avoids the consequences of antibody-mediated defense.

Not only is death of the host a mechanism of aggressive attack, death of phagocytes and lymphocytes would result in evasion of antibody synthesis as well as opsonization. *Pasteurella haemolytica* is a prime example of this mechanism, since it secretes a leukotoxin (LKT) that kills bovine phagocytes and lymphocytes. This results in persistence of the organism in the lung and can result in severe pneumonia. Gene cloning and sequencing revealed that LKT is part of a family of RTX toxins (47). Other members of the family contribute to evasion of antibody defenses as well.

COATS OF ARMOR
Another evasion mechanism of antibody defenses relies on the outer surface of the bacterium being resistant to the antibody-mediated defense. Some of these "coats of armor" have been known for decades. For example, the thick peptidoglycan layer in the cell wall of gram-positive bacteria protects them from antibody-mediated C killing. Capsules and long O side chains of lipopolysaccharide (LPS) provide protection against antibody-mediated complement killing and opsonization to many gram-negative pathogens (31, 38). Other "coats of armor" associated with resistance to C killing include bacterial S layers (44, 65) and surface fibrillar networks (17). The mechanisms of resistance to serum C killing (serum resistance) are complex; several will be discussed below. However, the above physical barriers may play a role in preventing the formation of the membrane attack complex (MAC) on the gram-negative bacterial surface. For example, O antigens sterically inhibit the accessibility of the MAC to hydrophobic domains in the outer membrane. Capsules and LPS O side chains and other structures preventing MAC insertion are often very antigenic. Therefore, although they prevent C killing by nonimmune serum, antibody directed against the capsule, LPS, etc., with activation of the classical C pathway may initiate killing of some resistant organisms.

HIDING FROM HOST DEFENSE
The first two strategies resemble either bold attack or a wall of protection against host defense. Alternatively, some bacteria evade these defenses by hiding themselves. This may result either in immune defenses not being stimulated or in the organism being inaccessible to immune mechanisms, if present. An example

of the latter is the protected site in bacterial endocarditis (6). In this case, bacteria normally sensitive to opsonization and killing by phagocytes are able to hide in endocardial vegetations to escape host defense. Other organisms escape into intracellular locations. *Listeria monocytogenes* invades many types of mammalian cells including epithelial cells, endothelial cells, fibroblasts, hepatocytes, and macrophages (45). The initial step is mediated by two *L. monocytogenes* proteins, internalin A (InlA) and InlB. The InlA receptor is the cell-adhesion molecule, E-cadherin (46), but the receptor for InlB is unknown. After uptake of *L. monocytogenes* via phagocytosis, the organism escapes the phagosome by secretion of listeriolysin (Hly). The bacteria then replicate in the cytoplasm and induce polymerization of cometlike actin tails at one pole of their surface, which rapidly propels them through the cytoplasm (25). When the propelled bacteria reach the cell border, protrusions of the double membrane extend into the adjacent cell, the membranes lyse, and the bacteria are released into the cytoplasm of the next cell (50). In this way, the organism spreads from cell to cell without exposure to blood or intercellular fluids containing antibody and complement. Similarly, macrophages infected with *L. monocytogenes* can transfer the organism to endothelial cells, including the brain microvascular endothelium (36), to penetrate the blood-brain barrier without exposure to antibody or C. This is truly a "Trojan horse" evasive mechanism.

STEALTH

Antigenic mimicry and masking of bacterial antigens with host proteins are two mechanisms closely related to the "Trojan horse" mechanism. These variations on a theme could be considered a "wolf in sheep's clothing" approach to evasion of host responses. Some organisms bind host proteins to their surfaces, resulting either in recognition of organisms as "self" by the host or in masking of antigens to avoid recognition. It has been well demonstrated that binding of albumin, fibrin-

ogen, or IgG to gram-positive cocci changes their surface characteristics (49). Binding of heparin to the surface of *Neisseria gonorrhoeae* may mask lipooligosaccharide (LOS) epitopes as well as increase serum resistance (13). Although trichomonads are not bacteria, this protozoal example will be used because it demonstrates in vivo relevance of IgG binding. Our early attempts to measure antibody to *Tritrichomonas foetus* in infected cattle by enzyme-linked immunosorbent assay (ELISA) resulted in very high backgrounds with organisms grown in media containing bovine serum (21). This suggested that the anti-Ig conjugate reacted with the bovine immunoglobulin on the surface of the organism. Subsequent studies showed that bovine IgG1 and IgG2 bound very tightly to the surface. In fact, rabbits immunized with these coated trichomonads mounted a greater antibody response to bovine IgG than to trichomonad antigens (21). This may explain, in part, the fact that little local antibody response to trichomonads is detectable until 4 to 6 weeks after experimental infection of the genital tract (12, 16, 39) and essentially no inflammation in the genital mucosa is detected until 8 weeks of infection (52). It appears that the organism evades recognition by donning a mask of host protein until it is well established. After the mucosal antibody response is detectable, covering the surface antigens with host protein may allow the organisms to escape host defense again. Because antigens are not exposed, no antigen–antibody complexes may form. Both mechanisms appear to play a role in the chronicity of this sexually transmitted disease.

Campylobacter jejuni LPS and streptococcal M protein are examples of antigenic mimicry that may be a factor in evasion of host responses. The structure of the outer core of some *C. jejuni* LPS serotypes mimics gangliosides prevalent in cell membranes of the peripheral nerves (51). This avoids recognition and thus prevents immune stimulation, but when antibody responses do occur, antibodies cross-react with nerve membranes. This is often associated with Guillain-Barré syndrome

(51). Similarly, streptococcal M proteins exhibit structural similarity to mammalian tropomyosin, and antibodies to muscle tissue, cardiac sarcolemma, and glomerular membranes result when animals are immunized with streptococcal M proteins (32). A similar situation exists with antibodies to streptococcal peptidoglycan–polysaccharide complexes. Monoclonal antibodies (MAbs) generated to these complexes cross-react strongly with keratin from human skin (62). These MAbs react with N-acetyl-β-D-glucosamine (GlcNAc), which is the immunodominant epitope of streptococcal group A-specific carbohydrate. The epitope may stimulate broad cross-reactivity with cytokeratine, possibly resulting in autoimmune disease (62). However, the mimicking of self-antigens may more often inhibit recognition of these epitopes by the host, resulting in evasion of host responses. The extent to which streptococcal antigenic mimicry results in evasion of host responses versus immune-mediated diseases is not clear. Another example of antigenic mimicry is provided by gonococcal LPS, which undergoes sialylation in vivo as has been demonstrated in urethal secretions of patients (3). Only strains with LOS having a Gal β 1-4-GlcNAc epitope are sialylated in vivo. This epitope is immunochemically similar to the precursor of human erythrocyte antigen i (3), which is sialylated in human cells. Therefore, the in vivo sialylation of gonococcal LOS is an example of both masking of LOS epitopes and antigenic mimicry (3). An even more complex interaction with host protein is demonstrated by "sulfated polysaccharide-directed recruitment of host proteins" (28). Many bacterial pathogens, including *Staphylococcus* spp., *Yersinia* spp., *Neisseria* spp., *Helicobacter pylori*, and *Streptococcus pyogenes*, specifically bind heparin and related sulfated polysaccharides. These in turn bind a multitude of host heparin-binding proteins. This provides for the potential of antigen masking and indirect mimicry, and it also imparts the function of the bound host protein. For example, binding of adhesive glycoproteins, such as vitronectin and fibronectin, mediated increased bacteria invasion of epithelial cells (28). This would result in intracellular escape from antibody, constituting another bacterial evasive mechanism.

BLOCKADE

Several gram-negative bacteria block the antibody activation of the C cascade, resulting in escape from C-mediated killing. Classic examples of blocking antibody are found with both *Neisseria meningitidis* and *N. gonorrhoeae*. For example, antibodies specific for terminal galactose-alpha 1,3-galactose residues (anti-Gal) react with pili of *N. meningitidis*. IgA anti-Gal antibodies in serum or colostrum block C-mediated killing of meningococci, whereas IgG anti-Gal antibodies do not (37). IgA antibodies to the capsule of group C *N. meningitidis* also block IgG-initiated C killing (42), as has been shown for IgA-blocking antibodies with other organisms (37). Therefore, IgA antibodies may promote bacterial survival in the host. In contrast, blocking antibodies for *N. gonorrhoeae* are IgG. Antibodies to LPS or porin (protein I or PI) antigens mediate C killing, whereas IgG antibodies to outer membrane protein III (PIII) or reduction modifiable protein (Rmp) block killing of gonococci by fresh normal human serum (11, 60). Serum-resistant strains also inactivate more C3b to iC3b and generate less C5a (60). Significantly, an increased rate of transmission of gonococci has been demonstrated in the presence of increased levels of antibody to Rmp or PIII (60), suggesting that blocking antibody does facilitate infection.

Blocking antibodies also appear to play a role in bovine brucellosis. Fresh normal bovine serum or fresh agammaglobulinemic serum from colostrum-deprived calves killed virulent *Brucella abortus* in vitro (18). Smooth strains (LPS with O side chains) were much less serum sensitive than rough vaccine strains (with no O antigen), again demonstrating the role of LPS O side chains in protecting against C-mediated killing. Although *B. abortus* could be killed by C in the absence of immunoglobulin, antibody increased killing of the rough strains. However, killing of the smooth virulent strain was significantly reduced in the presence of specific antibody. In infected animals, the earliest serum with IgM antibody did not decrease killing, whereas later IgG responses did. Affinity-purified IgG1 or IgG2 anti-smooth LPS blocked killing, but IgM and IgA antibodies to smooth LPS did not. The data from infected animals showed that serum from time zero (with no specific antibody) and serum from early in the primary response (with predominantly IgM antibody) were able to kill the organism. Thus, during the early extracellular phase, the host could kill the organism by IgM antibody and C or by C alone. Later, when the organism was intracellular and IgG antibodies predominated, blocking antibodies prevented extracellular killing (18). Thus, *B. abortus* evades antibody defenses both by its intracellular location and by stimulating blocking antibody.

DECOYS

Bacteria lacking a protective coat such as a capsule may avoid antibody-mediated defenses by shedding antigen into the external milieu. This cloud of antigen reacts with antibody, resulting in deposition of C components away from the surface of the organism. Therefore, the MAC may not be inserted in the membrane and killing would not occur. Similarly, opsonization would be evaded. Outer membrane vesicles (MVs) from the surface of many gram-negative bacteria are common means of discharging antigen from the surface (10). Whether this is associated with decreased C killing and opsonization is unclear. However, fibrils released from the surface of *Haemophilus somnus* (17) are associated with both serum resistance and reduced phagocytosis in the presence of immunoglobulin (C. Alexander-Handley and P. Widders, Abstr. Gen. Meet. Am. Soc. Microbiol. 1989, abstr. B-174, 1989). This fibrillar network, which is shed into the supernatant (17), is present in serum-resistant virulent isolates but not in some serum-sensitive isolates from asymptomatic carriers (14). Recent preliminary data

show that if the fibrils in concentrated supernatant from virulent strains are added back to these serum-sensitive carrier isolates, serum resistance results (Corbeil, unpublished data). This could be explained simply by antibody–antigen complexes activating C away from the surface as described above. However, the situation may be much more complex. The fibrillar network of *H. somnus* has been shown to bind bovine IgG2 by the Fc portion (6, 17, 72, 75). The role of these immunoglobulin-binding proteins will be discussed at the end of this chapter.

SHIFTING TARGETS

Antigenic variation is a mechanism of evading a vigorous antibody response. Early studies with pathogenic protozoa defined this complex mechanism in detail. The first bacteria shown to undergo antigenic variation in the face of an immune response were *Campylobacter fetus* (23, 27, 61) and *Borrelia hermsii* (the cause of relapsing fever) (7). For *B. hermsii*, this antigenic variation was shown to be associated with DNA rearrangements with resulting expression of different serotypes (53). One of the best-studied examples of bacterial antigenic variation involves surface antigens of *Neisseria* spp. Both *N. gonorrhoeae* and *N. meningitidis* have a variety of mechanisms of phase variation or antigenic variation of pilin and outer-membrane proteins (26, 31). The genetic mechanisms of antigenic variation in bacteria, protozoa, and fungal pathogens were reviewed recently (26) and will be discussed in chapter 10.

Some years ago, we noted that the LOS of *H. somnus* isolated from calf lungs in a chronic pneumonia study had LOS with different migration patterns in sodium dodecyl sulfate-polyacrylamide gel electrophoresis (SDS-PAGE) than the LOS of the infecting strain. We hypothesized that this was due to selection of a minor population in the intrabronchial inoculum. The experimental pneumonia study was repeated using an inoculum of *H. somnus* cloned three times by subculturing a single colony each day for 3 days (40).

Weekly isolates from bronchial lavage of 3 calves demonstrated a high rate of phenotypic variation in the LOS. Antigenic variation was detected, since rabbit antiserum to the LOS of the inoculated cloned strain did not react with any of the postchallenge isolates (40). Inzana et al. (41) later showed that the phase-variable epitopes were encoded by a gene with a tandem 5′-CAAT-3′ repeat. It was thought that variation in the number of 5′-CAAT-3′ sequences may occur via slip-strand repair mechanism as in *Haemophilus influenzae* LOS (41). Now researchers (24) have further elucidated the structure of the phase-variable LOS of *H. somnus*, so that the mechanism of variation can be better addressed. Whether this phase and antigenic variation is significant in vivo was studied in the natural host. Calves were inoculated intrabronchially with a cloned isolate of *H. somnus*, and the organism was reisolated from bronchioalveolar lavage fluids weekly until clearance (40). Although prechallenge serum did not react with the LOS of the challenge strain, convalescent serum taken at 35 days postchallenge reacted with the challenge-strain LOS and that of 6 of 31 subsequent isolates from the 3 calves. Convalescent-phase serum taken 74 days after challenge reacted with LOS of 15 of 31 subsequent isolates, indicating that the antibody response was catching up with the variant antigens. The suggestion of evasion of host response early in the infection with later antibody responses to all variants was strengthened by high counts of colony-forming units early in infection, but clearing of infection in the two calves whose antibody response at 74 days recognized the LOS of the last isolate at 63 days. Lavage fluid from the other calf still had $>10^4$ CFU/ml by 70 days, and its 74-day convalescent-phase serum did not react with the LOS of the isolate taken at 63 days (40). Antibody titers to *H. somnus* were high by 14 days, indicating persistence in the face of immune responses (35). This is a good demonstration of evasion of antibody-mediated defense coupled with antigenic var-

iation. It also shows that, in time, the host may overcome the evasion.

SUBVERSION OF ANTIBODY RESPONSES

Microbes divert immune responses away from productive, protective immunity in several ways. Some bacteria, including streptococci and staphylococci, release superantigens that stimulate T or B cells nonspecifically. T-cell superantigens bind both to the MHC class II molecules and to the variable domain of the β subunit (V β) of the T-cell antigen receptor (TCR). This prevents antigen stimulation of T cells for a specific T-helper response, but induces proliferation of a much greater polyclonal T-cell population than antigen-induced specific responses (67, 71). The result is a lack of specific protective response, but polyclonal stimulation with cytokine release that may be pathological. Similarly, B-cell superantigens are polyclonal stimulators of a large subpopulation of B cells (64), but without antigen-specific responses. Other microbes subvert the immune response away from protection by regulation of specific responses. Perturbation of cytokine networks by bacteria can modulate the balance between Th1 and Th2 responses (74). The predominant response depends both on the release of immunoregulatory molecules by the pathogen and on the genetic makeup of the host. For example, in human leprosy, polarized Th1 or Th2 responses result in relatively contained tuberculoid disease or disseminated lepromatous disease, respectively (63). Recent work shows that CD1 expression by dendritic cells (correlated with controlled disease) was due to factors at the site of infection rather than the genetic ability of patient cells to express CD1 antigens (63). A similar polarized response is seen in patients with tuberculosis, in that some patients have well-contained disease (or no overt disease), whereas others develop disseminated infection. Activated macrophages are critical to control of infection. Again, CD1-restricted T cells contribute to this defense. Stenger et al. (66) recently showed that infection of CD1-positive antigen-presenting cells with virulent *Mycobacterium tuberculosis* resulted in down-regulation of CD1 expression. Thus, both *M. leprae* and *M. tuberculosis* may evade host response by down-regulating protective CD1-restricted T-cell activation of macrophages. Other studies show modulation of Th1 versus Th2 responses with various bacterial infections. Immunization of cattle with DNA expressing a major surface antigen of *Anaplasma marginale* revealed a bias for IgG1 antibody responses (4) previously shown to be IL4 dependent and therefore Th2-like. Our own studies have shown that *H. somnus* antigens also induce biased responses in cattle. Immunization with one purified outer membrane protein (OMP p78) induced only detectable IgG1 antibody to p78, whereas another OMP (p40) induced almost equal amounts of IgG1 and IgG2 p40-specific antibody (17, 34). The former antibody was not passively protective, whereas the latter did protect in experimental *H. somnus* pneumonia in calves (34). Since IgG2 antibody was correlated with protection against this infection (34), modulation toward an IgG1 response may be an evasive mechanism. A similar bias was detected in our studies with the protective surface antigen (TF1.17) of *T. foetus*. Systemic immunization with immunoaffinity-purified TF1.17 antigen results in almost entirely IgG1 antibody to this antigen (16). A high IgG1 response does result in clearance of infection earlier than in nonimmunized controls (16), and IgG1 antibody was shown to be more effective in preventing adherence of *T. foetus* to bovine vaginal epithelial cells than IgG2 antibody (20). However, IgG2 antibody to *T. foetus* is most effective in opsonizing and intracellular killing by neutrophils (5). Therefore, it is not clear whether preferential IgG1 responses are an advantage to the host or an evasive mechanism of the parasite.

DESTRUCTION OF ANTIBODY DEFENSES

Thus far, we have seen that pathogens evade host defense (i) at the beginning of infection by killing the host or the phagocyte before an immune response occurs, (ii) by hiding from protective responses in cells or in vegetations, (iii) by avoiding recognition (masking or mimicry), (iv) by varying antigenic targets, (v) by blocking protective antibody functions (blocking antibodies or immune complex formation away from the surface), or (vi) by subverting immune responses. All of these "tactics" avoid antibody recognition. Another strategy used by many pathogens is the destruction of antibody and/or complement by proteinase cleavage of the molecule to render it nonfunctional. Many human mucosal pathogens produce proteinases that cleave IgA1 (59). Pathogenic species of both *Neisseria* and *Haemophilus* produce proteinases that cleave human IgA1 at the hinge. Human challenge demonstrated that *N. gonorrhoeae* does not require IgA1 proteinase to cause experimental urethritis in men (43). However, it was not determined whether the wild-type, IgA1-proteinase-positive strain had a longer duration of infection due to evasion of the IgA response than the IgA1-proteinase-negative mutant in these 5-day experiments. Obviously, evasion of a specific immune response does not provide a selective advantage until the immune response occurs. Some studies on duration of infection and IgA1 proteinases have been done with nontypable *H. influenzae* in children (48). The strains that remained in the respiratory tract were shown to undergo antigenic variation of the IgA1 proteinase (48), demonstrating a double evasive mechanism. Although many bacteria secrete proteinases (56), in some cases proteinases of parasites provide better examples of the role of proteinase in immune evasion. *Entamoeba histolytica* has been shown to secrete cysteine proteinase that cleaves C3 to produce C3a and C3b (57). This activates the alternative C pathway, resulting in lysis of nonpathogenic but not pathogenic *E. histolytica*. Further investigations of the relationship between extracellular cysteine proteinase and pathogenesis demonstrated cleavage of anaphylatoxins C3a and C5a to small inactive peptides (58). Thus, although *E.*

histolytica extracellular cysteine proteinase activates the C cascade, it also provides a mechanism of evading proinflammatory factors C3a and C5a. This same cysteine proteinase was later shown to degrade human IgG, which correlates with the inability of certain levels of specific IgG antibodies to prevent invasive amebiasis or recurrent infection (70). Another study provided additional evidence that cysteine proteinases are important in pathogenesis in vivo. Ankri et al. (2) showed that virulent *E. histolytica* transfected with the antisense gene-encoding cysteine proteinase 5 had only 10% of the cysteine proteinase activity and was incapable of forming liver abscesses in hamsters, whereas the wild-type and the control transfectant both produced liver abscesses. This result was not due to differences in cytopathic activity in mammalian cell monolayers (2), so evasion of host defense may be a factor. Similarly, trichomonads have been shown to produce cysteine proteinases that degrade immunoglobulins (54, 69). *Trichomonas vaginalis*, the human pathogen, degrades human IgG, IgM, and IgA (54), whereas *T. foetus* degrades primarily bovine IgG1 and IgG2 (69). Interestingly, recent studies show that the two bovine IgG2 allotypes are not equally susceptible to cleavage by *T. foetus* extracellular cysteine proteinase in that IgG2^a is cleaved at a significantly faster rate than IgG2^b (8). This implies that *T. foetus* may evade IgG2 responses of animals with the IgG2^a allotype better than those of animals with the IgG2^b allotype. In vivo challenge studies of animals of known allotype will need to be done to test this hypothesis. It is clear, however, that *T. foetus* causes a chronic genital infection in cattle with a robust local immune response. The persistence of infection may be due to digestion of antibody, in addition to binding of IgG to mask antigen and biased IgG1 responses.

CRIPPLING ANTIBODY DEFENSES

This strategy could be labeled "If you can't beat them, join them." Rather than degrading immunoglobulins as described above, some bacteria hold the antigen combining site "at arms' length" by binding antibody Fc domains to specific immunoglobulin-binding proteins (IgBPs) on the bacterial surface. The best known IgBPs are Protein A of *Staphylococcus aureus* (33) and Protein G of Groups C and G streptococci (1), although streptococci of other groups also have IgBPs (55). Protein A coating of *S. aureus* with immunoglobulin has been shown to decrease opsonization (33), and Group A streptococci expressing high levels of IgG-binding protein after eight passages in human blood were more resistant to opsonophagocytosis (55). Studies with Protein H, another IgBP from *S. pyogenes*, demonstrated that soluble IgBP–IgG complexes activate C, but that protein H inhibits C activation by IgG-coated targets (9). This suggests that the released IgBP–IgG complexes would activate C away from the bacteria surface, but that C activation on the bacterial surface would be inhibited. Both mechanisms would contribute to evasion of host defenses.

Studies in our own laboratories demonstrated IgBPs released from and on the surface of *H. somnus* (75). These IgBPs were shown to be associated with serum resistance of *H. somnus* (18, 73) and thus evasion of host responses. Molecular analysis revealed that serum-sensitive isolates from asymptomatic preputial carriers were missing a 13.4 kb sequence of DNA that encoded high-molecular-weight (HMW) IgBPs as well as a 76-kDa protein (14). Sequencing of the DNA insert encoding p76 showed that it contained 1.2 kb tandem direct repeats similar to insertion sequences (15). Whether the presence of insertion elements was related to acquiring virulence by isolates from disease was unknown. However, later studies showed that p76 was also an IgBP and was probably a peripheral membrane protein (17). The HMW IgBPs, which often are shed into the medium, were shown to consist of a very fine fibrillar network on the surface of *H. somnus* (17). These easily shed fibrils bind bovine IgG2 by the Fc fragment (17). As noted above, activation of C away from the surface of *H. somnus* by shed fibrils may protect the organism from insertion of the MAC in the bacterial outer membrane, resulting in serum resistance. Like *S. aureus* and *S. pyogenes*, *H. somnus* IgBPs were also shown to be associated with resistance to phagocytosis (C. Alexander-Handley and P. Widders, Abstr. Gen. Meet. Am. Soc. Microbiol. 1989, abstr. B-174, 1989), so these IgBPs contribute to evasion of several antibody-mediated defenses. Sequencing of the gene encoding the *H. somnus* IgBPs revealed that both HMW IgBPs and the p76 IgBP are encoded by one very large open reading frame (Y. Tagawa, J. D. Sanders, I. Uchida, F. D. Bastida-Corcuera, and L. B. Corbeil, Abstr. Vir. Mech. Bact. Pathog., abstr. 14, 1999). Motifs discovered in the sequence analysis provided a basis for functional studies of the protein to aid in understanding its relevance to evasion of antibody-mediated defense. These should help to explain the persistence of *H. somnus* infection in calf lungs (35) for at least 10 weeks in the presence of specific antibody.

REFERENCES

1. **Akerstrom, B., T. Brodin, K. Reis, and L. Bjorck.** 1985. Protein G: a powerful tool for binding and detection of monoclonal and polyclonal antibodies. *J. Immunol.* **135**:2589–2592.
2. **Ankri, S., T. Stolarsky, R. Bracha, F. Padilla-Vaca, and D. Mirelman.** 1999. Antisense inhibition of expression of cysteine proteinases affects *Entamoeba histolytica*-induced formation of liver abscess in hamsters. *Infect. Immun.* **67**:421–422.
3. **Apicella, M. A., R. E. Mandrell, M. Shero, M. E. Wilson, J. M. Griffiss, G. F. Brooks, C. Lammel, J. F. Breen, and P. A. Rice.** 1990. Modification by sialic acid of *Neisseria gonorrhoeae* lipooligosaccharide epitope expression in human urethral exudates: an immunoelectron microscopic analysis. *J. Infect. Dis.* **162**:506–512.
4. **Arulkanthan, A., W. C. Brown, T. C. McGuire, and D. P. Knowles.** 1999. Biased immunoglobulin G1 isotype responses induced in cattle with DNA expressing *msp*1a of *Anaplasma marginale*. *Infect. Immun.* **67**:3481–3487.
5. **Aydintug, M. K., P. R. Widders, and R. W. Leid.** 1993. Bovine polymorphonuclear leukocyte killing of *Tritrichomonas foetus*. *Infect. Immun.* **61**:2995–3002.
6. **Baddour, L. M.** 1994. Virulence factors among gram-positive bacteria in experimental endocarditis. *Infect. Immun.* **62**:2143–2148.
7. **Barbour, A. G., S. L. Tessier, and H. G. Stoenner.** 1982. Variable major proteins of *Borrelia hermsii*. *J. Exp. Med.* **156**:1312–1324.
8. **Bastida-Corcuera, F., J. E. Butler, H. Heyermann, J. W. Thomford, and L. B. Corbeil.** *Tritrichomonas foetus* extracellular cysteine proteinase cleavage of bovine IgG2 allotypes. *J. Parasitol.*, in press.
9. **Berge, A., B. Kihlberg, A. G. Sjoholm, and L. Bjorck.** 1997. Streptococcal protein H forms soluble complement-activating complexes with IgG, but inhibits complement activation by IgG-coated targets. *J. Biol. Chem.* **272**:20774–20781.
10. **Beveridge, T. J.** 1999. Structures of gram-negative cell walls and their derived membrane vesicles. *J. Bacteriol.* **181**:4725–4733.
11. **Blake, M. S., L. M. Wetzler, E. C. Gotschlich, and P. A. Rice.** 1989. Protein III: structure, function, and genetics. *Clin. Microbiol. Rev.* **2**:S60–S63.
12. **BonDurant, R. H., R. R. Corbeil, and L. B. Corbeil.** 1993. Immunization of virgin cows with surface antigen TF1.17 of *Tritrichomonas foetus*. *Infect. Immun.* **61**:1385–1394.
13. **Chen, T., J. Swanson, J. Wilson, and R. J. Belland.** 1995. Heparin protects Opa *Neisseria gonorrhoeae* from the bactericidal action of normal human serum. *Infect. Immun.* **63**:1790–1795.
14. **Cole, S. P., D. G. Guiney, and L. B. Corbeil.** 1992. Two linked genes for outer membrane proteins are absent in four non-disease strains of *Haemophilus somnus*. *Mol. Microbiol.* **6**:1895–1902.
15. **Cole, S. P., D. G. Guiney, and L. B. Corbeil.** 1993. Molecular analysis of a gene encoding a serum-resistance-associated 76 kDa surface antigen of *Haemophilus somnus*. *J. Gen. Microbiol.* **139**:2135–2143.
16. **Corbeil, L. B., M. L. Anderson, R. R. Corbeil, J. M. Eddow, and R. H. BonDurant.** 1998. Female reproductive tract immunity in bovine trichomoniasis. *Am. J. Reprod. Immunol.* **39**:189–198.
17. **Corbeil, L. B., F. D. Bastida-Corcuera, and T. J. Beveridge.** 1997. *Haemophilus somnus* immunoglobulin binding proteins and surface fibrils. *Infect. Immun.* **65**:4250–4257.
18. **Corbeil, L. B., K. Blau, T. J. Inzana, K. H. Nielsen, R. H. Jacobson, R. R. Corbeil, and A. J. Winter.** 1988. Killing of *Brucella abortus* by bovine serum. *Infect. Immun.* **56**:3251–3261.
19. **Corbeil, L. B., R. P. Gogolewski, I. Kacskovics, K. H. Nielsen, R. R. Corbeil, J. L. Morrill, R. Greenwood, and J. E. Butler.**

1997. Bovine IgG2a antibodies to *Haemophilus somnus* and allotype expression. *Can. J. Vet. Res.* **61:**207–213.

20. **Corbeil, L. B., J. L. Hodgson, D. W. Jones, R. R. Corbeil, P. R. Widders, and L. R. Stephens.** 1989. Adherence of *Tritrichomonas foetus* to bovine vaginal epithelial cells. *Infect. Immun.* **57:**2158–2165.

21. **Corbeil, L. B., J. L. Hodgson, and P. R. Widders.** 1991. Immunoglobulin binding by *Tritrichomonas foetus. J. Clin. Microbiol.* **29:**2710–2714.

22. **Corbeil, L. B., S. A. Kania, and R. P. Gogolewski.** 1991. Characterization of immunodominant surface antigens of *Haemophilus somnus. Infect. Immun.* **59:**4295–4301.

23. **Corbeil, L. B., G. G. D. Schurig, P. J. Bier, and A. J. Winter.** 1975. Bovine venereal vibriosis: antigenic variation of the bacterium during infection. *Infect. Immun.* **11:**240–244.

24. **Cox, A. D., M. D. Howard, J.-R. Brisson, M. Van Der Zwan, P. Thibault, M. B. Perry, and T. J. Inzana.** 1998. Structural analysis of the phase-variable lipooligosaccharide from *Haemophilus somnus* strain 738. *Eur. J. Biochem.* **253:**507–516.

25. **Dabiri, G. A., J. M. Sanger, D. A. Portnoy, and F. S. Southwick.** 1990. *Listeria monocytogenes* moves rapidly through the host-cell cytoplasm by inducing directional actin assembly. *Proc. Natl. Acad. Sci. USA* **87:**6068–6072.

26. **Deitsch, K. W., E. R. Moxon, and T. E. Wellems.** 1997. Shared themes of antigenic variation and virulence in bacterial, protozoal, and fungal infections. *Microbiol. Mol. Biol. Rev.* **61:**281–293.

27. **Dubreuil, J., Daniel, M. Kostrzynska, J. W. Austin, and T. J. Trust.** 1990. Antigenic differences among *Campylobacter fetus* S-layer proteins. *J. Bacteriol.* **172:**5035–5043.

28. **Duensing, T. D., J. S. Wing, and J. P. M. Van Putten.** 1999. Sulfated polysaccharide-directed recruitment of mammalian host proteins: a novel strategy in microbial pathogenesis. *Infect. Immun.* **67:**4463–4468.

29. **Duesbery, N. S., C. P. Webb, S. H. Leppla, V. M. Gordon, K. R. Klimpel, T. D. Copeland, N. G. Ahn, M. K. Oskarsson, K. Fukasawa, K. D. Paull, and G. F. Wande Woude.** 1998. Proteolytic inactivation of MAP-kinase-kinase by anthrax lethal factor. *Science* **280:**734–737.

30. **Feldmann, R. C., B. Henrich, V. Kolb-Bachofen, and U. Hadding.** 1992. Decreased metabolism and viability of *Mycoplasma hominis* induced by monoclonal antibody-meditated agglutination. *Infect. Immun.* **60:**166–174.

31. **Findlay, B. B., and S. Falkow.** 1997. Common themes in microbial pathogenicity revisited. *Microbiol. Mol. Biol. Rev.* **61:**136–169.

32. **Fischetti, V. A.** 1989. Streptococcal M. protein: molecular design and biological behavior. *Clin. Microbiol. Rev.* **2:**285–314.

33. **Foster, T. J., and D. McDevitt.** 1994. Surface-associated proteins of *Staphylococcus aureus*: their possible roles in virulence. *FEMS Microbiol. Lett.* **118:**199–206.

34. **Gogolewski, R. P., S. A. Kania, D. Liggitt, and L. B. Corbeil.** 1988. Protective ability of antibodies against 78- and 40-kilodalton outer membrane antigens of *Haemophilus somnus. Infect. Immun.* **56:**2307–2316.

35. **Gogolewski, R. P., D. C. Schaefer, S. K. Wasson, R. R. Corbeil, and L. B. Corbeil.** 1989. Pulmonary persistence of *Haemophilus somnus* in the presence of specific antibody. *J. Clin. Microbiol.* **27:**1767–1774.

36. **Greiffenberg, L., W. Goebel, K. S. Kim, I. Weiglein, A. Bubert, F. Engelbrecht, M. Stins, and M. Kuhn.** 1998. Interaction of *Listeria monocytogenes* with human brain microvascular endothelial cells: InlB-dependent invasion, long-term intracellular growth, and spread from macrophages to endothelial cells. *Infect. Immun.* **66:**5260–5267.

37. **Hamadeh, R. M., M. M. Estabrook, P. Zhou, G. A. Jarvis, and J. M. Griffiss.** 1995. Anti-gal binds to pili of *Neisseria meningitidis*: the immunoglobulin A isotype blocks complement-mediated killing. *Infect. Immun.* **63:**4900–4906.

38. **Hoffman, J. A., C. Wass, M. F. Stins, and K. S. Kim.** 1999. The capsule supports survival but not traversal of *Escherichia coli* K1 across the blood-brain barrier. *Infect. Immun.* **67:**3566–3570.

39. **Ikeda, J. S., R. H. BonDurant, and L. B. Corbeil.** 1995. Bovine vaginal antibody responses to immunoaffinity-purified surface antigen of *Tritrichomonas foetus. J. Clin. Microbiol.* **33:**1158–1163.

40. **Inzana, T. J., R. P. Gogolewski, and L. B. Corbeil.** 1992. Phenotypic phase variation in *Haemophilus somnus* lipooligosaccharide during bovine pneumonia and after in vitro passage. *Infect. Immun.* **60:**2943–2951.

41. **Inzana, T. J., J. Hensley, J. McQuiston, A. J. Lesse, A. A. Campagnari, S. M. Boyle, and M. Apicella.** 1997. Phase variation and conservation of lipooligosaccharide epitopes in *Haemophilus somnus. Infect. Immun.* **65:**4675–4681.

42. **Jarvis, G. A., and J. M. Griffiss.** 1991. Human IgA2 blockade of IgG-initiated lysis of *Neisseria meningitidis* is a function of antigen-binding frag-

ment binding to the polysaccharide capsule. *J. Immunol.* **147:**1962–1967.

43. **Johannsen, D. B., D. M. Johnston, H. O. Koymen, M. S. Cohen, and J. G. Cannon.** 1999. A *Neisseria gonorrhoeae* immunoglobulin A1 protease mutant is infectious in the human challenge model of urethral infection. *Infect. Immun.* **67:**3009–3013.

44. **Kay, W. W., and T. J. Trust.** 1991. Virulence factors. Form and functions of the regular surface array (S-layer) of *Aeromonas salmonicida. Experientia* **47:**412–414.

45. **Krull, M., R. Nost, S. Hippenstiel, E. Domann, T. Chakraborty, and N. Suttorp.** *Listeria monocytogenes* potently induces up-regulation of endothelial adhesion molecules and neutrophil adhesion to cultured human endothelial cells. *J. Immunol.* **159:**1970–1976.

46. **Lecuit, M., H. Ohayon, L. Braun, J. Mengaud, and P. Cossart.** 1997. Internalin of *Listeria monocytogenes* with an intact leucine-rich repeat region is sufficient to promote internalization. *Infect. Immun.* **65:**5309–5319.

47. **Lo, R. Y. C., C. A. Strathdee, and P. E. Sheen.** 1987. Nucleotide sequence of the leukotoxin genes of *Pasteurella haemolytica* A1. *Infect. Immun.* **55:**1987–1996.

48. **Lomholt, H., L. van Alphen, and M. Kilian.** 1993. Antigenic variation of immunoglobulin A1 proteases among sequential isolates of *Haemophilus influenzae* from healthy children and patients with chronic obstructive pulmonary disease. *Infect. Immun.* **61:**4575–4581.

49. **Miorner, H., E. Myhre, L. Bjorck, and G. Kronvall.** 1980. Effect of specific binding of human albumin, fibrinogen, and immunoglobulin G on surface characteristics of bacterial strains as revealed by partition experiments in polymer phase systems. *Infect. Immun.* **29:**879–885.

50. **Mounier, J., A. Ryter, M. Coquis-Rondon, and P. Sansonetti.** 1990. Intracellular and cell-to-cell spread of *Listeria monocytogenes* involves interaction with F-actin in the enterocytelike cell line Caco-2. *Infect. Immun.* **58:**1048–1058.

51. **Nachamkin, I., B. M. Allos, and T. Ho.** 1998. *Campylobacter* species and Guillain-Barre Syndrome. *Clin. Microbiol. Rev.* **11:**555–567.

52. **Parsonson, I. M., B. L. Clark, and J. H. Dufty.** 1976. Early pathogenesis and pathology of *Tritrichomonas foetus* infection in virgin heifers. *J. Comp. Pathol.* **86:**59–66.

53. **Plasterk, R. H. A., M. L. Simon, and A. G. Barbour.** 1985. Transposition of structural genes to an expression sequence on a linear plasmid causes antigenic variation in the bacterium *Borrelia hermsii. Nature* **318:**257–263.

54. **Provenzano, D., and J. F. Alderete.** 1995. Analysis of human immunoglobulin-degrading cysteine proteinases of *Trichomonas vaginalis. Infect. Immun.* **63:**3388–3395.

55. **Raeder, R. M., and D. P. Boyle.** 1993. Association of type II immunoglobulin G-binding protein expression and survival of group A streptococci in human blood. *Infect. Immun.* **61:**3696–3702.

56. **Rao, M. B., A. M. Tanksale, M. S. Chatge, and V. V. Deshpande.** 1998. Molecular and biotechnological aspects of microbial proteases. *Microbiol. Mol. Biol. Rev.* **62:**597–635.

57. **Reed, S. L., W. E. Keene, J. H. McKerrow, and I. Gigli.** 1989. Cleavage of C3 by a neutral cysteine proteinase of *Entamoeba histolytica. J. Immunol.* **143:**189–195.

58. **Reed, S. L., J. A. Ember, D. S. Herdman, R. G. DiScipio, T. E. Hugli, and I. Gigli.** 1995. The extracellular neutral cysteine proteinase of *Entamoeba histolytica* degrades anaphylatoxins C3a and C5a. *J. Immunol.* **155:**266–274.

59. **Reinholdt, J., and M. Kilian.** 1997. Comparative analysis of immunoglobulin A1 protease activity among bacteria representing different genera, species, and strains. *Infect. Immun.* **65:**4452–4459.

60. **Rice, P. A., D. P. McQuillen, S. Gulati, D. B. Jani, L. M. Wetzler, M. S. Blake, and E. C. Gotschlich.** 1994. Serum resistance of *Neisseria gonorrhoeae. Ann. N. Y. Acad. Sci.* **730:**7–14.

61. **Schurig, G. D., C. E. Hall, K. Burda, L. B. Corbeil, J. R. Duncan, and A. J. Winter.** 1973. Persistent genital tract infection with *Vibrio fetus intestinalis* associated with serotypic alteration of the infecting strain. *Am. J. Vet. Res.* **34:**1399–1403.

62. **Shikhman, A. R., and M. W. Cunningham.** 1994. Immunological mimicry between N-acetyl-β-D-glucosamine and cytokeratin peptides. Evidence for a microbially driven anti-keratin antibody response. *J. Immunol.* **152:**4375–4387.

63. **Sieling, P. A., D. Jullien, M. Dahlem, T. F. Tedder, T. H. Rea, R. L. Modlin, and S. A. Porcelli.** 1999. CD1 expression by dendritic cells in human leprosy lesions: correlation with effective host immunity. *J. Immunol.* **162:**1851–1858.

64. **Silverman, G. J.** 1997. B-cell superantigens. *Immunol. Today* **18:**379–386.

65. **Sleytr, U. B., and P. Messner.** 1988. Crystalline surface layers in procaryotes. *J. Bacteriol.* **170:**2891–2897.

66. **Stenger, S., K. R. Niazi, and R. L. Modlin.** 1998. Down-regulation of CD1 on antigen-

presenting cells by infection with *Mycobacterium tuberculosis. J. Immunol.* **161:**3582–3588.

67. **Stohl, W., and J. E. Elliott.** 1995. Differential human T-cell dependent B cell differentiation induced by staphylococcal superantigens (SAG). *J. Immunol.* **155:**1838–1850.

68. **Strauss, E.** 1998. New clue to how anthrax kills. *Science* **280:**676.

69. **Talbot, J. A., K. Nielsen, and L. B. Corbeil.** 1991. Cleavage of proteins of reproductive secretions by extracellular proteinases of *Tritrichomonas foetus. Can. J. Microbiol.* **37:**384–390.

70. **Tran, V. Q., D. Scott Herdman, B. E. Torian, and S. L. Reed.** 1997. The neutral cysteine proteinase of *Entamoeba histolytica* degrades IgG and prevents its binding. *J. Infect. Dis.* **177:**508–511.

71. **Ulrich, R. G., M. A. Olson, and S. Bavari.** 1998. Development of engineered vaccines effec-

tive against structurally related bacterial superantigens. *Vaccine* **16:**1857–1864.

72. **Widders, P. R., J. W. Smith, M. Yarnall, T. C. McGuire, and L. B. Corbeil.** 1987. Non-immune immunoglobulin binding by *Haemophilus somnus. J. Med. Microbiol.* **26:**307–311.

73. **Widders, P. R., L. A. Dorrance, M. Yarnall, and L. B. Corbeil.** 1989. Immunoglobulin-binding activity among pathogenic and carrier isolates of *Haemophilus somnus. Infect. Immun.* **57:**639–642.

74. **Wilson, M., R. Seymour, and B. Henderson.** 1998. Bacterial perturbation of cytokine networks. *Infect. Immun.* **66:**2401–2409.

75. **Yarnall, M., P. R. Widders, and L. B. Corbeil.** 1988. Isolation and characterization of Fc receptors from *Haemophilus somnus. Scand. J. Immunol.* **28:**129–137.

MECHANISMS OF RESISTANCE TO NO-RELATED ANTIBACTERIAL ACTIVITY

Andrés Vazquez-Torres and Ferric C. Fang

9

Little more than a decade ago, it was discovered that an array of physiological processes ranging from vascular homeostasis to neurotransmission and immunity are regulated by endogenously produced nitric oxide (NO) (for reviews, see references 62, 70). NO is the enzymatic product of a group of hemoproteins collectively known as NO synthases that catalyze the oxidation of one of the guanidino groups of L-arginine, giving rise to equimolar amounts of citrulline and NO. Production of this small diatomic chemical species is a complicated process that requires L-arginine, molecular oxygen, and reducing equivalents provided by NADPH, along with the cofactors FAD, FMN, and tetrahydrobiopterin (45, 50, 73). Of the three isoforms of NO synthase and a variety of splice variants that can catalyze this reaction, the cytokine-inducible high output NO synthase (also designated iNOS or NOS2) has been most strongly associated with antimicrobial activity (20, 72). NO can also arise by the reduction of ingested nitrate to nitrite by commensal bacteria in the oral

cavity and the subsequent nonenzymatic reduction of nitrite to NO in the acidic environment of the stomach (6, 25).

NO has a broad range of antimicrobial activity toward pathogens encompassing viruses, bacteria, fungi, and parasites (for reviews, see references 20, 31). The exquisite diversity of microbial NO targets reflects a complicated chemistry resulting in good part from the reactivity of the unpaired electron present in the π antibonding orbital of the molecule (39). Although relatively stable, NO combines directly with metals, accounting for the activation of specific heme proteins and the slow inactivation of dehydratases (36; reviewed in reference 103). The microbicidal and microbiostatic properties of NO also derive from reactions with other radicals and oxygen. NO reacts with superoxide to form peroxynitrite (ONOO$^-$) with a rate constant of approximately 7×10^9 M^{-1} s^{-1}; peroxynitrite is a strong oxidant able to modify lipids, DNA, and proteins, such as bacterial aconitases (12, 47, 55). The auto-oxidation of NO can give rise to several reactive molecular species, ranging from the strong oxidant nitrogen dioxide to the N- and S-nitrosating agent dinitrogen trioxide (N$_2$O$_3$). In the presence of iron, S-nitrosothiols readily form dinitrosyl–iron complexes, nitrosating species little studied in

Andrés Vazquez-Torres and Ferric C. Fang Departments of Medicine, Pathology, and Microbiology, University of Colorado Health Sciences Center, 4200 E. 9th Ave., B168, Denver, CO 80262.

Virulence Mechanisms of Bacterial Pathogens, 3rd ed., Edited by K. A. Brogden et al.
©2000 ASM Press, Washington, D.C.

the context of antimicrobial action but potentially able to mediate specific damage. The relative abundance, diffusibility, and stability of specific NO congeners, together with specific targeting of microbial molecules, probably account for the broad antimicrobial actions of this radical.

Abundant evidence implicates endogenous NO production in host defense. Both humans and experimental animals produce dramatically increased quantities of NO during infection (2, 26, 29, 80, 101), and NOS expression can be directly demonstrated in infected tissues (40, 60, 75, 90). NOS inhibition exacerbates microbial proliferation during experimental infections (9, 29, 60; additional references are listed in reference 20) or phagocyte-killing assays (23, 43, 58, 77, 98; additional references are listed in reference 20), and a variety of chemical NO donors have been shown to inhibit or kill diverse microbial species in vitro (3, 21, 53, 71, 93; other references are listed in reference 20).

NO clearly plays diverse roles during infection, not only limiting microbial proliferation but also controlling vascular tone (83) and cell adhesion, mediating both cytoprotection (103) and cytotoxicity (42), and modulating immune system activation (65). This chapter focuses on some of the recently elucidated strategies by which bacteria cope with the cytotoxic effects of NO and its redox congeners (Fig. 1). Microbes, particularly pathogenic species, have coapted previously existing mechanisms and evolved novel strategies to avoid NO exposure or detoxify nitrogen oxides.

AVOIDANCE OR INHIBITION OF NO SYNTHESIS

Lipopolysaccharide (LPS) from gram-negative bacteria is one of the most potent inducers of iNOS transcription (92), although repeated activation of murine macrophages with trace quantities of LPS actually results in inhibition of iNOS expression (8, 19, 86, 99), possibly by a mechanism that involves prolonged elevation of intracellular cyclic AMP synthesis

(11). Some pathogens may modify their LPS to avoid triggering iNOS expression. For example, LPS phase variants of the intracellular pathogen *Francisella tularensis* fail to stimulate macrophage NO production and consequently have enhanced intracellular survival (16). Virulence-associated modifications of lipid A in *Salmonella enterica* serovar Typhimurium result in reduced tumor necrosis factor alpha (TNF-α) production (46), which suggests that NO production may be reduced as well (34).

Antiphagocytic factors such as capsular polysaccharides (84) can allow pathogenic bacteria to avoid exposure to toxic products of iNOS. An alternative avoidance strategy is employed by *Listeria monocytogenes*, which escapes from the phagosome to a more hospitable cytosolic environment (95). YopH, a tyrosine phosphatase secreted by pathogenic *Yersinia* species, can subvert multiple functions of macrophages (7), which are likely to include iNOS expression (24).

STRESS REGULONS

The SoxRS and OxyR regulons are well characterized regarding their roles in resistance of enteric bacteria to oxidative stress, but both appear to be involved in resistance to reactive nitrogen species as well.

SoxRS

The SoxRS two-component regulatory system coordinates the expression of at least 16 genes in response to superoxide-generating redox-cycling agents (51). Oxidation of the (2Fe-2S) cluster of SoxR promotes binding of this regulator to the promoter of *soxS*, whose gene product in turn triggers the transcription of other genes such as *sodA* (encoding Mn-superoxide dismutase), *nfo* (encoding endonuclease IV), and *zwf* (encoding glucose 6-phosphate dehydrogenase [G6PD] required for NADPH synthesis) (41). In addition to responding to redox-cycling agents, the SoxRS regulon can be activated by NO gas (78). By analogy to other (Fe-S) containing proteins, activation of SoxR by NO may result from

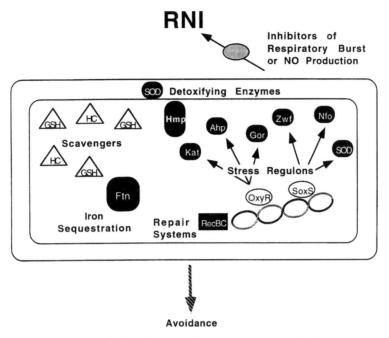

FIGURE 1 Microbial defenses against reactive oxygen and nitrogen intermediates. Abbreviations: SOD, superoxide dismutase; Hmp, flavohemoprotein; GSH, glutathione; HC, homocysteine; FumC, resistant fumarase; Ftn, ferritin; Kat, catalase; Ahp, alkyl hydroperoxide reductase; Gor, glutathione reductase; Zwf, glucose-6-phosphate dehydrogenase; Nfo, endonuclease IV; OxyR, H_2O_2 response regulator; SoxS, superoxide response regulator; RecBC, exonuclease V. Adapted from reference 31 with copyright permission by the *Journal of Clinical Investigation*.

disruption of the (Fe-S) cluster. NO-dependent activation of *soxS* transcription appears to enhance the ability of *Escherichia coli* to withstand NO-related cytotoxicity following ingestion by murine macrophages (79).

OxyR

The OxyR transcription factor is activated in response to hydrogen peroxide, thus promoting expression of an antioxidant regulon that includes genes encoding catalase and glutathione reductase (91). Hydrogen peroxide is believed to activate the formation of an intramolecular disulfide bridge by oxidizing the thiol groups of Cys199 and Cys208 (105). As described above, S-nitrosation is a favored reaction of NO in biological systems. Therefore, it is perhaps not surprising that one of the free thiol groups of OxyR can be nitrosylated by S-nitrosothiols (particularly during conditions of glutathione depletion), stimulating transcription of the OxyR-regulated *katG* gene (48). OxyR-deficient *E. coli* is slightly more susceptible to bacteriostatic actions of S-nitrosocysteine when compared with wild-type controls, an effect possibly related to the down-regulation of catalase.

The general role of SoxR and OxyR in the adaptive response of pathogenic bacteria to oxidative and nitrosative stress has been called into question by some experimental observations. For instance, the transcriptional regulator SoxS appears to be dispensable for resistance of serovar Typhimurium to NO donors or killing by macrophages in vitro, as well as for virulence for mice in vivo (32), even though the production of reactive nitrogen and oxygen species has been shown to play an important role in experimental salmonellosis (87). Similarly, OxyR does not seem to be required for virulence of serovar Typhimurium or *Mycobacterium tuberculosis* (35, 94). Certain genes belonging to the SoxRS or OxyR regulons such as *zwf* (glucose-6-phosphate dehydrogenase) and *ahpC* (alkyl hydroperoxide reductase) do appear to play a role in the resistance of *Salmonella* and *M. tuberculosis* to NO-dependent cytotoxicity (see below for details), but this may reflect the incomplete dependence of these loci on SoxRS or OxyR for expression. It is very likely that additional microbial oxidative and nitrosative stress responses have yet to be fully identified, and these may account for the dispensability of SoxRS and OxyR for virulence. For example, induction of the iron-repressed *Salmonella hmp* gene (see below) by NO (17) suggests that Fe-NO interactions and the subsequent activation of iron-regulated genes may play an important role in the adaptive response to NO.

DETOXIFYING ENZYMES

A few recently described enzymatic pathways appear to ameliorate NO-mediated cytotoxicity directly by metabolizing NO congeners or indirectly by scavenging reactive oxygen intermediates that synergize with NO to produce cell damage.

Primary NO Detoxification

FLAVOHEMOPROTEIN (Hmp)
Recent observations have revealed a novel NO-detoxifying role for the *hmp*-encoded bacterial flavohemoprotein (flavohemoglobin), in addition to its known oxygen and ferric-reducing functions (28, 66). The C-terminal domain of Hmp contains a single heme group, whereas the N-terminal domain has homology to ferredoxin-NADP$^+$ reductase and contains conserved binding sites for FAD and NAD(P)H (1). The first indication

that flavohemoprotein might play a role in NO metabolism came from the observation that expression of an *hmp-lacZ* gene fusion was induced by nitrate, nitrite, and NO (82). Induction of *hmp* expression by NO-derived species has now been observed in serovar Typhimurium and *M. tuberculosis* as well (17, 52). *E. coli* or serovar Typhimurium strains deficient in *hmp* are hypersusceptible to NO and the NO-generating compounds GSNO (*S*-nitrosoglutathione), sodium nitroprusside, acidified nitrite, and *S*-nitrosocysteine (18, 37, 49). Hausladen and colleagues have proposed a model in which the NO group of *S*-nitrosothiols is homolytically released by a currently uncharacterized lyase activity (49). The liberated NO is converted to nitrate by the dioxygenase function of the heme group of flavohemoprotein, in a reaction that consumes NADPH, FAD, and O_2 (37, 38, 49). Although the flavohemoprotein can reduce molecular oxygen to $O_2^{\cdot-}$, the Hmp-dependent accumulation of NO_3^- does not seem to involve the isomerization of ONOO$^{\cdot}$, a product of the reaction of NO and superoxide, but rather on the 2e$^-$ oxidation of NO by a heme-Fe3+$-O_2^-$ or a heme-Fe2+$-O_2$ intermediate (37, 49). For this reaction, detoxification of NO by the dioxygenase activity of Hmp relies on the presence of O_2. However, the ability of serovar Typhimurium flavohemoprotein to confer resistance to acidified nitrite and *S*-nitrosothiols under both aerobic and anaerobic conditions (18) implies that Hmp also possesses a denitrosolase function responsible for anoxic reduction of NO to nitrous oxide (49, 56).

Along with mediating protection against nitrosative stress, *E. coli* Hmp has been reported to confer resistance to the redox cycling agent, paraquat (69), a phenotype not found in association with its serovar Typhimurium homolog (18). Divergence between *E. coli* and serovar Typhimurium flavohemoproteins is evident at the regulatory level as well. Expression of a serovar Typhimurium *hmp-lacZ* operon fusion in response to NO donors is dependent on the iron uptake regulator Fur, but independent of the Fnr, OxyR, and SoxRS transcriptional regulators (17). In contrast, NO-inducible *hmp* expression in *E. coli* is independent of SoxRS and Fur, but negatively regulated by Fnr and positively regulated by the stationary-phase alternative sigma factor σ^S (67, 82). The latter observation is somewhat unanticipated since stationary-phase bacteria are relatively susceptible (21, 38) and σ^S-deficient bacteria are relatively resistant (21) to at least some nitrogen oxides. Expression of *hmp* in *M. tuberculosis* may also be controlled by an alternative sigma factor (52). Recent evidence additionally indicates that NO-induced *hmp* expression in *E. coli* involves the transcriptional regulator MetR (68). The intergenic region between *glyA* and *hmp* contains two MetR binding sites. According to a model proposed by Poole and coworkers, binding of MetR to the coactivator homocysteine favors *glyA* transcription. However, *S*-nitrosation of homocysteine prevents its interaction with MetR, leading to a weak interaction of MetR with promoter site 1 and the preferential expression of *hmp*. Thus, multiple independent pathways appear to have convergently evolved to ensure the induction of *hmp* by nitrogen oxides. The importance of *hmp* for bacterial virulence has not yet been established.

THIOREDOXIN

Whereas NO_x are generally considered to be terminal nontoxic oxidation products of NO, GSNO is an NO congener with antimicrobial activity that is formed under multiple physiological conditions involving aerobic NO metabolism. Thioredoxin reductases of mammalian cells and *E. coli* have been observed to homolytically metabolize GSNO to GSH and NO (76), by a reaction in which the intramolecular cysteines of the enzyme are oxidized to form a disulfide bond. The reducing potential of thioredoxin reductase is regenerated at the expense of thioredoxin and NADPH. Although a theoretically attractive

defense mechanism, the role of thioredoxin reductase in the resistance of pathogenic organisms to NO–related cytotoxicity remains to be tested experimentally.

G6PD

As noted above, the detoxifying properties of flavohemoglobin and thioredoxin reductase require the contribution of reducing equivalents in the form of NADPH. It is therefore not surprising that *zwf* mutant serovar Typhimurium, lacking the enzyme G6PD (which provides NADPH reducing equivalents) is hypersusceptible to GSNO in vitro and to the cytotoxic effects of NO in vivo (61). It is likely that *zwf* plays a similar role in *E. coli* (79).

Secondary NO Detoxification

SodC

An alternative strategy to resist NO cytotoxicity is to limit reactions of NO with other molecular species that produce potent oxidants. Serovar Typhimurium mutants deficient in periplasmic copper, zinc-cofactored superoxide dismutase are attenuated for virulence (33) and exhibit increased susceptibility to macrophage killing; both phenotypes are antagonized by inhibition of NO synthesis or superoxide production (23). These observations can be rationalized by the efficient metabolism of superoxide by superoxide dismutase, thereby preventing the synergistic toxicity of superoxide and NO. These two radicals rapidly combine to form peroxynitrite, a potent oxidizing species capable of damaging a variety of biological molecules. Since superoxide dismutation indirectly reduces levels of hydrogen peroxide as well (59), superoxide dismutase might also limit synergistic cytotoxic interactions of hydrogen peroxide and NO (81).

HYDROPEROXIDASES

Alkyl hydroperoxide reductase encoded by the *ahpC* gene is induced in response to hydrogen peroxide-mediated oxidative stress under the positive control of OxyR. Recently, alkyl hydroperoxide reductase has been associated with resistance of serovar Typhimurium and *M. tuberculosis* to GSNO and acidified nitrite in vitro (13), in a reaction that is independent of the reducing equivalents provided by the AhpF protein. GSNO induction of *E. coli katG* expression further suggests an antinitrosative function for hydroperoxidases (48), and *E. coli* deficient in catalase in fact are slightly hypersusceptible to S-nitrosocysteine (48). The mechanism of protection of hydroperoxidases against NO-related congeners is not clear, but does not seem to be due to increased RSNO (S-nitrosothiol) catabolism. By analogy to SodC, catalases may ameliorate cytotoxic effects of NO by limiting synergistic interactions of NO and H_2O_2 (81). However, the significance of hydroperoxidases in bacterial pathogenesis remains unclear, especially since both *ahpC* and *kat* genes seem to be dispensable for *Salmonella* virulence (10, 94).

SCAVENGERS

Low-molecular-weight thiols are abundant in organisms adapted to aerobic life, serving a crucial role in the prevention of oxidative damage. Glutathione (GSH) is the most abundant thiol in gram-negative bacteria. Exposure of *E. coli* to the NO donor 2-(N,N-diethylamino)-diazenolate-2-oxide/NO results in a 75 to 85% decrease in the intracellular content of reduced GSH (81), suggesting that this small thiol may also play a role in defense against nitrosative stress. In fact, mutation of the *gshA* gene required for GSH synthesis enhances the susceptibility of serovar Typhimurium to the toxic effects of NO, peroxynitrite, or RSNO in vitro and to the cytotoxicity of NO-producing macrophages in vivo (F. C. Fang, M. A. DeGroote, and A. Vazquez-Torres, unpublished data). The complex chemistry of GSH interactions with reactive nitrogen intermediates may explain the protective role of this small thiol protein against NO-associated cytotoxicity. For example, the interaction of GSH with

GSNO produces oxidized glutathione (GSSG), nitrite, nitrous oxide, and ammonia, with the proportion of reaction products dependent upon the concentrations of GSH and oxygen (88). Moreover, GSH can detoxify peroxynitrite to NO by a process that involves S-nitroglutathione ($GSNO_2$) or GSNO as intermediates (4, 97). Many bacterial species lack GSH (30), but other low-molecular-weight thiols such as mycothiols in actinobacteria and nocardioforms (including *M. tuberculosis*) (74) may serve analogous functions. The role of these thiols in pathogenesis remains to be investigated.

Screening of a transposon library has revealed that a mutation in the *metL* gene, encoding the bifunctional enzyme aspartokinase II-homoserine dehydrogenase II of the threonine and methionine biosynthetic pathway, diminishes the virulence of serovar Typhimurium in mice and enhances its susceptibility to several S-nitroso donors (22, 68). Moreover, *Salmonella* lacking *metL* is killed more efficiently by NO-dependent systems of peritoneal macrophages (A. Vazquez-Torres and F. C. Fang, unpublished observations). A mutation in *thrA*, encoding the related enzyme aspartokinase I-homoserine dehydrogenase I, has an identical RSNO-susceptible phenotype that can be reversed by supplementation with exogenous homocysteine. These observations suggest that homocysteine, a low-molecular-weight thiol intermediate in the biosynthesis of methionine, may be an endogenous NO scavenger. Transnitrosation reactions from S-nitrosothiols such as GSNO to homocysteine appear to be favored reactions, and it has been suggested that this might "shift the physiological pattern of S-nitroso compounds to a less biologically active NO pool," since S-nitrosohomocysteine is a poor NO donor and S-nitrosylating compound (96). It is intriguing to consider that antagonistic relationships between S-nitrosothiols and homocysteine might play important roles in both host-pathogen interactions and the pathogenesis of vascular disease (5).

A less well-characterized NO scavenger is pyocyanin of *Pseudomonas aeruginosa* (100), which reacts with NO to form a nitroso derivative. The involvement of this pigment in the pathogenesis of *Pseudomonas* infections awaits further investigation.

REPAIR SYSTEMS

Early in 1985, Kosaka and collaborators described the induction of the serovar Typhimurium *umuC* gene in response to the radical nitrogen dioxide (57), implicating DNA repair in the adaptive response to oxidative stress generated by NO congeners. In fact, DNA repair must constitute part of the normal microbial response to NO since reactive nitrogen intermediates are genotoxic. This is exemplified by the capacity of NO or some of the species generated from it to deaminate deoxynucleosides, deoxynucleotides, or intact DNA (102). NO delivered in the form of nitroglycerine or spermine-NO is a mutagen for *Salmonella*, an activity overtly revealed in DNA repair-deficient (*uvrB*) strains (64, 102). Other DNA repair systems of *Salmonella* are also involved in antagonizing the genotoxic nature of NO congeners, as demonstrated by the hypersusceptibility of strains deficient in RecBC-dependent homologous recombination to GSNO (21). The attenuation of *recBC* mutant bacteria is reversed in double immunodeficient mice lacking both the gp91 subunit of the phagocyte oxidase and the inducible NO synthase (87), suggesting that the RecBC system mediates repair of insults resulting from a coordinated action of NO and superoxide metabolism. Peroxynitrite is likely to be one of the mediators of NO-related DNA damage with its hydroxyl-like ability to induce double-stranded DNA breaks (54). This reactivity of peroxynitrite may contribute to the hypersusceptibility of DNA repair-deficient *nfo* mutant *E. coli* to macrophage killing (78).

OTHER STRATEGIES

Alveolar macrophages from patients with active tuberculosis, a pandemic infection that currently afflicts more than one third of the world's population, demonstrate exuberant expression of inducible NO synthase (75). Furthermore, NO-dependent cytostasis within granulomata may contribute to the latency of tuberculosis (44). Mice lacking iNOS are unable to contain experimental *M. tuberculosis* infections (63), illustrating the importance of nitrogen oxides in antituberculous host defense. Mechanisms of mycobacterial resistance to NO are poorly characterized, but recent attempts to screen a mycobacterial gene library expressed in *Salmonella* have identified several candidate NO-resistance genes, designated *noxR1* and *noxR3*, that increase the resistance of *E. coli*, serovar Typhimurium, or *Mycobacterium smegmatis* to GSNO, acidified nitrite, and macrophage-derived NO, as well as to hydrogen peroxide (27, 85). The mechanism of action of these genes is unknown at present. In addition, 2D-electrophoretic studies of cell extracts obtained from NO-treated *M. tuberculosis* have led to the identification of two novel NO-induced genes encoding a small heat shock protein and a ferritin homolog (35).

CONCLUSIONS

Microbial adaptation to nitrosative stress is likely to date back to primordial times. NO generated during denitrification and nitrate respiration (104) or by other microbial NO synthetic pathways (14, 15, 89) can impose stresses similar to those derived from NO synthase-derived products. The diverse regulatory networks, detoxifying enzymes, scavengers, and repair systems described in this chapter are used by bacteria to prevent or repair cytotoxicity inherent to reactive oxygen and nitrogen species. Functional overlap between antioxidant and antinitrosative defenses represents an economic use of resources and suggests redundancy in the toxic actions of reactive oxygen and nitrogen intermediates. Nevertheless, some defenses, such as NO metabolism by bacterial flavohemoprotein and scavenging by homocysteine, appear to be unique to the detoxification of nitrogen oxides. Additional microbial strategies for resistance to NO-dependent cytotoxicity remain to be elucidated. A more complete comprehension of mechanisms of resistance to NO-related antibacterial activities will shed new light in understanding infectious processes and lead to the discovery and application of new therapies.

ACKNOWLEDGMENTS

We are grateful to the National Institutes of Health, the U.S. Department of Agriculture, and the James Biundo Foundation for their support.

REFERENCES

1. **Andrews, S. C., D. Shipley, J. N. Keen, J. B. Findlay, P. M. Harrison, and J. R. Guest.** 1992. The haemoglobin-like protein (HMP) of *Escherichia coli* has ferrisiderophore reductase activity and its C-terminal domain shares homology with ferredoxin NADP$^+$ reductases. *FEBS Lett.* **302:**247–252.

2. **Anstey, N. M., J. B. Weinberg, M. Y. Hassanali, E. D. Mwaikambo, D. Manyenga, M. A. Misukonis, D. R. Arnelle, D. Hollis, M. I. McDonald, and D. L. Granger.** 1996. Nitric oxide in Tanzanian children with malaria: inverse relationship between malaria severity and nitric oxide production/nitric oxide synthase type 2 expression. *J. Exp. Med.* **184:**557–567.

3. **Assreuy, J., F. Q. Cunha, M. Epperlein, A. Noronha-Dutra, C. A. O'Donnell, F. Y. Liew, and S. Moncada.** 1994. Production of nitric oxide and superoxide by activated macrophages and killing of *Leishmania major*. *Eur. J. Immunol.* **24:**672–676.

4. **Balazy, M., P. M. Kamiski, K. Mao, J. Tan, and M. S. Wolin.** 1998. S-nitroglutathione, a product of the reaction between peroxynitrite and glutathione that generates nitric oxide. *J. Biol. Chem.* **273:**32009–32015.

5. **Bellamy, M. F., and I. F. McDowell.** 1997. Putative mechanisms for vascular damage by homocysteine. *J. Inher. Metab. Dis.* **20:**307–315.

6. **Benjamin, N., and R. Dykhuizen.** 1999. Nitric oxide and epithelial host defense, p. 215–230. *In* F. C. Fang (ed.), *Nitric Oxide and Infection*. Kluwer Academic/Plenum, New York, N.Y.

7. **Bliska, J. B., K. L. Guan, J. E. Dixon, and S. Falkow.** 1991. Tyrosine phosphate hydrolysis of host proteins by an essential *Yersinia* virulence determinant. *Proc. Natl. Acad. Sci. USA* **88:**1187–1191.

8. **Bogdan, C., Y. Vodovotz, J. Paik, Q.-W. Xie, and C. Nathan.** 1993. Traces of bacterial lipolysaccharide suppress IFN-γ-induced nitric oxide synthase gene expression in primary mouse macrophages. *J. Immunol.* **151:**301–309.

9. **Boockvar, K. S., D. L. Granger, R. M. Poston, M. Maybodi, M. K. Washington, J. B. Hibbs, Jr., and R. L. Kurlander.** 1994. Nitric oxide produced during murine listeriosis is protective. *Infect. Immun.* **62:**1089–1100.

10. **Buchmeier, N. A., S. J. Libby, Y. Xu, P. C. Loewen, J. Switala, D. G. Guiney, and F. C. Fang.** 1995. DNA repair is more important than catalase for *Salmonella* virulence in mice. *J. Clin. Invest.* **95:**1047–1053.

11. **Bulut, V., A. Severn, and F. Y. Liew.** 1993. Nitric oxide production by murine macrophages is inhibited by prolonged elevation of cyclic AMP. *Biochem. Biophys. Res. Commun.* **195:**1134–1138.

12. **Castro, L., M. Rodriguez, and R. Radi.** 1994. Aconitase is readily inactivated by peroxynitrite, but not by its precursor, nitric oxide. *J. Biol. Chem.* **269:**29409–29415.

13. **Chen, L., Q.-W. Xie, and C. Nathan.** 1998. Alkyl hydroperoxidase reductase subunit C (AhpC) protects bacterial and human cells against reactive nitrogen intermediates. *Mol. Cell* **1:**795–805.

14. **Chen, Y., and J. P. Rosazza.** 1995. Purification and characterization of nitric oxide synthase (NOSNoc) from a *Nocardia* species. *J. Bacteriol.* **177:**5122–5128.

15. **Choi, W.-S., J. W. Chang, S.-Y. Han, S.-Y Hong, and H.-W. Lee.** 1997. Identification of nitric oxide synthase in *Staphylococcus aureus*. *Biochem. Biophys. Res. Commun.* **237:**554–558.

16. **Cowley, S. C., S. V. Myltseva, and F. E. Nano.** 1996. Phase variation in *Francisella tularensis* affecting intracellular growth, lipopolysaccharide antigenicity and nitric oxide production. *Mol. Microbiol.* **20:**867–874.

17. **Crawford, M. J., and D. E. Goldberg.** 1998. Regulation of the *Salmonella typhimurium* flavohemoglobin gene. A new pathway for bacterial gene expression in response to nitric oxide. *J. Biol. Chem.* **273:**34028–34032.

18. **Crawford, M. J., and D. E. Goldberg.** 1998. Role for the *Salmonella* flavohemoglobin in protection from nitric oxide. *J. Biol. Chem.* **273:**12543–12547.

19. **Cunha, F. Q., J. Assreuy, D. Xu, I. Charles, F. Y. Liew, and S. Moncada.** 1993. Repeated induction of nitric oxide synthase and leishmanicidal activity in murine macrophages. *Eur. J. Immunol.* **23:**1385–1388.

20. **DeGroote, M. A., and F. C. Fang.** 1999. Antimicrobial properties of nitric oxide, p. 231–261. *In* F. C. Fang (ed.), *Nitric Oxide and Infection.* Kluwer Academic/Plenum, New York, N.Y.

21. **DeGroote, M. A., D. Granger, Y. Xu, G. Campbell, R. Prince, and F. C. Fang.** 1995. Genetic and redox determinants of nitric oxide cytotoxicity in a *Salmonella typhimurium* model. *Proc. Natl. Acad. Sci. USA* **92:**6399–6403.

22. **DeGroote, M. A., T. Testerman, Y. Xu, G. Stauffer, and F. C. Fang.** 1996. Homocysteine antagonism of nitric oxide-related cytostasis in *Salmonella typhimurium. Science* **272:**414–417.

23. **DeGroote, M. A., U. A Ochsner, M. U. Shiloh, C. Nathan, J. M. McCord, M. C. Dinauer, S. J. Libby, A. Vazquez-Torres, Y. Xu, and F. C. Fang.** 1997. Periplasmic superoxide dismutase protects *Salmonella* from products of phagocyte NADPH-oxidase and nitric oxide synthase. *Proc. Natl. Acad. Sci. USA* **94:**13997–14001.

24. **Diaz-Guerra, M. J., A. Castrillo, P. Martin-Sanz, and L. Bosca.** 1999. Negative regulation by protein tyrosine phosphatase of IFN-gamma-dependent expression of inducible nitric oxide synthase. *J. Immunol.* **162:**6776–6783.

25. **Duncan, C., H. Dougall, P. Johnston, S. Green, R. Brogan, C. Leifert, L. Smith, M. Golden, and N. Benjamin.** 1995. Chemical generation of nitric oxide in the mouth from the enterosalivary circulation of dietary nitrate. *Nat. Med.* **1:**546–551.

26. **Dykhuizen, R. S., J. Masson, G. McKnight, A. N. Mowat, C. C. Smith, L. M. Smith, and N. Benjamin.** 1996. Plasma nitrate concentration in infective gastroenteritis and inflammatory bowel disease. *Gut* **39:**393–395.

27. **Ehrt, S., M. U. Shiloh, J. Ruan, M. Choi, S. Gunzburg, C. Nathan, Q.-W. Xie, and L. W. Riley.** 1997. A novel antioxidant gene from *Mycobacterium tuberculosis. J. Exp. Med.* **186:**1885–1896.

28. **Eschenbrenner, M., J. Coves, and M. Fontecave.** 1994. Ferric reductases in *Escherichia coli*: the contribution of the haemoglobin-like protein. *Biochem. Biophys. Res. Commun.* **198:**127–131.

29. **Evans, T. G., L. Thai, D. L. Granger, and J. B. Hibbs, Jr.** 1993. Effect of *in vivo* inhibition of nitric oxide production in murine leishmaniasis. *J. Immunol.* **151:**907–915.

30. **Fahey, R. C., W. C. Brown, W. B. Adams, and M. B. Worsham.** 1978. Occurrence of glutathione in bacteria. *J. Bacteriol.* **133:**1126–1129.

31. **Fang, F. C.** 1997. Perspectives series: host/pathogen interactions. Mechanisms of nitric oxide–related antimicrobial activity. *J. Clin. Invest.* **99:**2818–2825.

32. **Fang, F. C., A. Vazquez-Torres, and Y. Xu.** 1997. The transcriptional regulator SoxS is required for resistance of *Salmonella typhimurium* to paraquat but not for virulence in mice. *Infect. Immun.* **65:**5371–5375.

33. **Fang, F. C., M. A. DeGroote, J. W. Foster, A. J. Baumler, U. Ochsner, T. Testerman, S. Bearson, J. C. Giard, Y. Xu, G. Campbell, and T. A. Laessig.** 1999. Virulent *Salmonella typhimurium* has two periplasmic Cu, Zn-superoxide dismutases. *Proc. Natl. Acad. Sci. USA* **96:**7502–7507.

34. **Funatogawa, K., M. Matsuura, M. Nakano, M. Kiso, and A. Hasegawa.** 1998. Relationship of structure and biological activity of monosaccharide lipid A analogues to induction of nitric oxide production by murine macrophage RAW264.7 cells. *Infect. Immun.* **66:**5792–5798.

35. **Garbe, T. R., N. S. Hibler, and V. Deretic.** 1999. Response to reactive nitrogen intermediates in *Mycobacterium tuberculosis*: induction of the 16-kilodalton alpha-crystallin homologue by exposure to nitric oxide donors. *Infect. Immun.* **67:** 460–465.

36. **Gardner, P. R., G. Costantino, C. Szabo, and A. L. Salzman.** 1997. Nitric oxide sensitivity of the aconitases. *J. Biol. Chem.* **272:**25071–25076.

37. **Gardner, P. R., A. M. Gardner, L. A. Martin, and A. L. Salzman.** 1998. Nitric oxide dioxygenase: an enzymic function for flavohemoglobin. *Proc. Natl. Acad. Sci. USA* **95:**10378–10383.

38. **Gardner, P. R., G. Costantino, and A. L. Salzman.** 1998. Constitutive and adaptive detoxification of nitric oxide in *Escherichia coli*. Role of nitric-oxide dioxygenase in the protection of aconitase. *J. Biol. Chem.* **273:**26528–26533.

39. **Gaston, B., and J. S. Stamler.** 1999. Biochemistry of nitric oxide, p. 37–55. *In* F. C. Fang (ed.), *Nitric Oxide and Infection.* Kluwer Academic/Plenum, New York, N.Y.

40. **Gazzinelli, R. T., I. Eltoum, T. A. Wynn, and A. Sher.** 1993. Acute cerebral toxoplasmosis is induced by *in vivo* neutralization of TNF-α and correlates with the down-regulated expression of inducible nitric oxide synthase and other markers of macrophage activation. *J. Immunol.* **151:**3672–3681.

41. **Gonzalez-Flecha, B., and B. Demple.** 1999. Biochemistry of redox signaling in the activation of oxidative stress genes, p. 133–153. *In* D. L. Gilbert and C. A. Colton (ed.), *Reactive Oxygen Species in Biological Systems.* Kluwer Academic/ Plenum, New York, N.Y.

42. **Gow, A. J., R. Foust III, S. Malcolm, M. Gole, and H. Ischiropoulos.** 1999. Biochem-

ical regulation of nitric oxide cytotoxicity, p. 175–186. *In* F. C. Fang (ed.), *Nitric Oxide and Infection.* Kluwer Academic/Plenum, New York, N.Y.

43. **Granger, D. L., J. B. Hibbs, Jr., J. R. Perfect, and D. T. Durack.** 1988. Specific amino acid (L-arginine) requirement for the microbiostatic activity of murine macrophages. *J. Clin. Invest.* **81:**1129–1136.

44. **Granger, D. L., M. L. Cameron, K. Lee-See, and J. B. Hibbs, Jr.** 1993. Role of macrophage-derived nitrogen oxides in antimicrobial function, p. 7–30. *In* G. Lopez Berenstein and J. Klostergaard (ed.), *Mononuclear Phagocytes in Cell Biology.* CRC Press, Boca Raton, Fla.

45. **Griffith, O. W., and D. J. Stuehr.** 1995. Nitric oxide synthases: properties and catalytic mechanism. *Annu. Rev. Physiol.* **57:**707–736.

46. **Guo, L., K. B. Kim, J. S. Gunn, B. Bainbridge, R. P. Darveau, M. Hackett, and S. I. Miller.** 1997. Regulation of lipid A modifications by *Salmonella typhimurium* virulence genes *phoP-phoQ. Science* **276:**250–253.

47. **Hausladen, A., and I. Fridovich.** 1994. Superoxide and peroxynitrite inactivate aconitases, but nitric oxide does not. *J. Biol. Chem.* **269:** 29405–29408.

48. **Hausladen, A., C. T. Privalle, T. Keng, J. DeAngelo, and J. S. Stamler.** 1996. Nitrosative stress: activation of the transcription factor OxyR. *Cell* **86:**719–729.

49. **Hausladen, A., A. J. Gow, and J. S. Stamler.** 1998. Nitrosative stress: metabolic pathway involving the flavohemoglobin. *Proc. Natl. Acad. Sci. USA* **95:**14100–14105.

50. **Hemmens, B., and B. Mayer.** 1999. Enzymology of nitric oxide biosynthesis, p. 57–76. *In* F. C. Fang (ed.), *Nitric Oxide and Infection.* Kluwer Academic/Plenum, New York, N.Y.

51. **Hidalgo, E., H. Ding, and B. Demple.** 1997. Redox signal transduction via iron-sulfur clusters in the SoxR transcription activator. *Trends Biochem. Sci.* **22:**207–210.

52. **Hu, Y., P. D. Butcher, J. A. Mangan, M. A. Rajandream, and A. R. Coates.** 1999. Regulation of *hmp* gene transcription in *Mycobacterium tuberculosis*: effects of oxygen limitation and nitrosative and oxidative stress. *J. Bacteriol.* **181:**3486–3493.

53. **Incze, K., J. Farkas, V. Mihalys, and E. Zukal.** 1974. Antibacterial effect of cysteine-nitrosothiol and possible precursors thereof. *Appl. Microbiol.* **27:**202–205.

54. **Juedes, M. J., and G. N. Wogan.** 1996. Peroxynitrite-induced mutation spectra of pSP189 following replication in bacteria and in human cells. *Mutat. Res.* **349:**51–61.

55. **Keyer, K., and J. A. Imlay.** 1997. Inactivation of dehydratase (4Fe-4S) clusters and disruption of iron homeostasis upon cell exposure to peroxynitrite. *J. Biol. Chem.* **272:**27652–27659.

56. **Kim, S. O., Y. Orii, D. Lloyd, M. N. Hughes, and R. K. Poole.** 1999. Anoxic function for the *Escherichia coli* flavohaemoglobin (Hmp): reversible binding of nitric oxide and reduction to nitrous oxide. *FEBS Lett.* **445:**389–394.

57. **Kosaka, H., Y. Oda, and M. Uozumi.** 1985. Induction of *umuC* gene expression by nitrogen dioxide in *Salmonella typhimurium. Mutat. Res.* **142:**99–102.

58. **Liew, F. Y., S. Millott, C. Parkinson, R. M. Palmer, and S. Moncada.** 1990. Macrophage killing of *Leishmania* parasite *in vivo* is mediated by nitric oxide from L-arginine. *J. Immunol.* **144:** 4794–4797.

59. **Liochev, S. I., and I. Fridovich.** 1994. The role of $O_2\cdot^-$ in the production of HO· *in vitro* and *in vivo. Free Radical Biol. Med.* **16:**29–33.

60. **Lowenstein, C. J., S. L. Hill, A. Lafond-Walker, J. Wu, G. Allen, M. Landavere, N. R. Rose, and A. Herskowitz.** 1996. Nitric oxide inhibits viral replication in murine myocarditis. *J. Clin. Invest.* **97:**1837–1843.

61. **Lundberg, B. E., R. E. Wolf, Jr., M. C. Dinauer, Y. Xu, and F. C. Fang.** 1999. Glucose 6-phosphate dehydrogenase is required for *Salmonella typhimurium* virulence and resistance to reactive oxygen and nitrogen intermediates. *Infect. Immun.* **67:**436–438.

62. **MacMicking, J., Q.-W. Xie, and C. Nathan.** 1997. Nitric oxide and macrophage function. *Annu. Rev. Immunol.* **15:**323–350.

63. **MacMicking, J. D., R. J. North, R. LaCourse, J. S. Mudgett, S. K. Shah, and C. F. Nathan.** 1997. Identification of nitric oxide synthase as a protective locus against tuberculosis. *Proc. Natl. Acad. Sci. USA* **94:**5243–5248.

64. **Maragos, C. M., A. W. Andrews, L. K. Keefer, and R. K. Elespuru.** 1993. Mutagenicity of glyceryl trinitrate (nitroglycerin) in *Salmonella typhimurium. Mutat. Res.* **298:**187–195.

65. **McInnes, I. B., and F. Y. Liew.** 1999. Immunomodulatory actions of nitric oxide, p. 199–213. *In* F. C. Fang (ed.), *Nitric Oxide and Infection.* Kluwer Academic/Plenum, New York, N.Y.

66. **Membrillo-Hernandez, J., N. Ioannidis, and R. K. Poole.** 1996. The flavohaemoglobin (HMP) of *Escherichia coli* generates superoxide *in vitro* and causes oxidative stress *in vivo. FEBS Lett.* **382:**141–144.

67. **Membrillo-Hernandez, J., S. O. Kim, G. M. Cook, and R. K. Poole.** 1997. Paraquat regulation of *hmp* (flavohemoglobin) gene expression

in *Escherichia coli* K-12 is SoxRS independent but modulated by sigma S. *J. Bacteriol.* **179:**3164–3170.

68. **Membrillo-Hernandez, J., M. D. Coopamah, A. Channa, M. N. Hughes, and R. K. Poole.** 1998. A novel mechanism for upregulation of the *Escherichia coli* K-12 *hmp* (flavohaemoglobin) gene by the 'NO releaser', S-nitrosoglutathione: nitrosation of homocysteine and modulation of MetR binding to the *glyA-hmp* intergenic region. *Mol. Microbiol.* **29:**1101–1112.

69. **Membrillo-Hernandez, J, M. D. Coopamah, M. F. Anjum, T. M. Stevanin, A. Kelly, M. N. Hughes, and R. K. Poole.** 1999. The flavohemoglobin of *Escherichia coli* confers resistance to a nitrosating agent, a "nitric oxide releaser," and paraquat and is essential for transcriptional responses to oxidative stress. *J. Biol. Chem.* **274:** 748–754.

70. **Moncada, S., R. M. J. Palmer, and E. A. Higgs.** 1991. Nitric oxide—physiology, pathophysiology, and pharmacology. *Pharmacol. Rev.* **43:**109–142.

71. **Morris, S. L., and J. N. Hansen.** 1981. Inhibition of *Bacillus cereus* spore outgrowth by covalent modification of a sulfhydryl group by nitrosothiol and iodoacetate. *J. Bacteriol.* **148:** 465–471.

72. **Mühl, H., and C. A. Dinarello.** 1999. Cytokine regulation of nitric oxide production, p. 77–94. *In* F. C. Fang (ed.), *Nitric Oxide and Infection.* Kluwer Academic/Plenum, New York, N.Y.

73. **Nathan, C. F., and Q. W. Xie.** 1994. Nitric oxide synthase: roles, tolls, and controls. *Cell* **79:** 915–918.

74. **Newton, G. L., K. Arnold, M. S. Price, C. Sherrill, S. B. Delcardayre, Y. Aharonowitz, G. Cohen, J. Davies, R. C. Fahey, and C. E. Davis.** 1996. Distribution of thiols in microorganisms: mycothiol is a major thiol in most actinomycetes. *J. Bacteriol.* **178:**1990–1995.

75. **Nicholson, S., M. da G. Bonecini-Almeida, J. R. Lapa e Silva, C. Nathan, Q.-W. Xie, R. Mumford, J. R. Weidner, J. Calaycay, J. Geng, N. Boechat, C. Linhares, W. Rom, and J. L. Ho.** 1996. Inducible nitric oxide synthase in pulmonary alveolar macrophages from patients with tuberculosis. *J. Exp. Med.* **183:** 2293–2302.

76. **Nikitovic, D., and A. Holmgren.** 1996. S-nitrosoglutatione is cleaved by the thioredoxin system with liberation of glutathione and redox regulating nitric oxide. *J. Biol Chem.* **271:**19180–19185.

77. **Nozaki, Y., Y. Hasegawa, S. Ichiyama, I. Nakashima, and K. Shimokata.** 1997. Mechanism of nitric oxide-dependent killing of *Mycobacterium bovis* BCG in human alveolar macrophages. *Infect. Immun.* **65:**3644–3647.

78. **Nunoshiba, T., T. deRojas-Walker, J. S. Wishnok, S. R. Tannenbaum, and B. Demple.** 1993. Activation by nitric oxide of an oxidative-stress response that defends *Escherichia coli* against activated macrophages. *Proc. Natl. Acad. Sci. USA* **90:**9993–9997.

79. **Nunoshiba, T, T. deRojas-Walker, S. R. Tannenbaum, and B. Demple.** 1995. Roles of nitric oxide in inducible resistance of *Escherichia coli* to activated murine macrophages. *Infect. Immun.* **63:**794–798.

80. **Ochoa, J. B., A. O. Udekwu, T. R. Billiar, R. D. Curran, F. B. Cerra, R. L. Simmons, and A. B. Peitzman.** 1991. Nitrogen oxide levels in patients after trauma and during sepsis. *Ann. Surg.* **214:**621–626.

81. **Pacelli, R., D. A. Wink, J. A. Cook, M. C. Krishna, W. DeGraff, N. Friedman, M. Tsokos, A. Samuni, and J. B. Mitchell.** 1995. Nitric oxide potentiates hydrogen peroxide-induced killing of *Escherichia coli. J. Exp. Med.* **182:**1469–1479.

82. **Poole, R. K., M. F. Anjum, J. Membrillo-Hernandez, S. O. Kim, M. N. Hughes, and V. Stewart.** 1996. Nitric oxide, nitrite, and Fnr regulation of *hmp* (flavohemoglobin) gene expression in *Escherichia coli* K-12. *J. Bacteriol.* **178:** 5487–5492.

83. **Rees, D.** 1999. Cardiovascular actions of nitric oxide, p. 151–174. *In* F. C. Fang (ed.), *Nitric Oxide and Infection.* Kluwer Academic/Plenum, New York, N.Y.

84. **Robbins, J. D., and J. B. Robbins.** 1984. Reexamination of the protective role of the capsular polysaccharide (Vi antigen) of *Salmonella typhi. J. Infect. Dis.* **150:**436–449.

85. **Ruan, J., G. St. John, S. Ehrt, L. Riley, and C. Nathan.** 1999. *noxR3*, a novel gene from *Mycobacterium tuberculosis*, protects *Salmonella typhimurium* from nitrosative and oxidative stress. *Infect. Immun.* **67:**3276–3283.

86. **Severn, A., D. Xu, J. Koyle, L. M. C. Leal, C. A. O'Donnell, S. J. Brett, D. W. Moss, and F. Y. Liew.** 1993. Pre-exposure of murine macrophages to lipopolysaccharide inhibits the induction of nitric oxide synthase and reduces leishmanicidal activity. *Eur. J. Immunol.* **23:**1711–1714.

87. **Shiloh, M. U., J. D. MacMicking, S. Nicholson, J. E. Brause, S. Potter, M. Marino, F. Fang, M. Dinauer, and C. Nathan.** 1999. Phenotype of mice and macrophages deficient in both phagocyte oxidase and inducible nitric oxide synthase. *Immunity* **10:**29–38.

88. **Singh, S. P., J. S. Wishnok, M. Keshive, W. M. Deen, and S. R. Tannenbaum.** 1996. The chemistry of the S-nitroglutathione/glutathione system. *Proc. Natl. Acad. Sci. USA* **93:** 14428–14433.

89. **Stachura, J., J. W. Konturek, A. Karczewska, W. Domschke, T. Popiela, and S. J. Konturek.** 1996. *Helicobacter pylori* from duodenal ulcer patients expresses inducible nitric oxide synthase immunoreactivity *in vivo* and *in vitro. J. Physiol. Pharmacol.* **47:**131–135.

90. **Stenger, S., N. Donhauser, H. Thuring, M. Rollinghoff, and C. Bogdan.** 1996. Reactivation of latent leishmaniasis by inhibition of inducible nitric oxide synthase. *J. Exp. Med.* **183:** 1501–1514.

91. **Storz, G., and J. A. Imlay.** 1999. Oxidative stress. *Curr. Opin. Microbiol.* **2:**188–194.

92. **Stuehr, D. J., and M. A. Marletta.** 1985. Mammalian nitrate biosynthesis: mouse macrophages produce nitrite and nitrate in response to *Escherichia coli* lipopolysaccharide. *Proc. Natl. Acad. Sci. USA* **82:**7738–7742.

93. **Tarr, H. L. A.** 1941. Bacteriostatic action of nitrates. *Nature* **147:**417–418.

94. **Taylor, P. D., C. J. Inchley, and M. P. Gallagher.** 1998. The *Salmonella typhimurium* AhpC polypeptide is not essential for virulence in BALB/c mice but is recognized as an antigen during infection. *Infect. Immun.* **66:**3208–3217.

95. **Tilney, L. G., and D. A. Portnoy.** 1989. Actin filaments and the growth, movement, and spread of the intracellular bacterial parasite, *Listeria monocytogenes. J. Cell Biol.* **109:**1597–1608.

96. **Tsikas, D., J. Sandmann, S. Rossa, F. M. Gutzki, and J. C. Frolich.** 1999. Investigations of S-transnitrosylation reactions between low- and high-molecular-weight S-nitroso compounds and their thiols by high-performance liquid chromatography-mass spectrometry. *Anal. Biochem.* **270:**231–241.

97. **Van der Vliet, A., P. A. Chr.'t Hoen, P. S.-Y. Wong, A. Bast, and C. E. Cross.** 1998. Formation of S-nitrosothiols via direct nucleophilic nitrosation of thiols by peroxynitrite with elimination of hydrogen peroxide. *J. Biol. Chem.* **273:**30155–30162.

98. **Vazquez-Torres, A., J. Jones-Carson, and E. Balish.** 1996. Peroxynitrite contributes to the candidacidal activity of nitric oxide-producing macrophages. *Infect. Immun.* **64:** 3127–3133.

99. **Vodovotz, Y., N. S. Kwon, M. Pospischil, J. Manning, J. Paik, and C. Nathan.** 1994. Inactivation of nitric oxide synthase after prolonged incubation of mouse macrophages with IFN-γ and bacterial lipopolysaccharide. *J. Immunol.* **152:**4110–4118.

100. **Warren, J. B., R. Loi, N. B. Rendell, and G. W. Taylor.** 1990. Nitric oxide is inactivated by the bacterial pigment pyocyanin. *Biochem. J.* **266:**921–923.

101. **Weinberg, J. B.** 1999. Human mononuclear phagocyte nitric oxide production and inducible nitric oxide synthase expression, p. 95–150. *In* F. C. Fang (ed.), *Nitric Oxide and Infection.* Kluwer Academic/Plenum, New York, N.Y.

102. **Wink, D. A., K. S. Kasprzak, C. M. Maragos, R. K. Elespuru, M. Misra, T. M. Dunams, T. A. Cebula, W. H. Koch, A. W. Andrews, J. S. Allen, and L. K. Keefer.** 1991. DNA deaminating ability and genotoxicity of nitric oxide and its progenitors. *Science* **254:**1001–1003.

103. **Wink, D. A., M. Feelisch, Y. Vodovotz, J. Fukuto, and M. B. Grisham.** 1999. The chemical biology of nitric oxide, p. 245–291. *In* D. L. Gilbert and C. A. Colton (ed.), *Reactive Oxygen Species in Biological Systems.* Kluwer Academic/Plenum, New York, N.Y.

104. **Xiao-Bing, J., and T. C. Hollocher.** 1988. Reduction of nitrite to nitric oxide by enteric bacteria. *Biochem. Biophys. Res. Commun.* **157:** 106–108.

105. **Zheng, M., F. Aslund, and G. Storz.** 1998. Activation of the OxyR transcription factor by reversible disulfide bond formation. *Science* **279:** 1718–1721.

DNA REPAIR AND MUTATORS: EFFECTS ON ANTIGENIC VARIATION AND VIRULENCE OF BACTERIAL PATHOGENS

Thomas A. Cebula and J. Eugene LeClerc

10

"A slow sort of country!" said the Queen. "Now, HERE, you see, it takes all the running YOU can do, to keep in the same place. If you want to get somewhere else, you must run at least twice as fast as that."
—L. Carroll, *Through the Looking Glass*

PROLOGUE

A favorable mutation occurs within the genome at the rate of 1×10^{-8} per generation. Yet a deleterious mutation is spawned about 20,000 times more frequently, and one in ten of these events leads to extinction because it is lethal (Table 1). Alice's encounter with the Red Queen in *Through the Looking Glass* seems then a most apropos metaphor to describe the constant need for adaptation within the bacterial community (97). We are reminded though that "the objective in population genetics is to reconstruct the possible, or more rarely the actual, sequence of events in the evolution of organisms in terms of changes resulting from the interplay of mutation and

selection" (4). However, if that sequence is to be delineated correctly, we must travel beyond metaphor and completely assess the genetic mechanisms involved to which Atwood et al. (3, 4) briefly alluded.

INTRODUCTION

The genetic diversity inherent in the bacterial population and the selection landscape that works upon that diversity are the ingredients for evolution. In describing periodic selection within bacterial populations, where a rare mutant founds a population only to be "swept" periodically by another mutation as selection pressures change, Atwood et al. (3, 4) tacitly assumed that the nonsexual nature of bacteria was unquestionable. All genetic variability upon which selection acted, they reasoned, was due solely to the emergence of mutants and not to recombinants. Although the clonal selection theory has been modified somewhat to accommodate an incidental horizontal event (25, 50, 64), more and more we are appreciating the significance of recombination among bacteria (9, 15, 36, 58, 63, 74, 80). Comparative genomics has helped sway a new appreciation for the considerable genetic exchange that has taken place among prokaryotic organisms (33, 39, 41, 42, 43, 69). Comparisons of the 18 complete genomes from 17

Thomas A. Cebula Division of Molecular Biological Research and Evaluation (HFS-235), Center for Food Safety and Applied Nutrition, Food and Drug Administration, Washington, DC 20204. *J. Eugene LeClerc* Molecular Biology Branch (HFS-237), Center for Food Safety and Applied Nutrition, Food and Drug Administration, Washington, DC 20204.

Virulence Mechanisms of Bacterial Pathogens, 3rd ed., Edited by K. A. Brogden et al.
©2000 ASM Press, Washington, D.C.

bacterial species (complete genomes as of June 1999) depict a much more fluid and mosaic organization of the bacterial chromosome than previously imagined—one that is formed by genetic exchange (recombination) as well as by change (mutation). These analyses have underscored that horizontal transfer has been underestimated and also have revealed how commonly deletions, duplications, and gene rearrangements occur. An assessment of the relative contribution of such events is essential, then, for our understanding of the evolution of bacterial populations.

Elsewhere, we have discussed how particular mutator strains of bacteria, namely, those defective in methyl-directed mismatch repair (MMR), contribute significantly to the antibiotic resistance conundrum (12). In this chapter, we shall discuss MMR mutators in the context of the evolution of bacterial pathogens. In particular, we will explore the many and disparate effects elicted by MMR defects and how these defects may aid in antigenic diversification and virulence of a pathogen.

MUTATION RATE AND BACTERIAL EVOLUTION

It is inevitable that any discussion of bacterial evolution ultimately focuses on mutation rate. Witness, for example, the discussion that ensued following Atwood's classic delivery of the periodic selection theory (4). In describing antigenic switching in *Salmonella enterica* serovar Typhimurium (work referred to by Atwood), Stocker remarked that the mutation rates that he had calculated for antigenic switching were much greater than those described by Atwood for the occurrence of histidine mutations in *Escherichia coli*. He reasoned that, in his case, the transmission into a new host may sometimes be caused by a very small number of cells. Atwood duly noted that differences in the two systems may lie in the high rates of mutation quoted by Stocker, given that the time to establish equilibrium among sublines would be drastically shortened for "rapidly mutating characters than for those with mutation rates of the usual order of mag-

nitude" (4). Ironically, Stocker utilized serovar Typhimurium LT7 strains in his studies. As discovered years later, LT7 strains have high mutation rates due to a defect in *mutL* (38, 84), one component of the MMR system (see below).

Considering the choice of words, Atwood assumed that bacteria necessarily possessed a constant and low mutation rate. This seemed obvious; over time, the more frequent mutations spawned by higher mutation rates would likely be deleterious more often than not. Thus, bacteria with high mutation rates would be at a long-term disadvantage in a stable environment. From data based on 12 microbial genes, Drake calculated the mutation rate for *E. coli* and other microbes and found a rather low and constant rate (22); the mutation rate (μ_g) for *E. coli* was recently recalculated to be 2.5×10^{-3} mutations per genome per replication (24). As there was only a 2.5-fold variance in mutation rates among bacteriophages, bacteria, and microbial eukaryotes, this might suggest that the mutation rate evolved—the residuum of natural selection. Yet Drake cautions (22) that there may be times when an enhanced mutation rate could be advantageous.

For example, Ninio (73) hypothesizes that a "transient mutator" could contribute significantly to complex mutations arising spontaneously, and Leigh (48, 49) reminds us that in bacteria an optimal rate, rather than a low mutation rate, has been selected. Indeed, many have reasoned that an enhanced mutation rate, especially in an unstable environment, should be beneficial (9, 20, 46, 48, 77, 88, 90, 91, 92). In contemplating the importance of mutation rate, it is therefore essential to ask how often bacteria find themselves in a stable environment, other than in the coddled environment of the laboratory. Most often, it seems that pathogenic bacteria spend their time trying to conform to and obey a "survive or die" dictum.

The question then is—can a bacterium with an increased mutation rate exist, persist, and compete successfully with bacteria that possess "the more normal mutation rates" to which Atwood intimated (4)? Lytic RNA phages are able to do so, and they possess an exceptionally high mutation rate, $\mu_g \sim 1$ mutation per genome per replication (23). These phages are constantly pushing the envelope, considering that a μ_g only a few-fold higher would mean extinction of the species (21, 23). Notably, treatment of RNA phages with mutagens that cause even modest increases in mutant yields (2.5-fold) profoundly affects viability (23), emphasizing the precariousness of a mutation rate set too high. Unlike the RNA phages, treatment of E. coli with various exogenous mutagens can lead to a >1,000-fold increase in mutant frequencies. This is not too surprising since μ_g for E. coli is 0.0025 and, in theory, the mutation rate could be increased several thousand-fold before risking extinction. For example, even when μ_g is enhanced 10,000-fold genetically in E. coli, cells remain viable (at least for short periods of time) (85). Dead cells obviously accumulate in such cultures because, with a $\mu_g \sim 25$, the population is likely to be extinguished due to the deleterious and lethal mutations rapidly accumulating in the genome. Clearly, this is the Red Queen's dilemma—a mutator phenotype, by spawning more mutations, has more of a chance to sport a right and favorable mutation; by the same virtue, it has an increased probability of causing extinction.

DNA REPLICATION, DNA REPAIR, AND MUTATION RATE

With the cacophonies of endogenous metabolism and the exogenous insults of an ever-changing environment, bacteria constantly face the daunting challenge of preserving their genetic identity (Fig. 1). In their feral settings, bacteria experience a steady barrage of natural and anthropogenic mutagenic insults, ranging from UVB of the sun's irradiation to the mutagenic byproducts of disinfection or of natural products. So too does the endogenous metabolism of bacteria (or that of their host) deliver many insults to the bacterial genome. Together, endogenous and exogenous insults

produce about 3,000 to 5,000 DNA lesions per generation in every E. coli cell, with most of these lesions the result of oxidative damage (17, 82).

Bacteria weather such onslaughts by actively repairing DNA damage and replicating DNA in a near error-free manner. The genetic heritage of a species depends upon the fidelity of DNA replication and upon the various repair pathways that recognize and correct DNA mismatches and lesions before heritable mutations become fixed into the genome. DNA mismatches and lesions present as formidable problems to bacterial populations. If not repaired, most will lead to deleterious consequences, but if all are repaired, that too has a catastrophic outcome—ultimate extinction in the event of a changing environment. It is not surprising then that bacteria, over evolutionary time, have evolved an optimal mutation rate, one that allows the sporting of sufficient mutants to ensure the genetic variation necessary for survival. However, as discussed below, optimal may not always mean minimal (48, 56).

The unique chemistry of DNA (Watson-Crick base pairing) and enzymology of DNA polymerase are crucial, but not sufficient, in maintaining the integrity of the genome (Fig. 2). For example, if left to its own devices, the synthetic property of DNA polymerase III would introduce 1 to 40 errors into the genome every time the E. coli chromosome was replicated. Yet it is apparent that the mutation rate of E. coli is such that only about one error occurs in every 400 genomes replicated (Table 1).

Errors introduced by the DNA polymerase are circumvented both by its proofreading function and by the various DNA repair enzymes, primarily the MMR system (Fig. 2), which monitor and repair lesions in the DNA. This underscores why DNA repair is essential to any discussion of bacterial evolution. Although beyond the scope of this chapter, we must mention the various genetic loci that, in mutant form, increase the mutation rate (Fig. 3). Mutations in any of these loci could lead

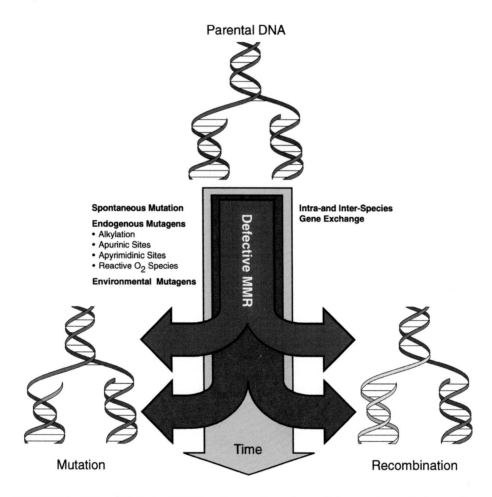

Parental DNA

Spontaneous Mutation

Endogenous Mutagens
• Alkylation
• Apurinic Sites
• Apyrimidinic Sites
• Reactive O_2 Species

Environmental Mutagens

Intra- and Inter-Species Gene Exchange

Defective MMR

Time

Mutation   Recombination

FIGURE 1   Effect of defective MMR on the time frame for evolutionary change. Defects in the MMR pathway increase mutations from errors in DNA replication and repair and enhance recombination of diverged DNA among and between bacterial species. The increased rates of spontaneous mutation and homeologous recombination observed in MMR-defective strains suggest that MMR mutators may act to cause rapid evolutionary change.

to aberrant replication or repair. Such mutants would be conspicuous among their siblings because of their mutator phenotypes. Whereas tens of loci can give rise to a mutator phenotype, it will become clear that certain mutators, namely those defective in MMR, may be of special importance in the evolution of bacterial populations.

MMR

From Fig. 2, it is obvious that if the chemistry of the DNA structure alone were to preserve

genetic integrity, then 1 in 10 base pairs would be mismatched. The cellular machinery must be such to contend with these mismatches, accurately discarding the wrong bases in order to fix the correct bases into the chromosome. Even with the editing function of the 3   5 exonuclease, DNA polymerase III is ill equipped to handle this high rate of mismatches. Thus, if there were no further discrimination, 50% of the mismatches at this stage would be fixed into the chromosome as mutations. MMR is a repair system that helps

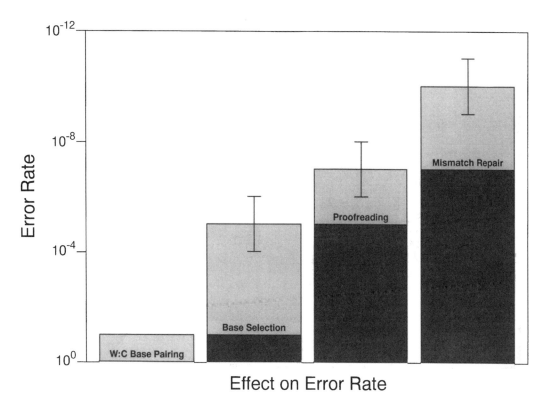

FIGURE 2 Estimates of fidelity during the course of DNA replication. Bar graphs represent the additive effect of each step that controls the fidelity of DNA replication and show the decrease in error rate, given as error per DNA nucleotide replicated.

govern ultimate fidelity of the DNA replication process by correcting base-pairing errors in newly synthesized DNA (see references 67 and 68 for reviews). Exquisite monitoring of the DNA by MMR ensures the ultimate fidelity of the replication process. By following closely behind the replication machinery, MMR recognizes and repairs mismatched base pairs and local distortions (one to three unmatched base pairs) within the DNA.

MMR is guided by an asymmetric cue of methylation within the DNA. Bacterial DNA is methylated by Dam, the gene product of *dam*, an adenine methylase that modifies the

TABLE 1 Number of total, deleterious, lethal, and favorable mutations per genome, gene, and base pair replication

| Mutation | No. of mutations: | | | Reference |
	per genome per replication	per gene per replication	per base pair per replication	
Total	2.5×10^{-3}	5.4×10^{-7}	5.4×10^{-10}	24
Deleterious	2×10^{-4}	4.3×10^{-8}	4.3×10^{-11}	37
Lethal	2×10^{-5}	4.3×10^{-9}	4.3×10^{-12}	37
Favorable	1×10^{-8}	2.2×10^{-12}	2.2×10^{-15}	92

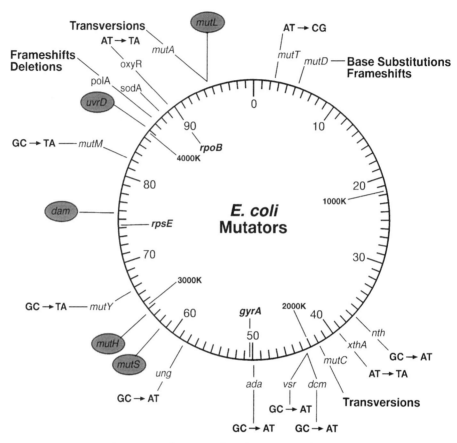

FIGURE 3 Mutator loci in *E. coli*. Mutant loci that increase the spontaneous mutation rate are shown on the outside of the genetic map of *E. coli*. Preferential mutations induced in each mutant strain are given, except for the pleiotropic MMR mutators, which are highlighted in gray. Genetic loci shown inside the map are antibiotic resistance determinants used to screen the mutator phenotype (see text). The circular genetic map is shown with centisome intervals indicated for the 100-minute map and 1,000-kb (K) intervals indicated for the 4,639-kb genome.

DNA specifically at 5′GATC3′ sequences within the DNA. Because bacterial DNA is methylated at the adenine positions of 5′GATC3′ sequences after replication, it ensures a transient hemimethylated state, of which the cellular machinery can (and does) take advantage. Whenever a mismatch is encountered, because the bases of 5′GATC3′ sequences of the template strand of DNA are methylated and the 5′GATC3′ sequences of the nascent strand briefly are not, there is a natural asymmetry that can be recognized biochemically. MutS binds to the mismatch and, in the presence of ATP and MutL, a confor-mational change within the DNA occurs, bringing into juxtaposition the mismatched sequence and the nearest 5′GATC3′ sequence. Activated MutH then incises at the 5′GATC3′ sequence of nascent DNA, i.e., the unmethylated strand of DNA. UvrD, DNA helicase II, unwinds the DNA. Specific nucleases, depending upon whether the closest GATC site was 5′ or 3′ to the mismatch (68), digest the DNA from the point of incision past the mismatch, degrading as many as 1,000 base pairs. DNA polymerase I provides a gap-filling function, and DNA ligase seals the ultimate nick to complete the repair process. From the enzymology of MMR, it is clear that a defect in any component of MMR will reduce the fidelity of repair, causing a greater number of errant bases to be fixed into the chromosome as mutations. Thus, bacteria harboring a defective MMR system exhibit a mutator phenotype; mutation rates for MMR-defective strains are typically about 100-fold greater than those of MMR-proficient strains (47, 57, 65), although this can vary depending on the locus under investigation.

In addition to its role in managing the fidelity of DNA replication, MMR is integral to the monitoring and maintenance of the integrity of the bacterial genome. MMR is the major gatekeeper in the host bacterium that prevents foreign DNA from infiltrating the genome. Indeed, a single base pair mismatch is sufficient to abort recombination when MMR is operating, though in its absence, sequences that are 30% disparate are able to recombine (59, 60). Thus, for example, in conjugational and transductional crosses between *E. coli* K-12 and serovar Typhimurium LT2, recipients mutant in *mutS* or *mutL* had recombination frequencies up to three orders of magnitude greater than mut^+ strains (80).

Therefore, MMR gene defects relax recombination barriers between species that normally do not mate. The impact on bacterial populations is apparent and has led many to conjecture how MMR might control the rate of bacterial evolution (9, 10, 15, 44, 59, 60, 80, 98). Recent studies using laboratory-evolved strains have demonstrated how MMR can influence speciation by simultaneously affecting both mutation and recombination (99). From a practical standpoint, in order to rapidly produce a genetic map for serovar Typhi, Zahrt and Maloy (103, 104) have taken advantage of this biological quirk to shuttle marker genes from serovar Typhimurium to serovar Typhi using P22 transduction. How does MMR protect against the promiscuous nature of homeologous recombination? Biochemical studies have shown that MutS protein blocks strand transfer in response to mispairs in heteroduplex DNA, an inhibition

enhanced by MutL (101). Although MutL enhances this process, the mismatch recognition by MutS alone stalls strand displacement sufficiently to inhibit RecA reassociation, thus making recombination in such areas unlikely (100).

MMR is the major paladin that monitors and prevents promiscuous encounters. Thus, an MMR mutant is poised to evolve quickly, perhaps spawning and inheriting a favorable mutation vertically. Or, an MMR mutant can recombine with other bacterial species, thus acquiring the useful variation laterally. Such puckish behavior should be beneficial and should therefore have an adaptive advantage, to a pathogen infecting an intended host. Facing the inhospitable environs of the host and armed with such promiscuity, the invading pathogen opportunistically can sample and/or acquire new sequence elements from similar or disparate genomes, thus expanding its repertoire for adaptation with one genetic step. This reasoning has led us to suggest that particular MMR mutators may play a vital role during the course of a bacterial infection (9, 10, 11, 12, 44, 46).

MUTATORS IN LABORATORY EXPERIMENTS AND NATURAL POPULATIONS

Cox and coworkers studied growth of *E. coli* mutators (and nonmutators) in long-term chemostat experiments and showed that the mutator phenotype enjoyed an adaptive advantage over its nonmutator counterpart (13, 14, 16, 72). This advantage was frequency-dependent, however. At a starting ratio $>5 \times 10^{-5}$, a mutator consistently outstripped the growth of its nonmutator counterpart, whereas below this starting inoculum the nonmutator phenotype invariably was favored. This selective advantage exhibited in relatively short time frames, manifesting within 60 to 130 generations. When a mutator is outcompeted, it is destined then either to be part of a stable subpopulation carrying beneficial and neutral mutations or, more likely, to become extinct because of deleterious mutations introduced in vital genes.

The chemostat experiments also showed that high mutation rates persisted for relatively short periods of time, eventually being supplanted by populations that displayed lower mutation rates. For example, when a *mutT* point mutant was used, causing AT → CG transversions and exhibiting a mutation rate 100 to 500 fold greater than that of wild type, it consistently won out in competition experiments. However, when assessed after 2,200 generations, the mutation rate was only about 2- to 12-fold different than that of the original nonmutator utilized in the experiment. The lower mutation rate observed at later times was not due to loss of the mutator genotype, but was probably the result of a secondary mutation in the background that suppressed the mutator phenotype (93). The presence of a suppressor was inferred; i.e., crossing the *mutT* allele from isolates displaying the attenuated mutation rate into a naïve strain showed that a high mutation rate (akin to the original *mutT* strain) could be restored. Simply stated, the chemostat experiments show that the success of a bacterial mutator rests on timing. It must spawn a beneficial mutation sooner than its nonmutator sibling or be lost. Even when it wins, its longevity will depend both on the inherent strength of the mutator and on the relative stability of the niche in which the mutator resides.

The chemostat studies, as well as two independent reports that detected mutators among hospital isolates of *E. coli* (29, 34), suggested that mutators could persist in natural populations of bacteria. The conventional thinking of the time, however, was that a mutator phenotype, because of its initial low numbers and its propensity for spawning lethal mutations more frequently, could not rise to prominence in the population (78). This incongruence prompted us to ask in 1996 whether mutators existed among pathogenic isolates of *E. coli* and *Salmonella enterica* (46).

We review the criteria that we used to assess a mutator phenotype because the conditions of selection help dictate what one ultimately finds in the screen. In our studies, mutators had to show at least a 50-fold increase in mutant frequencies at three distinct loci, compared with their nonmutator counterparts. In screening for the hypermutable phenotype, we selected strains with mutations that made them resistant to either rifampin, spectinomycin, or nalidixic acid. The β-subunit of RNA polymerase, the S5 protein of the 30S ribosomal subunit, and DNA gyrase, the targets for rifampin, spectinomycin, and nalidixic acid, respectively, are encoded by *rpoB* (90 min), *rpsE* (74.2 min), and *gyrA* (50.3 min), respectively, in *E. coli* (Fig. 3). These three resistances are due to different modes of action that affect different chromosomal determinants and locations. Thus, this screen should enrich for strains that were not only strong mutators but general mutators as well. With such a screen, we expected to capture most of the known mutants of *E. coli* that are responsible for conferring a mutator phenotype (Fig. 3). Using these criteria, we found that a relatively high number (>1%) of the isolates examined displayed a mutator phenotype. Notably, all mutators isolated by these criteria were found defective in MMR. Although as with other deleterious mutations, those in mutator alleles are expected to be selected against in the population, our studies demonstrated that MMR mutators were maintained in feral populations at frequencies 10 to 1,000-fold greater than expected. While it might be reasoned that a hypermutable phenotype could confer a selective advantage because it could spawn a favorable mutation more quickly, so too could any of twenty or so other mutators scattered throughout the chromosome—yet, we found only MMR mutators. Thus, we reasoned that it was not just the hypermutable phenotype but also the promiscuous phenotype of MMR mutators that provided an adaptive advantage to the bacteria (9, 12, 44, 46, 47).

Before proceeding, it is worth emphasizing the importance of carefully analyzing the populations in question for a mutator phenotype.

As each individual antibiotic resistance is established with a frequency of about 1×10^{-5}, it is imperative to distinguish between a mutator phenotype and a meroclone, one comprising each of the resistant subpopulations spawned by spontaneous mutation. For example, in our initial screen of 350 *E. coli* and *S. enterica* isolates (46), we found 26 "putative mutators," although only about one-third of these were truly mutators. Using sib selection, we showed that the remaining 17 nonmutator strains isolated as putative mutators were attributable to distinct subpopulations of mutants, each resistant to only one of the antibiotics (46). Thus, an essential control in confirming a mutator phenotype in studies such as these is to show that the high mutation rate segregates with the locus under selection (see below). Although the necessity is obvious, this obligate control is sometimes ignored. For example, although Matic et al. (61), in a report subsequent to ours, claim an even greater frequency of mutators among both pathogenic and commensal strains of *E. coli* based on mutant frequencies alone, the reader should be forewarned that the authors make no attempt to demonstrate a mutator allele in most of their isolates (see reference 44). This, compounded by the less stringent criteria employed, could help explain the high frequency of "putative mutators" (i.e., not mutators) that they report (61, 90).

Soon after our report, the work of Sniegowski et al. (88) demonstrated that in long-term bacterial cultures under selection pressure, a mutator cell arising spontaneously could survive, propagate, and successfully invade a population of nonmutator bacteria. That is, unlike the chemostat studies, 3 of 12 cultures founded from a single cell and propagated for 10,000 generations under limiting glucose conditions became mutators due to defects in MMR (88). Although this study did not address the frequency at which an MMR⁻ phenotype arises, more recent experiments have. Thus, for example, the frequency of spontaneously arising mutator cells in unselected clones of nonmutator cells was determined to be approximately 1×10^{-5} for *E. coli* (57, 65) and about 1×10^{-6} for serovar Typhimurium (47). Once again, all mutators isolated from these studies were MMR⁻. More recently, Lenski and coworkers have reported that in long-term continuous cultures (approximately 25,000 generations), most of the phenotypic diversity occurs within the first 2,000 generations (2), reminiscent of the chemostat studies discussed earlier. The phylogenies obtained from such cultures are formed because of beneficial mutations and become ancestral quickly, for "no clone left any descendants after 500 generations" (2). The spawning and silencing of mutators in such cultures reinforce the dynamism of the process. Such has bolstered our contention that there are strong selection pressures operating to maintain an MMR⁻ phenotype in natural populations (9, 10, 11, 12, 47). We believe that the MMR⁻ subpopulations serve as mixing pools for both enhanced genetic change and exchange (11, 47).

Computer simulations also have corroborated that mutators could persist and sweep populations (91, 92), though as Miller astutely notes, in silico experiments are a good reinforcement "that mutators are selected for in continued growth in chemostats" (65). Moreover, in silico experiments suffer in not being able to distinguish the individuality of a particular mutator. That is, once the various parameters enumerated in Table 1 are assigned, the computer runs on, completely oblivious to whether particular mutators are biologically capable of affecting even some of the processes that are blindly subsumed into the tabular data entered.

One must pause, therefore, to appreciate an important nuance in analyzing the data from natural populations and laboratory strains. Although in both cases, 100% of the mutator phenotypes were due to defects in MMR, the plurality of defects mapped to *mutS* (~73%; 11/15) in the natural isolates (9), whereas *mutS* defects were comprised in a minority class (18%; 12/65) among *E. coli* K-12 and serovar Typhimurium LT2 (47, 57, 65). We earlier pointed to this biological dichotomy (9), and further studies have reinforced this incongruity. Whether in silico approaches can distinguish these subtleties must await more sophisticated modeling that must incorporate the nuances of this intricate repair system.

DEFECTS IN MMR AND THEIR EFFECTS ON ANTIGENIC VARIATION AND VIRULENCE

Figure 4 summarizes the many and varied effects that mutant MMR alleles have on DNA change and exchange within the bacterial chromosome. The effects of MMR defects on spontaneous base-substitution and frameshift mutations are well documented (18, 31, 47, 79, 86, 87). In MutS-, MutH-, MutL-, or UvrD-deficient cells, mutational spectra show that base-substitution mutations are recovered more frequently than are frameshift mutations and, among the base substitutions, transition mutations predominate over transversion mutations (86). However, at particular loci containing homopolymeric tracts or oligonucleotide repeats (either mono-, di-, or trinucleotide repeats), frameshift mutations may

prevail, their incidence dependent on both the number (96) and length of repeats (51, 52, 96). Slipped mispairing at such sites during replication can lead to expansion or contraction of these repeats. The distribution of tandem repeats of particular bases within bacterial genomes prompted Moxon and coworkers to suggest that these represent contingency loci–loci intended for rapid adaptation to changing environments (19, 27, 70, 71). Genes controlled by the on-off switches of slipped-strand mispairing and polymerase slippage—such as those involved in fimbriae phase and lipopolysaccharide antigenic variation (19, 27, 56, 105)—contain the very sequences most sensitive to modulation by the MMR⁻ phenotype.

Moreover, slipped mispairing can occur by sister chromatid exchange (SCE), a process normally monitored by MMR (8). Because MMR aborts SCE between homeologous sequences, deletion formation between these sequences is enhanced in MMR⁻ strains (8). Defects in MMR also enhance the frequency of duplications (75) and transposition events (53, 54). One wonders if such mechanisms

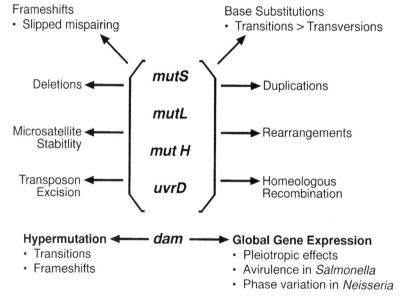

FIGURE 4 Phenotypes associated with mutant strains carrying defects in the MMR pathway.

provide the bases for pathoadaptive deletions in *Shigella* (62) and other pathogens (89) as well. Aside from the SCE process, *recA*-dependent recombinogenic effects of MMR mutants are seen as increased intraspecies exchange in *mutH* and *uvrD* mutants, as well as the homeologous recombination events promoted by *mutS* and *mutL* (see reference 67 and references therein).

The concept of contingency genes, or highly mutable loci, as a mechanism for rapid adaptation of virulence determinants (19, 27, 70, 71) can be expanded to envelop preferential targets for homeologous recombination after horizontal transfer events. We have proposed that the persistence of mutator alleles in nature is the consequence of selection for new gene functions gained from promiscuous exchange (homeologous recombination), since the mutators observed are notably MMR⁻ phenotypes (9). The available evidence further suggests a preference for *mutS* alleles among MMR mutators in nature. Our analysis of the *mutS* region of the chromosome offers a clue to explain this dichotomy. The *mutS* region is one of extensive genetic variability, as assessed by the heterogeneity of sequences found between the *mutS* gene and its neighboring *rpoS* gene in closely related enterics (45). *E. coli*, *Shigella*, and *Salmonella* species contain intergenic regions ranging from 3.7 to 12.6 kb, each comprising multiple sequence elements. These contrast to minimal regions of 88 bp in *Yersinia enterocolitica* (32) and 92 bp in *Vibrio cholerae* (102). The polymorphic character of the entire region is extended by considering the 40-kb pathogenicity island in serovar Typhimurium, which lies to the 5′ side of *mutS* and in fact encompasses the coding sequence for the first six or so amino acids of the MutS protein (66). And in more distantly related *Haemophilus influenzae*, a 3.1-kb tryptophanase gene cluster is inserted to the 3′ side of the *mutS* gene, primarily in pathogenic strains (58). The latter examples are ones in which sequence elements derived from horizontal transfer have selective value for pathogenesis, where they may enhance cell survival or proliferation in a particular niche of the infected host.

Significantly, such local sequence variations earmark the *mutS* region as a hot spot for recombination. In this respect, we would identify the region as a "bastion of polymorphism" in the chromosome, much as the *hsd* restriction and modification loci and the *rfb* complex for O-antigen biosynthesis (64). In particular, the polymorphic region between the *mutS* and *rpoS* genes of related enteric bacteria is reminiscent of the *rfb* gene cluster—a set of highly variable genes comprising 10 to 20 kb of sequence anchored within the conserved genes of related species (1, 5, 6, 35, 76, 81). We speculate that selection for new gene functions, aided by the promiscuous nature of MMR-defective cells, may itself generate the polymorphic DNA region found associated with the *mutS*⁺ gene. That is, since there is likely a long-term disadvantage to maintaining a high mutation rate, it would be favorable if a new sequence were linked to a functional *mut*⁺ gene, to correct the MMR defect and thereby quiet the mutator phenotype. This scenario, which would stabilize new gene combinations through long-term evolution, may lead to the genetically varied *mutS* regions observed in both closely and distantly related species.

E. coli dam mutants that are deficient in their ability to methylate 5′GATC3′ sites or strains that overexpress Dam are also mutators, evidence that subsequent steps in MMR rely on the state of Dam methylation. In addition to the hypermutable phenotype, mutations in *dam* make *E. coli* hyper-recombinogenic, enhance chromosomal segregation, and induce the SOS repair system moderately (18, 67). While the hypermutability of *dam* mutants is predictable—adenine methylation controls the specificity of MMR—the pleiotropy of Dam deficiency demonstrates yet again the intricacies of MMR repair.

In *Neisseria meningitidis*, a particular consequence of the mutator phenotype caused by a *dam* defect substantially favors pathogenicity (7). The regulation of capsular synthesis helps to define the virulence of meningococci. In group B meningococci, the *siaD* gene, encoding a polysialyltransferase that synthesizes a polymer of α-2,8-N-acetyl-neuraminic acid as its capsule, plays a pivotal role in this process. Several group B meningococci undergo capsular phase variation mediated by $+1/-1$ frameshifting within a $(dC)_7$ tract that alters the reading frame of the *siaD* gene. It is slipped-strand mispairing promoted by Dam deficiency that alters the reading frame within the poly-dC tract. This on-off switch thus changes gene expression to enable the pathogen to enter the epithelial cell in the nonencapsulated form and invade the cell, escaping host defenses; or to survive the extracellular environment in the capsular form. The reversible change in expression of capsular polysaccharide occurs at frequencies of 10^{-3} to 10^{-4} in *dam* meningococci, high enough to ensure the transitions required for a successful course of infection. Dam deficiency was found associated with 50% of commensal strains surveyed, whereas all pathogenic isolates were Dam$^-$. This partitioning of the hypermutability phenotype favoring the pathogen state is the likely outcome of the mechanism of Dam inactivation: transposition of a sequence specifying a GmeATC-endonuclease into the *dam*$^+$ gene both obliterates Dam activity and restricts further genetic exchange with Dam$^+$ cells. The mutator biotype of these meningococci is then locked in place, poised for the phase variations that promote successful infection.

Unlike the case in meningococci, mutations in *dam* render *Salmonella* pathogens avirulent (28, 30). The role that Dam methylation plays in the transcriptional regulation of particular genes may help explain why the Dam$^+$ phenotype is essential for virulence. Dam regulates the expression of more than 20 *ivi* (in vivo-induced) genes in serovar Typhimurium, including a subset controlled by the PhoP global virulence regulator (30). The effects of faulty regulation include the reduced ability to invade nonphagocytic cells and failure to invade enterocytes and cause M cell cytotoxicity

(28). Such deficiencies make the Dam methylase a candidate for targeting vaccines or antimicrobials against *Salmonella* and, potentially, other pathogenic bacteria (30).

EPILOGUE

The opposing effects of the Dam$^-$ phenotype in meningococci and salmonellae, favoring or disfavoring pathogenesis, emphasize the biological vagaries and subtleties contained within commensal and pathogenic strains of bacteria. The roles that MutS, MutH, MutL, or UvrD deficiency play in the development of antigenic variation and virulence may be even more varied and subtle, due to the multiple ways these mutator phenotypes are expressed and act to affect gene structure. That is:

- Whether mutator alleles exist in small subpopulations of a pathogen "clone" or have swept the entire population will determine the availability of genetic variants upon which selection acts in changing environments.

- We suspect that each mutator allele affects repair and recombination pathways differently, as evidenced by the increased ability of *mutS* strains, but not *mutH* or *uvrD*, to carry out homeologous recombination.

- A heightened recombinogenic potential among pathogens may amplify the MMR mutator effect in gene exchange. Such intergeneric exchange may result in chromosomal mosaicisms such as those found in naturally transformable *Neisseria* and *Haemophilus* (40).

- While we have confined our attention to stable mutators, the MMR$^-$ phenotype may be transiently expressed (46, 83), perhaps as a general strategy for responding to the stresses of unpredictable environments. Down-regulation of the levels of MutS and MutH has been demonstrated in stationary-phase cells (26), and even sequestering of MutS at base mismatches simulates MMR defects in mutation and recombination (55). That MutS and MutH levels are under the control of the Hfq and

RpoS global regulators (94) and that *mutL* is part of a super operon and may be regulated by yet other stimuli (95) indicate the complex regulation of MMR.

- Finally, the susceptibility of particular genes to MMR control, like the Dam⁻ effect on the *siaD* gene in group B meningococci, applies an added level of regulation for adaptation during pathogenesis (70, 71).

Using logic akin to the multistep model for carcinogenesis, Thaler and coworkers (27, 56, 70) illustrated nicely the tremendous potential of a mutator phenotype—that of making likely a most improbable event. For illustration they assumed that only five genes (gene differences) ultimately determined whether an organism was a commensal or a pathogen. Using a binary switch model and assigning a frequency of occurrence of 10^{-3} for each of the events, the probability of finding one microbe with all five events is $(10^{-3})^5$, or 10^{-15}. In other words, it would take sifting through a population of 10^{15} bacteria to find perhaps one time the improbable cell that had all five changes contained in it. However, in a mutator population, where the mutation rate is enhanced 100-fold, and assuming that each of the events is prone to the mutator effect, then the probability is $(10^{-1})^5$. Thus, in a population size 10^{10} smaller, it is probable that the improbable has been spawned and will be found. Of course, because of the manifold effects elicited by an MMR⁻ phenotype (some enumerated above), we expect MMR mutators, and not mutators in general, to play the pivotal role in this process.

REFERENCES

1. **Aoyama, K., A. M. Haase, and P. R. Reeves.** 1994. Evidence for effect of random genetic drift on G+C content after lateral transfer of fucose pathway genes to *Escherichia coli* K-12. *Mol. Biol. Evol.* **11:**829–838.
2. **Ash, C. P. J.** 1999. Mutation and adaption from the Great Lakes to the Rocky Mountains. *Trends Microbiol.* **7:**395–398.
3. **Atwood, K. C., L. K. Schneider, and F. J. Ryan.** 1951. Periodic selection of *Escherichia coli*. *Proc. Natl. Acad. Sci. USA* **37:**146–155.
4. **Atwood, K. C., L. K. Schneider, and F. J. Ryan.** 1951. Selective mechanisms in bacteria. *Cold Spring Harbor Symp. Quant. Biol.* **16:**345–355.
5. **Bastin, D. A., and P. R. Reeves.** 1995. Sequence and analysis of the O antigen gene (*rfb*) cluster of *Escherichia coli* O111. *Gene* **164:**17–23.
6. **Bisercic, M., J. Y. Feutier, and P. R. Reeves.** 1991. Nucleotide sequence of the *gnd* gene from nine natural isolates of *Escherichia coli*: evidence of intragenic recombination as contributing factor in the evolution of the polymorphic *gnd* locus. *J. Bacteriol.* **173:**3894–3900.
7. **Bucci, C., A. Lavitola, P. Salvatore, L. Del Giudici, D. R. Massardo, C. B. Bruni, and P. Alifano.** 1999. Hypermutation in pathogenic bacteria: frequent phase variation in meningococci is a phenotypic trait of a specialized mutator biotype. *Mol. Cell* **3:**435–445.
8. **Bzymek, M., C. J. Saveson, V. V. Feschenko, and S. T. Lovett.** 1999. Slipped misalignment mechanisms of deletion formation: in vivo susceptibility to nucleases. *J. Bacteriol.* **181:**477–482.
9. **Cebula, T. A., and J. E. LeClerc.** 1997. Hypermutability and homeologous recombination: ingredients for rapid evolution. *Bull. Inst. Pasteur* **95:**97–106.
10. **Cebula, T. A., and J. E. LeClerc.** 1997. To be a mutator, or how pathogenic and commensal bacteria can evolve rapidly (discussion). *Trends Microbiol.* **5:**428–429.
11. **Cebula, T. A., B. Li, W. L. Payne, and J. E. LeClerc.** 1998. Mutators among *Escherichia coli* and *Salmonella enterica*: adaptation and emergence of bacterial pathogens. *Am. Soc. Microbiol. Conference on Small Genomes*, abstr. SA-22, p. 16.
12. **Cebula, T. A., D. D. Levy, and J. E. LeClerc.** 1999. Mutator bacteria and resistance development. *In* D. Hughes and D. Andersson (ed.), *Antibiotic Resistance and Antibiotic Development*. Harwood, Amsterdam, in press.
13. **Chao, L., and E. C. Cox.** 1983. Competition between high and low mutating strains of *Escherichia coli*. *Evolution* **37:**125–134.
14. **Cox, E. C.** 1973. Mutator gene studies in *Escherichia coli*: the *mutT* gene. *Genetics* (suppl.) **73:**67–80.
15. **Cox, E. C.** 1995. Recombination, mutation and the origin of species. *Bioessays* **17:**747–749.
16. **Cox, E. C., and T. C. Gibson.** 1974. Selection for high mutation rates in chemostats. *Genetics* **77:**169–184.
17. **Cox, M. M.** 1997. Recombinational crossroads: eukaryotic enzymes and the limits of bacterial precedents. *Proc. Natl. Acad. Sci. USA* **94:**11764–11766.

18. **Craig, R. J., J. A. Araj, and M. G. Marinus.** 1984. Induction of damage inducible (SOS) repair in *dam* mutants of *Escherichia coli* exposed to 2-aminopurine. *Mol. Gen. Genet.* **194:**539–540.

19. **Deitsch, K. W., E. R. Moxon, and T. E. Wellems.** 1997. Shared themes of antigenic variation and virulence in bacterial, protozoal, and fungal infections. *Microbiol. Mol. Biol. Rev.* **61:** 281–293.

20. **de Visser, J. A. G. M., C. W. Zeyl, P. J. Gerrish, J. L. Blanchard, and R. E. Lenski.** 1999. Diminishing returns from mutation supply rate in asexual populations. *Science* **283:**404–406.

21. **Domingo, E., and J. J. Holland.** 1997. RNA virus mutations and fitness for survival. *Annu. Rev. Microbiol.* **51:**151–178.

22. **Drake, J. W.** 1991. A constant rate of spontaneous mutation in DNA-based microbes. *Proc. Natl. Acad. Sci. USA* **88:**7160–7164.

23. **Drake, J. W.** 1993. Rates of spontaneous mutation among RNA viruses. *Proc. Natl. Acad. Sci. USA* **90:**4171–4175.

24. **Drake, J. W., B. Charlesworth, D. Charlesworth, and J. F. Crow.** 1998. Rates of spontaneous mutation. *Genetics* **148:**1667–1686.

25. **Dykhuizen, D. E., and L. Green.** 1991. Recombination in *Escherichia coli* and the definition of biological species. *J. Bacteriol.* **173:**7257–7268.

26. **Feng, G., H. C. Tsui, and M. E. Winkler.** 1996. Depletion of the cellular amounts of the MutS and MutH methyl-directed mismatch repair proteins in stationary-phase *Escherichia coli* K-12 cells. *J. Bacteriol.* **178:**2388–2396.

27. **Field, D., M. O. Magnasco, E. R. Moxon, D. Metzgar, M. M. Tanaka, C. Wills, and D. S. Thaler.** 1999. Contingency loci, mutator alleles, and their interactions. Synergistic strategies for microbial evolution and adaptation in pathogenesis. *Ann. N. Y. Acad. Sci.* **870:**378–382.

28. **Garcia-Del Portillo, F., M. G. Pucciarelli, and J. Casadesus.** 1999. DNA adenine methylase mutants of *Salmonella typhimurium* show defects in protein secretion, cell invasion, and M cell cytotoxicity. *Proc. Natl. Acad. Sci. USA* **96:** 11578–11583.

29. **Gross, M. D., and E. C. Siegel.** 1981. Incidence of mutator strains in *Escherichia coli* and coliforms in nature. *Mutat. Res.* **91:**107–110.

30. **Heithoff, D. M., R. L. Sinsheimer, D. A. Low, and M. J. Mahan.** 1999. An essential role for DNA adenine methylation in bacterial virulence. *Science* **284:**967–970.

31. **Horst, J. P., T. H. Wu, and M. G. Marinus.** 1999. *Escherichia coli* mutator genes. *Trends Microbiol.* **7:**29–36.

32. **Iriarte, M., I. Stainier, and G. R. Cornelis.** 1995. The *rpoS* gene from *Yersinia enterocolitica*

and its influence on expression of virulence factors. *Infect. Immun.* **63:**1840–1847.

33. **Jain, R., M. C. Rivera, and J. A. Lake.** 1999. Horizontal gene transfer among genomes: the complexity hypothesis. *Proc. Natl. Acad. Sci. USA* **96:**3801–3806.

34. **Jyssum, K.** 1960. Observations on two types of genetic instability in *Escherichi coli*. *Acta Pathol. Microbiol. Scand.* **48:**113–120.

35. **Karaolis, D. K., R. Lan, and P. R. Reeves.** 1995. The sixth and seventh cholera pandemics are due to independent clones separately derived from environmental, nontoxigenic, non-O1 *Vibrio cholerae*. *J. Bacteriol.* **177:**191–198.

36. **Kehoe, M. A., V. Kapur, A. M. Whatmore, and J. M. Musser.** 1996. Horizontal gene transfer among group A streptococci: implications for pathogenesis and epidemiology. *Trends Microbiol.* **4:**436–443.

37. **Kibota, T. T., and M. Lynch.** 1996. Estimate of the genomic mutation rate deleterious to overall fitness in *E. coli*. *Nature* **381:**694–696.

38. **Kirchner, C. E. J., and M. J. Rudden.** 1966. Localization of a mutator gene in *Salmonella typhimurium* by cotransduction. *J. Bacteriol.* **92:** 1453–1456.

39. **Kolstø, A.-B.** 1999. Time for a fresh look at the bacterial chromosome. *Trends Microbiol.* **7:**223–226.

40. **Kroll, J. S., K. E. Wilks, J. L. Farrant, and P. R. Langford.** 1998. Natural genetic exchange between *Haemophilus* and *Neisseria*: intergeneric transfer of chromosomal genes between major human pathogens. *Proc. Natl. Acad. Sci. USA* **95:**12381–12385.

41. **Lawrence, J. G., and H. Ochman.** 1997. Amelioration of bacterial genomes: rates of change and exchange. *J. Mol. Evol.* **44:**383–397.

42. **Lawrence, J. G., and H. Ochman.** 1998. Molecular archaeology of the *Escherichia coli* genome. *Proc. Natl. Acad. Sci. USA* **95:**9413–9417.

43. **Lawrence, J. G., and J. R. Roth.** 1996. Selfish operons: horizontal transfer may drive the evolution of gene clusters. *Genetics* **143:**1843–1860.

44. **LeClerc, J. E., and T. A. Cebula.** 1997. Highly variable mutation rates in commensal and pathogenic *Escherichia coli*. *Science* **227:**1834. (Response.)

45. **LeClerc, J. E., B. Li, W. L. Payne, and T. A. Cebula.** 1999. Promiscuous origin of a chimeric sequence in the *Escherichia coli* O157:H7 genome. *J. Bacteriol.* **181:**7614–7617.

46. **LeClerc, J. E., B. Li, W. L. Payne, and T. A. Cebula.** 1996. High mutation frequencies among *Escherichia coli* and *Salmonella* pathogens. *Science* **274:**1208–1211.

47. **LeClerc, J. E., W. L. Payne, E. Kupchella, and T. A. Cebula.** 1998. Detection of mutator subpopulations in *Salmonella typhimurium* LT2 by reversion of *his* alleles. *Mutat. Res.* **400:**89–97.

48. **Leigh, E. G.** 1970. Natural selection and mutability. *Am. Nat.* **104:**301–305.

49. **Leigh, E. G.** 1973. The evolution of mutation rates. *Genetics* **73:**1–18.

50. **Levin, B. R., M. Lipsitch, and S. Bonhoeffer.** 1999. Population biology, evolution, and infectious disease: convergence and synthesis. *Science* **283:**806–809.

51. **Levinson, G., and G. A. Gutman.** 1987. High frequencies of short frameshifts in poly-CA/TG tandem repeats borne by bacteriophage M13 in *Escherichia coli* K-12. *Nucleic Acids Res.* **15:**5323–5338.

52. **Levy, D. D., and T. A. Cebula.** 1999. Mutagenesis patterns in a tRNA mutation marker gene altered to include repetitive sequence replicated in *mutS E. coli. Ann. N. Y. Acad. Sci.* **870:**392–395.

53. **Lundblad, V., and N. Kleckner.** 1982. Mutants of *Escherichia coli* K12 which affect excision of transposon Tn*10*. *Basic Life Sci.* **20:**245–258.

54. **Lundblad, V., and N. Kleckner.** 1985. Mismatch repair mutations of *Escherichia coli* K12 enhance transposon excision. *Genetics* **109:**3–19.

55. **Maas, W. K., C. Wang, T. Lima, A. Hach, and D. Lim.** 1996. Multicopy single-stranded DNA of *Escherichia coli* enhances mutation and recombination frequencies by titrating MutS protein. *Mol. Microbiol.* **19:**505–509.

56. **Magnasco, M. O., and D. S. Thaler.** 1996. Changing the pace of evolution. *Physics Lett.* **221:**287–292.

57. **Mao, E. F., L. Lane, J. Lee, and J. H. Miller.** 1997. Proliferation of mutators in a cell population. *J. Bacteriol.* **179:**417–422.

58. **Martin, K., G. Morlin, A. Smith, A. Nordyke, A. Eisenstark, and M. Golomb.** 1998. The tryptophanase gene cluster of *Haemophilus influenzae* type b: evidence for horizontal gene transfer. *J. Bacteriol.* **180:**107–118.

59. **Matic, I., C. Rayssiguier, and M. Radman.** 1995. Gene exchange in bacteria: the role of SOS and mismatch repair systems in evolution of species. *Cell* **80:**507–515.

60. **Matic, I., F. Taddei, and M. Radman.** 1996. Genetic barriers among bacteria. *Trends Microbiol.* **4:**69–73.

61. **Matic, I., M. Radman, F. Taddei, B. Picard, C. Doit, E. Bingen, E. Denamur, and J. Elison.** 1997. Highly variable mutation rates in commensal and pathogenic *Escherichia coli. Science* **227:**1833–1834.

62. **Maurelli, A. T., R. E. Fernandez, C. A. Bloch, C. K. Rode, and A. Fasano.** 1998. "Black holes" and bacterial pathogenicity: a large genomic deletion that enhances the virulence of *Shigella* spp. and enteroinvasive *Escherichia coli. Proc. Natl. Acad. Sci. USA* **95:**3943–3948.

63. **Médigue, C., T. Rouxel, P. Vigier, A. Hénaut, and A. Danchin.** 1991. Evidence for horizontal gene transfer in *Escherichia coli* speciation. *J. Mol. Biol.* **222:**851–856.

64. **Milkman, R.** 1997. Recombination and population structure in *Escherichia coli. Genetics* **146:**745–750.

65. **Miller, J. H., A. Suthar, J. Tai, A. Yeung, C. Truong, and J. L. Stewart.** 1999. Direct selection for mutators in *Escherichia coli. J. Bacteriol.* **181:**1576–1584.

66. **Mills, D. M., V. Bajaj, and C. A. Lee.** 1995. A 40 kb chromosomal fragment encoding *Salmonella typhimurium* invasion genes is absent from the corresponding region of the *Escherichia coli* K-12 chromosome. *Mol. Microbiol.* **15:**749–759.

67. **Modrich, P.** 1991. Mechanisms and biological effects of mismatch repair. *Annu. Rev. Genet.* **25:**229–253.

68. **Modrich, P., and R. Lahue.** 1996. Mismatch repair in replication fidelity, genetic recombination, and cancer biology. *Annu. Rev. Biochem.* **65:**101–133.

69. **Moxon, E. R.** 1995. Whole genome sequencing of pathogens: a new era in microbiology. *Trends Microbiol.* **3:**335–337.

70. **Moxon, E. R., and D. S. Thaler.** 1997. Microbial genetics. The tinkerer's evolving toolbox. *Nature* **387:**659–662.

71. **Moxon, E. R., R. B. Rainey, M. A. Nowak, and R. E. Lenski.** 1994. Adaptive evolution of highly mutable loci in pathogenic bacteria. *Curr. Biol.* **4:**24–33.

72. **Nestman, E. R., and R. F. Hill.** 1973. Population changes in continuously growing mutator cultures of *Escherichia coli. Genetics* **73**(Suppl.)**:**41–44.

73. **Ninio, J.** 1991. Transient mutators: a semiquantitative analysis of the influence of translation and transcription errors on mutation rates. *Genetics* **129:**957–962.

74. **Pang T.** 1998. Genetic dynamics of *Salmonella typhi*—diversity in clonality. *Trends Microbiol.* **6:**339–342.

75. **Petit, M. A., J. Dimpfl, M. Radman, and H. Echols.** 1991. Control of large chromosomal duplications in Escherichia coli by the mismatch repair system. *Genetics* **129:**327–332.

76. **Pupo, G. M., D. K. Karaolis, R. Lan, and P. R. Reeves.** 1997. Evolutionary relationships among pathogenic and nonpathogenic *Escherichia*

coli strains inferred from multilocus enzyme electrophoresis and *mdh* sequence studies. *Infect. Immun.* **65:**2685–2692.

77. **Radman, M., I. Matic, and F. Taddei.** 1999. Evolution of evolvability. *Ann. N. Y. Acad. Sci.* **870:**146–155.

78. **Rainey, P. B.** 1999. The economics of mutation. *Curr. Biol.* **9:**R371–R373.

79. **Raposa, S., and M. S. Fox.** 1987. Some features of base pair mismatch and heterology repair in *Escherichia coli. Genetics* **117:**381–390.

80. **Rayssiguier, C., D. S. Thaler, and M. Radman.** 1989. The barrier to recombination between *Escherichia coli* and *Salmonella typhimurium* is disrupted in mismatch-repair mutants. *Nature* **342:**396–401.

81. **Reeves, P. R.** 1992. Variation in O-antigens, niche-specific selection and bacterial populations. *FEMS Microbiol. Lett.* **79:**509–516.

82. **Roca, A. I., and M. M. Cox.** 1997. RecA protein: structure, function, and role in recombinational DNA repair. *Prog. Nucleic Acid Res. Mol. Biol.* **56:**129–223.

83. **Rosenberg, S. M., C. Thulin, and R. S. Harris.** 1998. Transient and heritable mutators in adaptive evolution in the lab and in nature. *Genetics* **148:**1559–1566.

84. **Sanderson, K. E., A. Hessel, and B. A. D. Stocker.** 1987. Strains of *Salmonella typhimurium* and other *Salmonella* species used in genetic analysis, p. 2496–2503. *In* F. C. Neidhardt, J. L. Ingraham, and H. E. Umbarger (ed.), *Escherichia coli and Salmonella Typhimurium: Cellular and Molecular Biology.* American Society for Microbiology, Washington, D.C.

85. **Schaaper, R. M.** 1993. Base selection, proofreading, and mismatch repair during DNA replication in *Escherichia coli. J. Biol. Chem.* **268:**23762–23765.

86. **Schaaper, R. M., and R. L. Dunn.** 1991. Spontaneous mutation in the *Escherichia coli lacI* gene. *Genetics* **129:**317–326.

87. **Siegel, E. C., and F. Kamel.** 1974. Reversion of frameshift mutations by mutator genes in *Escherichia coli. J. Bacteriol.* **117:**994–1001.

88. **Sniegowski, P. D., P. J. Gerrish, and R. E. Lenski.** 1997. Evolution of high mutation rates in experimental populations of *E. coli. Nature* **387:**703–705.

89. **Sokurenko, E. V., D. L. Hasty, and D. E. Dykhuizen.** 1999. Pathoadaptive mutations: gene loss and variation in bacterial pathogens. *Trends Microbiol.* **7:**191–195.

90. **Taddei, F., I. Matic, B. Godelle, and M. Radman.** 1997. To be a mutator, or how pathogenic and commensal bacteria can evolve rapidly. *Trends Microbiol.* **5:**427–428.

91. **Taddei, F., M. Radman, J. Maynard-Smith, B. Toupance, P. H. Gouyon, and B. Godelle.** 1997. Role of mutator alleles in adaptive evolution. *Nature* **387:**700–702.

92. **Tenaillon, O., B. Toupance, H. Le Nagard, F. Taddei, and B. Godelle.** 1999. Mutators, population size, adaptive landscape and the adaptation of asexual populations of bacteria. *Genetics* **152:**485–493.

93. **Tröbner, W., and R. Piechoki.** 1984. Selection against hypermutability in *Escherichia coli* during long term evolution. *Mol. Gen. Genet.* **198:**177–178.

94. **Tsui, H. C., G. Feng, and M. E. Winkler.** 1997. Negative regulation of *mutS* and *mutH* repair gene expression by the Hfq and RpoS global regulators of *Escherichia coli* K-12. *J. Bacteriol.* **179:**7476–7487.

95. **Tsui, H. C., G. Zhao, G. Feng, H. C. Leung, and M. E. Winkler.** 1994. The mutL repair gene of Escherichia coli K-12 forms a superoperon with a gene encoding a new cell-wall amidase. *Mol. Microbiol.* **11:**189–202.

96. **van Belkum, A., S. Scherer, L. van Alphen, and H. Verbrugh.** 1998. Short-sequence DNA repeats in prokaryotic genomes. *Microbiol. Mol. Biol. Rev.* **62:**275–293.

97. **Van Valen, L.** 1973. A new evolutionary law. *Evol. Theory* **1:**1–30.

98. **Vulic, M., F. Dionisio, F. Taddei, and M. Radman.** 1997. Molecular keys to speciation: DNA polymorphism and the control of genetic exchange in enterobacteria. *Proc. Natl. Acad. Sci. USA* **94:**9763–9767.

99. **Vulic, M., R. E. Lenski, and M. Radman.** 1999. Mutation, recombination, and incipient speciation of bacteria in the laboratory. *Proc. Natl. Acad. Sci. USA* **96:**7348–7351.

100. **Worth, L., Jr., T. Bader, J. Yang, and S. Clark.** 1998. Role of MutS ATPase activity in MutS,L-dependent block of in vitro strand transfer. *J. Biol. Chem.* **273:**23176–23182.

101. **Worth, L., S. Clark, M. Radman, and P. Modrich.** 1994. Mismatch repair proteins MutS and MutL inhibit RecA-catalyzed strand transfer between diverged DNAs. *Proc. Natl. Acad. Sci. USA* **91:**3238–3241.

102. **Yildiz, F. H., and G. K. Schoolnik.** 1998. Role of *rpoS* in stress survival and virulence of *Vibrio cholerae. J. Bacteriol.* **180:**773–784.

103. **Zahrt, T. C., and S. Maloy.** 1997. Barriers to recombination between closely related bacteria: MutS and RecBCD inhibit recombination between *Salmonella typhimurium* and *Salmonella typhi. Proc. Natl. Acad. Sci. USA* **94:**9786–9791.

104. **Zahrt, T. C., G. C. Mora, and S. Maloy.** 1994. Inactivation of mismatch repair overcomes the barrier to transduction between *Salmonella typhimurium* and *Salmonella typhi*. *J. Bacteriol.* **176:**1527–1529.

105. **Zhang, Q., and K. S. Wise.** 1997. Localized reversible frameshift mutation in an adhesin gene confers a phase-variable adherence phenotype in mycoplasma. *Mol. Microbiol.* **25:**859–869.

BACTERIAL EFFECTS ON HOST CELL FUNCTION

BACTERIAL TOXINS IN DISEASE PRODUCTION

Joseph T. Barbieri and Kristin J. Pederson

11

The determination that pathogenic bacteria may elicit disease through the action of a toxin was made in the 1880s, when culture filtrates of *Corynebacterium diphtheriae* were shown to evoke a toxic response in animals that was identical to the disease elicited by infection with the bacterium. The soluble component was later identified as diphtheria toxin and was shown to ADP-ribosylate elongation factor-2. This modification inactivated elongation factor-2, which inhibited protein synthesis and led to cell death. The pathologic changes elicited by diphtheria toxin were solely responsible for the disease associated with systemic diphtheria.

Progress to define the mechanism of action of toxins, such as diphtheria toxin, has paralleled advances in understanding the molecular and cell biology of the eukaryotic cell and advances in the physical and functional analysis of proteins. Our current understanding of the mechanism of action of bacterial toxins includes details of their genetic, physical, and enzymatic properties and insight into their translocation across the eukaryotic cell membrane to gain access to intracellular targets. This chapter will outline the molecular aspects of bacterial toxins.

Each bacterial toxin possesses a unique mechanism of action, which modifies a specific function within a eukaryotic cell. The disease elicited by a toxin-producing bacterium is often directly attributable to the action of the toxin, but also often requires the expression of factors that assist in host colonization and comprise host defense mechanisms. Toxin production is often the primary factor that differentiates a commensal bacterium from a disease-producing bacterial pathogen. Conversely, while toxins may elicit clinical disease, their production is often not sufficient to cause clinical disease. For example, anthrax toxin is considered to be the toxic component of *Bacillus anthracis*, but nonvirulent toxin-producing strains of *B. anthracis* have been isolated. These mutants fail to produce a polyglutamic acid capsule, which interferes with host clearance of the bacterium. An extension to this generalization is the intoxication elicited by botulinum neurotoxins and staphylococcal enterotoxins, where ingestion of preformed toxin is responsible for clinical disease. Food poisoning caused by botulinum neurotoxins and staphylococcal enterotoxins is an intoxication, rather than an infection, by

Joseph T. Barbieri and Kristin J. Pederson   Department of Microbiology and Molecular Genetics, Medical College of Wisconsin, Milwaukee, WI 53226.

Virulence Mechanisms of Bacterial Pathogens, 3rd ed., Edited by K. A. Brogden et al.
©2000 ASM Press, Washington, D.C.

toxin-producing *Clostridium botulinum* or *Staphylococcus aureus*, respectively.

The pathologic changes elicited by a bacterial toxin result from the catalytic covalent modification of a specific host cell component, which relates to the disease associated with that exotoxin. Although diphtheria toxin and cholera toxin are both ADP-ribosylating exotoxins, the pathogenesis elicited by each exotoxin is unique. This happens because diphtheria toxin ADP-ribosylates elongation factor-2, which results in the inhibition of protein synthesis, while cholera toxin ADP-ribosylates the Gsα component of Gs heterotrimeric protein, which stimulates the activity of host adenylate cyclase. Such stimulation results in the elevation of intracellular cyclic AMP (cAMP) and subsequent secretion of electrolytes and H_2O from the host cell, which is the clinical manifestation of cholera. The term "host cell" may apply to either vertebrate cells or lower eukaryotic cells, such as protozoa, since bacterial toxins intoxicate a broad range of host cells.

Bacterial toxins are classified into several families including exotoxins, pore-forming toxins, membrane-acting toxins, and type-III-secreted cytotoxins. Each family of toxins possesses features that differentiate that family from other families of toxins.

EXOTOXINS

Bacterial exotoxins, the most well-characterized family of toxins, are classified according to their mechanism of action. Exotoxins are secreted by the bacterium as soluble proteins, which enter eukaryotic cells primarily through receptor-mediated endocytosis. Early experiments that characterized the genetic organization of diphtheria toxin showed that the diphtheria toxin was encoded within the genome of the lysogenic β-phage (reviewed in reference 4). Because either nonlysogenic or lysogenic *C. diphtheriae* could establish an upper respiratory tract infection, diphtheria toxin was not required for the colonization. Only lysogenic *C. diphtheriae* caused systemic disease. This observation established a role for

diphtheria toxin in systemic disease production. In addition, a general theme for exotoxin-mediated infections was established that involved local colonization of the host and exotoxin-mediated systemic disease, which often occurred distal to the site of infection.

Genes encoding bacterial exotoxins may be located on the bacterial chromosome (exotoxin A of *Pseudomonas aeruginosa*), a bacteriophage (diphtheria toxin or cholera toxin), or plasmids (heat-labile enterotoxin of *Escherichia coli*). Genes encoding multisubunit toxins are often organized in operons to allow the coordinate expression of the subunit components. Regulation of gene expression of bacterial exotoxins varies with each exotoxin. Exotoxins are often produced during specific stages of growth. These differential expression profiles often reflect the complex regulation of transcription, which include responses to environmental conditions (such as iron concentration) in order to modulate transcription of the genes encoding the exotoxin. For example, transcription of the gene encoding diphtheria toxin is regulated through the diphtheria toxin repressor, which binds iron to form an active repressor. This complex binds to the diphtheria toxin promoter to inhibit transcription (35). As the concentration of iron decreases below its K_d for the diphtheria toxin repressor, the apo-diphtheria toxin repressor has a reduced affinity for the promoter and the gene-encoding diphtheria toxin is expressed.

AB ORGANIZATION OF EXOTOXINS

Bacterial exotoxins have been defined in terms of their AB structure-function properties, where the A domain comprises the catalytic domain and the B domain comprises the receptor-binding domain (Fig. 1) (26). The B domain was shown to be responsible for the delivery of the A domain across the eukaryotic cell membrane and into the cytosol. Recently, structural and biochemical studies (3) have extended the resolution of these exotoxins into three independent separate domains: the catalytic domain (also termed the C domain), the

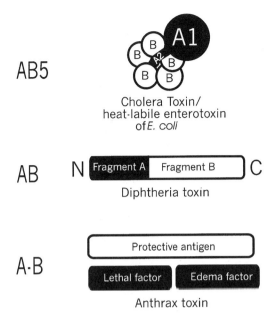

AB5

Cholera Toxin/
heat-labile enterotoxin
of *E. coli*

AB

N Fragment A   Fragment B   C

Diphtheria toxin

A·B

Protective antigen

Lethal factor   Edema factor

Anthrax toxin

FIGURE 1 AB organization of bacterial exotoxins. Indicated are the schematized representations of the three recognized organizations of bacterial exotoxins. The shaded area represents the catalytic domain, A, while the clear areas represent the translocation and binding domains, B.

binding (B) domain, and the translocation (T) domain (3). For simplicity, this review will describe the physical organization of the T and B domains as the B domain, because the T and B domains are often contiguous within the primary amino acid sequence of toxins. Each domain is responsible for a specific property of each exotoxin.

The simplest AB organization observed is when the A and B domains are included within a single protein. The prototype AB toxin is diphtheria toxin (3, 4). Diphtheria toxin is a 535-amino acid protein with the 1 to 193 amino-terminal residues comprising the A domain (the ADP-ribosyltransferase domain) and the 340 carboxyl-terminal residues comprising the B domain (the translocation and receptor-binding domains). The order of the AB domains may vary within a toxin. For example, exotoxin A of *P. aeruginosa* is also an AB exotoxin, with the B domain located within the amino terminus and the A domain located within the carboxyl terminus.

The AB5 exotoxins are composed of six noncovalently bound proteins. The prototype of the AB5 exotoxins is cholera toxin (1, 33). The A domain of cholera toxin, the ADP-ribosyltransferase domain, is organized into two functional regions termed A1, which includes the ADP-ribosyltransferase domain, and A2, which links A1 to the B5 component of the toxin. The B5 domain of cholera toxin is composed of 5 identical proteins, which form a pentamer that is organized into a ring structure upon which the A1 domain is positioned. The A2 domain inserts into the hole in the center of B5. The carboxyl terminus of A2 extends through the center of B5. Although the significance is not definitive, several models suggest that the carboxyl terminal-amino acids of the A2 domain contribute to the trafficking of the toxin within the eukaryotic cell. The five proteins that comprise the B domain may be identical, as is the case for cholera toxin or the heat-labile enterotoxin of *E. coli,* or composed of different proteins that form a nonsymmetrical ring structure (e.g., the B oligomer of pertussis toxin).

The third AB organization includes proteins that are not associated in solution, but associate upon binding to the host cell, termed A-B. The prototype for the A-B exotoxins is anthrax toxin (8, 18). Anthrax is a tripartite exotoxin composed of two proteins that encode different A domains, and a third protein that comprises the B domain. This B domain protein is called the protective antigen (PA), since antibodies directed against this protein are protective. PA interacts with either A domain protein. The mechanism of action of each A domain of anthrax has been determined. One A domain protein, edema factor, possesses adenylate cyclase activity. The other, called lethal factor, has been recently shown to possess an endoprotease activity. During the intoxication process, PA binds to sensitive cells and is cleaved by a eukaryotic protease, such as furin, which stimulates its oligomerization. The cleaved PA oligomer is now capable of binding either of the A domain proteins that are responsible for the pathology associated with anthrax.

Secretion and Activation of Exotoxins

Most bacteria utilize the general secretory pathway to secrete exotoxins across their cell membrane (also reviewed in chapter 14). Secretion of exotoxins by the general secretory pathway was predicted by the observation that the amino terminus of the mature exotoxins was processed relative to the predicted protein sequence of the gene encoding that exotoxin. The general secretory pathway involves the coordinate translation and secretion of the nascent protein across the cell membrane (6). During the translation of mRNA that encodes the leader sequence, the nascent polypeptide is targeted to the cell membrane. The translocated, nascent polypeptide folds into its native conformation and the leader sequence is cleaved by a periplasmic leader peptidase to yield a mature exotoxin. While some bacterial toxins such as the heat-labile enterotoxin of *E. coli* remain localized to the periplasmic space, other bacterial exotoxins, such as cholera toxin and pertussis toxin, are assembled in the periplasm and subsequently transported across the bacterial outer membrane and into the external environment. This secretion requires a complex export apparatus composed of multiple proteins and occurs by a mechanism that remains to be resolved (see chapter 14).

Bacterial exotoxins possess the ability to intoxicate sensitive cells or tissue. However, early biochemical studies observed that bacterial exotoxins had little intrinsic catalytic activity when assayed in vitro. Subsequent studies showed that many exotoxins were produced as proenzymes, which were activated to express catalytic activity. Because exotoxins intoxicate sensitive cells or tissue, it is apparent that the activation observed in vitro reflects an activation step that also occurs in vivo. Each exotoxin undergoes a specific activation process, which might include proteolysis, disul-fide bond reduction, or association with a nucleotide or eukaryotic accessory protein. Upon activation, the A domain may release from the B domain, or it may undergo a conformational change rendering it catalytic within the holotoxin. The activation process may be sequential. For example, diphtheria toxin is activated by proteolysis cleavage at an arginine that is located between the A and B domains, which is followed by disulfide bond reduction (4). This eliminates the covalent attachment between the A and B domains.

Structural and biochemical studies have demonstrated why exotoxins are secreted as proenzymes. Studies with pertussis toxin and diphtheria toxin have shown that the holotoxins are more resistant to proteolysis than their isolated respective A domain, the S1 subunit and fragment A, respectively. Prior to secretion and interaction with sensitive cells, the proenzyme (i.e., holotoxin) is present in a stable confirmation that is protease resistant as it is transported through the host.

The determination that the eukaryotic protein ARF (ADP-Ribosylation Factor) activated the ADP-ribosyltransferase activity of cholera toxin not only defined the activation mechanism of this bacterial exotoxin, but has also provided insight into eukaryotic cell physiology. ARF has been subsequently shown to play a central role in eukaryotic vesicle fusion (23). The ability of a host cell extract to activate cholera toxin is often used to implicate the presence of ARF. Likewise, elucidation of the mechanisms that pertussis toxin and cholera toxin use to intoxicate eukaryotic cells provided insight into the mechanism of signal transduction modulated by the heterotrimeric G proteins. The inhibition of a ligand-mediated signal transduction pathway by pertussis toxin is often used to implicate a role for G proteins in the signaling pathway.

ENTRY OF BACTERIAL EXOTOXINS INTO CELLS

Bacterial exotoxins utilize several mechanisms to enter host cells (Fig. 2). Receptor-mediated endocytosis is the most common eukaryotic

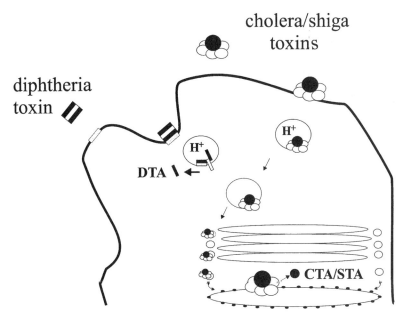

FIGURE 2 Entry of bacterial exotoxins into sensitive eukaryotic cells. Bacterial exotoxins associate with specific receptors on the cell surface or utilize more non-specific interactions to gain access into the eukaryotic cell by receptor-mediated endocytosis. Some exotoxins (diphtheria toxin and anthrax toxin) gain access into the intracellular environment at the level of the early endosome, while other exotoxins (cholera toxin, heat-labile enterotoxin, exotoxin A of *P. aeruginosa*, and shiga toxin) utilize the retrograde transport system of the eukaryotic cell to traffic to the Golgi and/or endoplasmic reticulum, where the translocation process is believed to occur.

process that exotoxins use to gain entry into the interior of the cell. The initial interaction of the exotoxin with the host cell may be through specific binding to a cell-surface receptor or through its association with a nonspecific component of the membrane. For example, diphtheria toxin binds productively to the epidermal growth factor precursor receptor, while exotoxin A of *P. aeruginosa* binds productively to the LDL-like receptor (15, 21). Upon binding to the cell surface, exotoxins may be nicked by host proteases, or proteolysis may occur in the early endosomal compartment. The actual translocation of the exotoxin across the eukaryotic cell membrane may occur within the early endosomal compartment or may occur following its retrograde transport into the Golgi or endoplasmic reticulum.

The A domains of diphtheria toxin and anthrax toxin utilize the intrinsic acidification of the early endosome to establish conditions favorable for the translocation of the A domain across the membrane of the endocytic vesicle and into the cytosol. Resolution of the crystal structure of diphtheria toxin, along with biochemical studies, contributed to development of a model to define the role of the T domain in the translocation of the A domain across the cell membrane. In this model, acidification of the lumen of the endosome results in the protonation of the carboxyl groups of two glutamic acids in the T domain, which are located at the termini of two hydrophobic alpha helices. This protonation initiates the translocation of these helices across the endosomal membrane. Upon translocation into the cytosol, the carboxylic groups are deprotonated, locking the two alpha helices in the membrane, which stabilizes the translocated helices. Biochemical studies have observed the formation of ion permeable channels upon formation of this structure. It is not clear what role the channel plays in the translocation event, but it appears that the A domain must unfold to some extent to effectively translocate across the endosomal cell membrane. Upon translocation across the endosomal membrane, the disulfide bond that covalently links the A and B domains is reduced by intracellular glutathione, allowing the release of the A domain into the cytosol. The translocation process of the A domains of anthrax toxin appears to utilize a similar mechanism as that proposed for diphtheria toxin.

Other exotoxins, including cholera toxin, pertussis toxin, shiga toxin, and exotoxin A of *P. aeruginosa*, appear to utilize the cell's retrograde transport system to traffic into the interior of the cell to be ultimately delivered to the Golgi or endoplasmic reticulum, where the translocation of the A domain into the cytosol occurs (Fig. 3) (29). Although the mechanism for their trafficking remains to be determined, many of these exotoxins possess a KDEL (Lys-Asp-Glu-Leu)-like retention signal on their carboxyl terminus. Eukaryotic cells retain proteins within the endoplasmic reticulum by retrograde transport upon recognition of the KDEL retention signal (22). Studies with chimeric proteins have shown that the introduction of a KDEL sequence onto proteins that normally traffic only through the early endosome pathway results in the retrograde transport of the chimeric protein through the Golgi apparatus and into the endoplasmic reticulum. Together, there appears to be a role for the KDEL-like sequences of exotoxins in intracellular trafficking. Upon delivery to the endoplasmic reticulum, the actual steps involved in the translocation of the A domain of these bacterial exotoxins across the endoplasmic reticulum cell membrane remain to be determined.

CATALYTIC PROPERTIES OF BACTERIAL EXOTOXINS

Bacterial exotoxins catalyze several covalent modifications to eukaryotic target proteins, which alter target protein function. These modifications include ADP-ribosylation, glucosylation, deamidation, endoproteolysis, and depurination. We often inherently consider that these chemical modifications inhibit the intrinsic activity of the targeted proteins, but should note that some modifications actually result in the activation of the intrinsic activity of the modified protein. For example, the ADP-ribosylation of $Gs\alpha$ by cholera toxin results in an activated form of the G protein (24). Likewise, the deamidation of Glutamine63 of RhoA by CNF1 of *E. coli* results in an activated form of the Rho protein (9, 31). We should also note that while most eukaryotic cellular targets of bacterial exotoxins are proteins, there are exceptions. Shiga toxin catalyzes the depurination of ribosomal RNA. Finally, exotoxins exhibit a range of target protein specificity. Diphtheria toxin possesses essentially an absolute specificity for its target protein elongation factor-2, such that within a crude eukaryotic cell extract, EF-2 is the only protein ADP-ribosylated by diphtheria toxin. One factor that contributes to this specificity is that diphtheria toxin ADP-ribosylates a posttranslationally modified histidine within EF-2. In contrast, within a crude eukaryotic cell extract, cholera toxin ADP-ribosylates several eukaryotic proteins, in addition to the $Gs\alpha$ protein. Likewise, CNF1 of *E. coli* deamidates the three members of the Rho family of GTPases (Rho, Rac, and Cdc42).

The recent advances in defining the mechanism of action of the botulinum toxins and tetanus toxin (BT/TT) as zinc proteases deserve comment (30). BT/TT are single proteins, which are organized as AB exotoxins. Although alignment of BT/TT with known zinc proteases revealed little overall primary amino acid homology, BT/TT were observed to possess a zinc protease motif within their A domains. Subsequent experiments showed

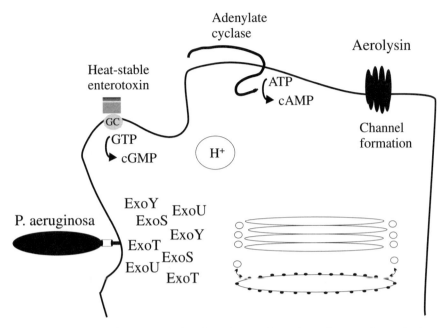

FIGURE 3 Entry of type III translocation of cytotoxins and membrane-acting toxins into sensitive eukaryotic cells. During type III translocation of cytotoxins into sensitive cells, bacteria bind to the surface of eukaryotic cells and utilize the type III secretion apparatus to translocate cytotoxins from the bacterial cytosol into the intracellular compartment of the eukaryotic cell. The membrane-active toxins bind directly to the cell surface and disrupt membrane function via the formation of a pore (aerolysin), stimulation of host guanylate cyclase (GC) through a signal transduction mechanism (heat-stable enterotoxin), or direct insertion of the catalytic domain into the eukaryotic cell cytosol (adenylate cyclase).

that BT/TT catalyzed specific cleavage of proteins involved in vesicle fusion.

Although earlier biochemical studies advanced the understanding of the structure-function properties of bacterial exotoxins, the development of molecular genetics has provided the technology to dissect their mechanism of action. Biochemical characterization indicated that catalytic activities of members of the same family of exotoxins were similar and predicted that there would be a conservation of genetic relatedness. For example, diphtheria toxin and exotoxin A of *P. aeruginosa* were shown to catalyze kinetically identical reactions. However, subsequent sequence analysis of the genes encoding diphtheria toxin and exotoxin A revealed that these toxins shared little overall primary amino acid ho-

mology. This apparent paradox was resolved by the analysis of the three-dimensional structures of exotoxin A and another ADP-ribosyltransferase, the heat-labile enterotoxin of *E. coli*. The three-dimensional structures of these exotoxins showed that the structure of the A domains was essentially superimposable, despite their little overall primary amino acid homology (34). For example, only 3 of the 43 amino acids comprising the active sites of exotoxin A and the heat-labile enterotoxin were conserved. Importantly, one of the conserved amino acids was a glutamic acid that had been shown earlier to be involved in catalysis for both exotoxins. These observations have proved to be a theme that is common among members of the family of bacterial ADP-ribosylating exotoxins, where despite their lack of primary amino acid homology, their three-dimensional structures are conserved and each active site includes a signature glutamic acid. This active site glutamic acid was first identified through a novel photochemical reaction between diphtheria toxin and NAD (2), which was subsequently shown to also occur in pertussis toxin, C3 cytotoxin of *Clostridium,* and exotoxin A of *P. aeruginosa.* Sequence alignment studies have allowed the identification of this active site glutamic acid in all members of the family of bacterial ADP-ribosylating exotoxins. It appears that the large glucosylating exotoxins of the clostridia also possess an analogous active site aspartic acid, which is present in each of the glucosylating exotoxins.

Although it is difficult to define all parameters that determine the absolute toxicity of exotoxins, the ability to bind to cell-surface receptors has a major impact on the sensitivity of a cell line or organism to that exotoxin. While some exotoxins such as exotoxin A possess similar toxicity among various species, other toxins such as diphtheria toxin show a preferred specificity among various species (20). The relative toxicity of diphtheria toxin has been shown to correlate to the presence of species-specific cell-surface receptors to which diphtheria toxin binds (21).

TOXOIDS OF BACTERIAL TOXINS

Many bacterial exotoxins can be chemically modified to toxoids, which are often effective vaccines. Examples of exotoxins that have been chemically toxoided include diphtheria toxin and tetanus toxin. In contrast, the chemical toxoiding of other exotoxins has proven more difficult. Recent studies have shown that bacterial toxins can also be genetically engineered to toxoids (28), presenting the possibility of development of a wider range of vaccine products with greater efficacy.

Bacterial exotoxins have also been utilized as therapeutic agents to correct various disorders; such uses include the treatment of muscle spasms (botulinum neurotoxins) (12), as carriers to deliver heterologous molecules to

elicit an immune response (diphtheria toxin) (32), and as agents to develop cell-specific immunotherapy (exotoxin A) (25). In addition, bacterial toxins have been used as research tools to help define various eukaryotic signaling pathways, such as G-protein mediated signal transduction (pertussis toxin) (14).

TYPE-III-SECRETED CYTOTOXINS

Relative to bacterial exotoxins, the type-III-secreted cytotoxins are a recently recognized family of bacterial toxins. The principal difference between exotoxins and type-III-secreted cytotoxins is that the latter are essentially noncytotoxic when incubated alone with cultured cells or in animal models. This observation was resolved with the determination that the type-III-secreted cytotoxins are delivered into eukaryotic cells directly by the bacterium (5). Thus, using a literal interpretation for the organization of bacterial exotoxins, the type-III-secreted cytotoxins could be considered A domains, and the bacterium is responsible for binding and translocation of the A domain into the interior of the eukaryotic cell. The translocation of the type-III-secreted cytotoxins is mediated by the action of a complex surface apparatus, termed the type-III secretion apparatus. The delivery of the type-III-secreted cytotoxins upon the direct binding of the bacterium to a eukaryotic cell suggests that there are intrinsic differences between the type-III-secreted cytotoxins and bacterial exotoxins in bacterial pathogenesis. While bacterial exotoxins disseminate from (and elicit pathology distal to) the site of infection, type-III-secreted cytotoxins appear to function at the site of infection, since the bacterium delivers the cytotoxin into the eukaryotic cell. Thus, these two classes of toxins would be predicted to function at different stages of the infectious process of a bacterial pathogen.

Type-III-secreted cytotoxins were recognized as being unique due to their secretion from bacteria via a mechanism that appeared to be different than type-I or type-II secretion pathways. Type-III-secreted cytotoxins were not processed at the amino terminus, a property of type-II secreted proteins, and deletion of the carboxyl terminus did not interfere with protein secretion, a property of type-I secreted proteins. Although initially considered to be a novel secretion pathway for cytotoxins produced by members of the genus *Yersinia*, several genera of bacteria have been shown to express type-III-secreted cytotoxins, including *Pseudomonas, Shigella, Bordetella, Salmonella, Escherichia,* and *Yersinia*. The type-III-secreted cytotoxins produced by these pathogens have been implicated in contributing to the bacterium's pathogenesis.

The secretion apparatus appears to be functionally conserved among these genera, since type-III-secreted cytotoxins can be translocated into eukaryotic cells by a type-III secretion apparatus of a heterologous bacterum. The translocation process appears to involve the translation of the mRNA in the cytosol; the nascent form of the protein is then complexed with a chaperon for stability. The protein is released into the type-III secretion apparatus upon binding of the bacterium to the eukaryotic cell membrane. Although the components of the secretion apparatus have been identified through genetic analysis, the mechanism for the delivery of type-III-secreted cytotoxins across the cell membrane remains to be resolved. Recent studies have implicated a role for the mRNA encoding the cytotoxins in the translocation process.

One of the common properties of the type-III secretion pathway is that the apparatus is capable of secreting several type-III-secreted cytotoxins that have different mechanisms of action. For example, the type-III secretion apparatus of *P. aeruginosa* can transport at least four different cytotoxins, including ExoS, ExoT, ExoU, and ExoY (10). One possible hypothesis for the presence of several different type-III-secreted cytotoxins is that each contributes to unique steps in bacterial pathogenesis. This is supported by the observation that a unique subset of the type-III-secreted cytotoxins of *P. aeruginosa* contributes to acute ocular pathogenesis. The type-III-secreted cytotoxins possess various catalytic activities, including an ADP-ribosyltransferase (ExoS of *P. aeruginosa*), a protein phosphatase (YopH of *Yersinia*), a GTPase activating protein for Rho GTPase (ExoS of *P. aeruginosa*), an adenylate cyclase (ExoY of *P. aeruginosa*), and a guanine nucleotide exchange factor (SopE of *Salmonella*).

Type-III-secreted cytotoxins also differ from exotoxins in the large quantity of toxin that the bacterium translocates into the host cell cytosol during a bacterial infection. This observation has made the type-III secretion apparatus a potential drug delivery mechanism or an agent for targeted gene therapy. The observation that bacteria can deliver chimeras of type-III-secreted cytotoxins coupled to heterologous proteins into the cytosol extends the potential for this delivery strategy.

MEMBRANE-ACTING TOXINS

Bacteria utilize several strategies to modulate eukaryotic cell physiology at the eukaryotic cell membrane. The common features of the membrane-acting toxins are that they elicit their effects without entering the interior of the cell, do not possess intrinsic catalytic activity, and lack A domains. There are several families of membrane-acting toxins, including the pore-forming toxins, the heat-stable enterotoxins, and the superantigens.

Pore-Forming Toxins

Pore-forming toxins lack a catalytic A domain and elicit their pathology on eukaryotic cells through their binding to sensitive cells and subsequent pore forming capabilities. There are three classes of pore-forming toxins: the alpha toxin of *S. aureus* (11), the aerolysin family (27), and the hemolysin family (19). The molecular models describing pore formation propose that the monomer initially binds to the plasma membrane. The membrane-bound monomers form an oligomer, which then inserts into the plasma membrane. Pathology associated with pore-forming toxins is due to the generation of ion-conducting channels.

Heat-Stable Enterotoxins

The heat-stable enterotoxins act by binding to the surface of eukaryotic cells, which elevates the concentration of intracellular cGMP. The heat-stable enterotoxin of *E. coli* is a low-molecular-weight peptide and is the prototype toxin for this family of toxins (13). The heat-stable enterotoxin of *E. coli* is produced as a protoxin that is processed to an 18- to 19-amino acid mature peptide. The mature peptide binds to the surface of sensitive cells and stimulates a signal transduction cascade, which results in an increase in the guanylate cyclase activity inside the cell. Thus, the heat-stable enterotoxin acts without entering the interior of the eukaryotic cell.

Superantigens

Both *Streptococcus pyogenes* and *S. aureus* produce soluble proteins of approximately 30 kDa, which act as superantigens (7). Superantigens bind to a component of the major histocompatibility complex of T lymphocytes. The interaction between the superantigen and the T lymphocyte is antigen independent and results in the stimulation of a large subset of lymphocytes.

ADENYLATE CYCLASE CYTOTOXINS

Three cytotoxins have been identified that express adenylate cyclase activity (A domains): CyaA produced by *Bordetella pertussis* (16), edema factor produced by *B. anthracis* (17), and ExoY produced by *P. aeruginosa* (36). These cytotoxins catalyze the conversion of ATP to cAMP. In addition, a eukaryotic protein activates the catalytic activity of each cytotoxin. CyaA and the edema factor are activated by calmodulin, while the eukaryotic protein that activates ExoY has not yet been determined, but does not appear to be calmodulin. The primary difference among the three cytotoxins is the mechanism for the delivery of the A domain into the eukaryotic cell. The edema factor is one of the A domains of anthrax toxin of *B. anthracis*. Edema factor is delivered into the cytosol through the action of the B domain of anthrax toxin (protective antigen) via receptor-mediated endocytosis. ExoY is delivered into the cytosol via the type-III secretion apparatus of *P. aeruginosa*. In contrast, the delivery of the A domain of CyaA into the cell cytosol appears unique, as it is translocated directly across the plasma membrane without the need for a separate intracellular trafficking pathway. The direct translocation of the A domain of CyaA across the plasma membrane is supported by several observations, including the determination that CyaA stimulates an immediate elevation of intracellular cAMP levels when added to sensitive cells. In addition, other studies showed that endosomal movement did not contribute to the delivery process. Intracellular delivery of these A domains into eukaryotic cells results in the elevation of intracellular cAMP concentrations by >1,000-fold, which in turn will interfere with numerous host cell signaling pathways.

CONCLUSIONS

Considerable progress has been made toward defining the mechanism of action of bacterial toxins and their role in bacterial pathogenesis. This progress has paralleled advances in our understanding of the molecular and cell biology of the eukaryotic cell and of the physical properties of proteins. Toxins possess defined structure-function relationships, which are common among toxins from quite diverse origins. It was unexpected that the active sites of the heat-labile enterotoxin of *E. coli* and exotoxin A would be essentially identical, despite little primary amino acid homology. Equally thematic was the determination that members of specific families of toxins possessed conserved catalytic motifs. Also impressive are the varied number of mechanisms that have evolved to deliver the A domains into the interior of the eukaryotic cell. A common, but not exclusive, feature of toxins is the utilization of the host cell trafficking pathways to move the A domain into a region of the host cell that can be used for translocation events.

Future studies on the role of toxins in bacterial pathogenesis will continue to unravel their mechanisms of action. Resolution of their molecular properties will allow an accurate interpretation of their role in the pathogenesis of each bacterial pathogen. Without this information, studies concerning these toxins in complex animal models will be descriptive and often speculative. Equally important to the study of the mode of action of these toxins are studies to provide a more complete understanding of transcriptional regulation of the genes that encode these toxins by environmental factors, in the environment and within the host. Defining the regulation of the expression of bacterial toxins may ultimately lead to an understanding of the regulation of toxin production during a bacterial infection. This could be an informative avenue of research to develop novel strategies to control the expression of toxin expression during bacterial infections, because many toxin-producing bacteria have acquired multiple drug resistance. There are two long-term goals for the study of bacterial toxins. The first is to develop strategies for the production of toxoids to prevent the clinical manifestation of disease that is elicited by a toxin-producing bacterium. The second is to utilize toxins or components of toxins for other scientific applications, including targeted therapies and as carrier molecules to deliver heterologous agents to the cell surface or into specific locations within the eukaryotic cell.

As we enter the era of genomics, we must anticipate future directions for medical microbiology. This will undoubtedly address the molecular aspects of the host, as well as the bacterial pathogen that contributes to clinical disease. Such research may ultimately define the factors that render an individual more susceptible to specific bacterial pathogens and help to develop new strategies to combat these pathogens.

REFERENCES

1. **Burnette, W. N.** 1994. AB5 ADP-ribosylating toxins: comparative anatomy and physiology. *Structure* **2**:151–158.

2. **Carroll, S. F., J. A. McCloskey, P. F. Crain, N. J. Oppenheimer, T. M. Marschner, and R. J. Collier.** 1985. Photoaffinity labeling of diphtheria toxin fragment A with NAD: structure of the photoproduct at position 148. *Proc. Natl. Acad. Sci. USA* **82**:7237–7241.

3. **Choe, S., M. J. Bennett, G. Fujii, P. M. Curmi, K. A. Kantardjieff, R. J. Collier, and D. Eisenberg.** 1992. The crystal structure of diphtheria toxin. *Nature* **357**:216–222.

4. **Collier, R. J.** 1975. Diphtheria toxin: mode of action and structure. *Bacteriol. Rev.* **39**:54–85.

5. **Cornelis, G. R., and H. Wolf-Watz.** 1997. The *Yersinia* Yop virulon: a bacterial system for subverting eukaryotic cells. *Mol. Microbiol.* **23**: 861–867.

6. **Duong, F., J. Eichler, A. Price, M. R. Leonard, and W. Wickner.** 1997. Biogenesis of the gram-negative bacterial envelope. *Cell* **91**:567–573.

7. **Fields, B. A., E. L. Malchiodi, H. Li, X. Ysern, C. V. Stauffacher, P. M. Schlievert, K. Karjalainen, and R. A. Mariuzza.** 1996. Crystal structure of a T-cell receptor beta-chain complexed with a superantigen [see comments]. *Nature* **384**:188–192.

8. **Finkelstein, A.** 1990. Channels formed in phospholipid bilayer membranes by diphtheria, tetanus, botulinum and anthrax toxin. *J. Physiol.* **84**: 188–190.

9. **Fiorentini, C., A. Fabbri, G. Flatau, G. Donelli, P. Matarrese, E. Lemichez, L. Falzano, and P. Boquet.** 1997. Escherichia coli cytotoxic necrotizing factor 1 (CNF1), a toxin that activates the Rho GTPase. *J. Biol. Chem.* **272**:19532–19537.

10. **Frank, D. W.** 1997. The exoenzyme S regulon of *Pseudomonas aeruginosa. Mol. Microbiol.* **26**:621–629.

11. **Gouaux, E.** 1998. Alpha-Hemolysin from *Staphylococcus aureus*: an archetype of beta-barrel, channel-forming toxins. *J. Struct. Biol.* **121**:110–122.

12. **Grandas, F.** 1995. Clinical application of botulinum toxin. *Neurologia* **10**:224–233.

13. **Guerrant, R. L., J. M. Hughes, B. Chang, D. C. Robertson, and F. Murad.** 1980. Activation of intestinal guanylate cyclase by heat-stable enterotoxin of Escherichia coli: studies of tissue specificity, potential receptors, and intermediates. *J. Infect. Dis.* **142**:220–228.

14. **Hazeki, K., T. Seya, O. Hazeki, and M. Ui.** 1994. Involvement of the pertussis toxin-sensitive GTP-binding protein in regulation of expression and function of granulocyte complement receptor type 1 and type 3. *Mol. Immunol.* **31**:511–518.

15. Kounnas, M. Z., R. E. Morris, M. R. Thompson, D. J. FitzGerald, D. K. Strickland, and C. B. Saelinger. 1992. The alpha 2-macroglobulin receptor/low density lipoprotein receptor-related protein binds and internalizes *Pseudomonas* exotoxin A. *J. Biol. Chem.* **267:**12420–12423.

16. Ladant, D., and A. Ullmann. 1999. Adenylate cyclase: a toxin with multiple talents. *Trends Microbiol.* **7:**172–176.

17. Leppla, S. H. 1984. Bacillus anthracis calmodulin-dependent adenylate cyclase: chemical and enzymatic properties and interactions with eucaryotic cells. *Adv. Cyclic Nucl. Prot. Phosphor. Res.* **17:**189–198.

18. Leppla, S. H. 1988. Production and purification of anthrax toxin. *Methods Enzymol.* **165:**103–116.

19. Menestrina, G., C. Moser, S. Pellet, and R. Welch. 1994. Pore-formation by *Escherichia coli* hemolysin (HlyA) and other members of the RTX toxins family. *Toxicology* **87:**249–267.

20. Middlebrook, J. L., and R. B. Dorland. 1977. Response of cultured mammalian cells to the exotoxins of *Pseudomonas aeruginosa* and *Corynebacterium diphtheriae*: differential cytotoxicity. *Can. J. Microbiol.* **23:**183–189.

21. Naglich, J. G., J. E. Metherall, D. W. Russell, and L. Eidels. 1992. Expression cloning of a diphtheria toxin receptor: identity with a heparin-binding EGF-like growth factor precursor. *Cell* **69:**1051–1061.

22. Nickel, W., B. Brugger, and F. T. Wieland. 1998. Protein and lipid sorting between the endoplasmic reticulum and the Golgi complex. *Semin. Cell Dev. Biol.* **9:**493–501.

23. Orcl, L., D. J. Palmer, M. Amherdt, and J. E. Rothman. 1993. Coated vesicle assembly in the Golgi requires only coatomer and ARF proteins from the cytosol. *Nature* **364:**732–734.

24. Owens, J. R., L. T. Frame, M. Ui, and D. M. Cooper. 1985. Cholera toxin ADP-ribosylates the islet-activating protein substrate in adipocyte membranes and alters its function. *J. Biol. Chem.* **260:**15946–15952.

25. Pai, L. H., and I. Pastan. 1998. Clinical trials with *Pseudomonas* exotoxin immunotoxins. *Curr. Top. Microbiol. Immunol.* **234:**83–96.

26. Pappenheimer, A. M., Jr. 1993. The story of a toxic protein, 1888-1992. *Protein Sci.* **2:**292–298.

27. Parker, M. W., J. T. Buckley, J. P. Postma, A. D. Tucker, K. Leonard, F. Pattus, and D. Tsernoglou. 1994. Structure of the *Aeromonas* toxin proaerolysin in its water-soluble and membrane-channel states. *Nature* **367:**292–295.

28. Pizza, M., A. Covacci, A. Bartoloni, M. Perugini, L. Nencioni, M. T. De Magistris, L. Villa, D. Nucci, R. Manetti, M. Bugnoli, et al. 1989. Mutants of pertussis toxin suitable for vaccine development. *Science* **246:**497–500.

29. Sandvig, K., and B. van Deurs. 1996. Endocytosis, intracellular transport, and cytotoxic action of Shiga toxin and ricin. *Physiol. Rev.* **76:**949–966.

30. Schiavo, G., F. Benfenati, B. Poulain, O. Rossetto, P. Polverino de Laureto, B. R. DasGupta, and C. Montecucco. 1992. Tetanus and botulinum-B neurotoxins block neurotransmitter release by proteolytic cleavage of synaptobrevin [see comments]. *Nature* **359:**832–835.

31. Schmidt, G., P. Sehr, J. Wilm, J. Selzer, M. Mann, and K. Aktories. 1997. Gln 63 of Rho is deamidated by *Escherichia coli* cytotoxic necrotizing factor-1. *Nature* **387:**725–729.

32. Seppala, I., H. Sarvas, O. Makela, P. Mattila, J. Eskola, and H. Kayhty. 1988. Human antibody responses to two conjugate vaccines of *Haemophilus influenzae* type B saccharides and diphtheria toxin. *Scand. J. Immunol.* **28:**471–479.

33. Sixma, T. K., K. H. Kalk, B. A. van Zanten, Z. Dauter, J. Kingma, B. Witholt, and W. G. Hol. 1993. Refined structure of *Escherichia coli* heat-labile enterotoxin, a close relative of cholera toxin. *J. Mol. Biol.* **230:**890–918.

34. Sixma, T. K., S. E. Pronk, K. H. Kalk, E. S. Wartna, B. A. van Zanten, B. Witholt, and W. G. Hol. 1991. Crystal structure of a cholera toxin-related heat-labile enterotoxin from *E. coli* [see comments]. *Nature* **351:**371–377.

35. White, A., X. Ding, J. C. vanderSpek, J. R. Murphy, and D. Ringe. 1998. Structure of the metal-ion-activated diphtheria toxin repressor/tox operator complex. *Nature* **394:**502–506.

36. Yahr, T. L., A. J. Vallis, M. K. Hancock, J. T. Barbieri, and D. W. Frank. 1998. ExoY, an adenylate cyclase secreted by the *Pseudomonas aeruginosa* type III system. *Proc. Natl. Acad. Sci. USA* **95:**13899–13904.

EXPLOITATION OF MAMMALIAN HOST CELL FUNCTION BY *SHIGELLA* SPP.

Larean D. Brandon and Marcia B. Goldberg

12

Shigella spp. are the causative agents of shigellosis (a form of bacillary dysentery) and lead to an estimated 1.1 million deaths per annum, most notably in developing world countries where nutrition and hygiene are suboptimal (27). A highly virulent serotype, *Shigella dysenteriae* serotype 1, is responsible for many of the epidemic cases, while less virulent strains, *S. flexneri, S. sonnei,* and *S. boydii* are responsible for endemic disease. Shigellosis is marked by symptoms of abdominal pain, fever, and a purulent diarrhea. In more severe cases, the disease is characterized by bloody stool and can lead to complications including sepsis subsequent to the ulceration and perforation of the bowel.

The natural habitat of shigellae is the gut of higher primates, most notably humans. Shigellae reach the gastrointestinal tract via the oral route by ingestion of contaminated food or water. From the intestinal lumen, the organisms become associated with the gut wall. The pathological process is primarily inflammatory and is localized to the mucosa of the terminal ileum (distal one-third of the small

intestine) and the colon (large intestine). Pathogenesis involves the invasion and destruction of colonic epithelial cells and the induction of an inflammatory response in the colonic mucosa and adjacent tissues. The incubation period is variable, but has been known to be as short as 24 h (14); therefore, *Shigella* spp., like many enteric pathogens, have developed an efficient mechanism for exploiting host cells to ensure their survival in nature.

The pathogenicity of *Shigella* depends upon a complex series of molecular events that are mediated by factors expressed on the large 220-kilobase virulence plasmid (53). Shigellae exploit M cells (membranous epithelial cells that sample antigens from the intestinal lumen) to gain access to the basolateral face of colonic epithelial cells, where they are taken up by bacterial-induced phagocytosis. They rapidly escape the phagocytic vacuole into the cell cytoplasm, where they divide and assemble actin filaments on one pole, which propels them through the cell cytoplasm. Actin assembly also generates sufficient force for them to push out against the cellular membrane such that they are taken up by adjacent cells. Upon being taken up in this manner, the bacteria are found enclosed within double membranes; they lyse these membranes and are thereby released into the cytoplasm of the adjacent cell,

Larean D. Brandon and Marcia B. Goldberg Infectious Disease Division, Massachusetts General Hospital, 55 Fruit St., Boston, MA 02114.

Virulence Mechanisms of Bacterial Pathogens, 3rd ed., Edited by K. A. Brogden et al.
©2000 ASM Press, Washington, D.C.

where the process of actin assembly and spread continues (Fig. 1). This review describes the detailed molecular events involved in these interactions of shigellae with the host cell.

EARLY STEPS: ADHERENCE TO AND INVASION OF COLONIC EPITHELIAL CELLS

Shigellae invade human colonic epithelial cells. The initial step in the pathogenesis of shigellosis most likely involves specific interactions of the organism, with the mucin associated with the apical surface of colonic epithelial cells. Rajkumar et al. (45) demonstrate in solid phase-binding assays using mu-

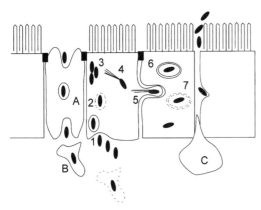

FIGURE 1 Simplified schematic of the pathogenesis of shigellosis. Shigellae (black ovals) are initially taken up by M cells (cell A) and transcytosed to the basolateral face of the epithelial cells, where they are engulfed by macrophages (cell B). The bacteria release factors that promote apoptosis of the macrophages (thereby releasing the bacteria) and that promote the release of interleukin 1-β. This release promotes an inflammatory response that induces the transmigration of PMNs (cell C) between the epithelial cells to the apical face of the epithelial cell layer. The inflammatory response destroys the integrity of the tight junctions (black rectangles), thus enhancing uptake of the bacteria. Once the bacteria reach the basolateral face of the epithelial cells they are endocytosed and rapidly escape the endocytic vesicle (steps 1 and 2). They are subsequently released into the host cell cytoplasm where they divide (3), recruit actin filaments (4), and invade adjacent epithelial cells via actin-mediated motility (5). In the new host cell, the bacteria are found associated with double-membrane vesicles, from which they rapidly escape (6 and 7).

cin isolated from various sources that the bacteria bind to the carbohydrate moieties associated with the mucin layer. Furthermore, the organisms only bind to mucin isolated from human colonic mucosa. The bacteria bind in a concentration-dependent manner that does not involve weak hydrophobic interactions since disruptive agents such as p-nitrophenol do not block binding. Periodate-mediated removal of the carbohydrates from the glycoprotein component of mucin significantly reduces the binding capabilities of *Shigella*; however, pretreatment of the bacterium with monosaccharides does not reduce binding, which suggests that the nature of binding may be complex.

In vitro studies demonstrate that shigellae are unable to enter polarized colonic epithelial cells at the apical pole, but rather can enter only at the basolateral surfaces (37). Access to the basolateral surfaces likely occurs by a complex series of events following uptake and passage of shigellae through M cells. Shigellae that enter M cells are transcytosed within the endocytic vacuole to the basolateral face of the cells, where they are rapidly phagocytosed by macrophages (41, 66). *Shigella* secretes a 62-kDa protein (IpaB) that binds to and activates caspase 1, thus triggering cell death (apoptosis) of macrophages and facilitating its release into the surrounding tissue (8, 35, 67). Furthermore, IpaB stimulates the release of interleukin-1-β, which leads to an acute inflammatory response in the intestinal mucosa (51). The proinflammatory cytokine IL-8, which is produced largely by epithelial cells, induces transmigration of polymorphonuclear leukocytes (PMNs) between epithelial cells into the intestinal lumen. The migration of the PMNs causes disruption of intercellular junctions sufficient to allow lumenal *Shigella* to penetrate the mucosa and gain access to the basolateral membranes of the epithelial cells (52, 66) (Fig. 1).

Entry of *Shigella* at the basolateral face of colonic epithelial cells involves a complex set of molecular reactions. The bacteria secrete factors that influence cellular signaling pathways in the host cell, while the host cell expresses factors that enhance secretion of these bacterial-derived factors (bacteria-host cell cross talk). *Shigella* secretes invasins upon contact with the host cell, which trigger a response that leads to endocytosis of the bacterium into a vacuole. Unlike some *Yersinia* spp. and *Listeria monocytogenes*, whose uptake involves direct interaction between the bacteria and the host cell via receptor-mediated endocytosis (25), *Shigella* does not form tight associations with the outer surface of the cell or with the endocytotic vacuole following uptake (1, 9). Instead, contact induces the secretion of invasins that promote rearrangements of the host cytoskeleton at the site of contact. The cell membranes form "membrane ruffles" that engulf the bacterium (1, 9). Uptake is mediated by secretion of the bacterial invasins, IpaA and the IpaB/C complex. These proteins are expressed by genes located on the *Shigella* virulence plasmid (32, 61). There are several candidates for factors expressed on the basolateral surface of host cells that might induce secretion; these possibly include the $\alpha_5\beta_1$ integrin (65). While the expression of the $\alpha_5\beta_1$ integrin on CHO (Chinese hamster ovary) cells mediates the binding of the Ipa proteins (65), the presence of this receptor on the basolateral surface of intestinal epithelial cells has not been demonstrated. Furthermore, Congo red, a dye whose molecular structure may resemble cell-surface adhesins, has been shown to induce the secretion of invasins (5). However, the function and nature of these factors remain unclear. Chapter 14 of this volume details the molecular mechanism of secretion of the bacterial invasins.

The Ipa proteins induce uptake by activating host cell signaling pathways that are involved in cell growth. Recent studies demonstrate that IpaA and IpaC are directly involved in promoting these changes in the cell cytoskeleton (11, 61). It is proposed that the IpaB–IpaC complex forms a pore in the membrane; thus, a channel is provided through which IpaA passes into the host cell cytoplasm. IpaB shares homology with a pore-forming hemolysin of *Escherichia coli* (10) and with the EspD protein of enteropathogenic *E. coli*, which is able to insert into HeLa cell membranes at the location of bacterial contact (63). However, EspD is not translocated into the cell (63). IpaC has been shown to disrupt and form stable interactions with lipid bilayers (11). It is unclear whether IpaC is involved in channel formation for the translocation of *Shigella*-derived factors into the host cells or whether IpaC is inserted into the membranes to directly elicit its effect on the host cell.

Upon bacterial contact with the host cell, rearrangements of the actin cytoskeleton occur around the site of contact. Several cytoskeletal proteins are recruited to the site of contact, including vinculin, which co-immunoprecipitates with IpaA, T-plastin, cortactin-P, ezrin, the tyrosine kinase pp60[c-src], and the small GTPase Rho (16, 57, 61). The role that Rho plays in actin cytoskeleton rearrangement at the entry site is unclear. Earlier studies demonstrate that cytoskeletal rearrangements are induced in a Rho-dependent fashion (2). However, recent studies suggest that cytoskeletal rearrangements are directly mediated by a small GTPase (&Rgr) and are indirectly mediated by IpaC, which activates Cdc42. This in turn activates Rac (31, 36, 62). During bacterial uptake, Src is activated and cortactin, an actin-associated protein and a substrate for Src kinase activity, is phosphorylated. pp60[c-src] is thought to be involved in phosphorylation of cortactin (12). Further indirect support for the possible role of Src and cortactin in uptake is the finding that overexpression of Src in transfected cells leads to membrane ruffling and uptake of avirulent *Shigella* mutants (16). IpaA interacts with vinculin, and IpaA mutants are impaired in the recruitment of α-actinin (61). However, IpaA does not appear to play a role in vinculin recruitment, suggesting that the role of IpaA may be in stabilizing vinculin after its recruitment (61).

Once *Shigella* is internalized, it rapidly escapes the vacuole, using a process that appears to involve IpaB (24) and IpaC (11). The molecular mechanism by which these proteins mediate the escape of *Shigella* from vacuoles is unknown. It seems likely that IpaB and IpaC may disrupt the lipid bilayer of the vacuolar membrane because they are able to mediate lysis of liposomes in a pH-dependent fashion (11, 50).

INTERMEDIATE STEPS: IcsA-MEDIATED INTRACELLULAR MOTILITY AND INTERCELLULAR SPREAD

Once *Shigella* gains entry into host cells, it uses components of the host cytoskeleton as a means to move within the host cytoplasm and invade adjacent cells. Movement within the cytoplasm is mediated by IcsA (VirG), a 120-kDa outer membrane protein that is sufficient and necessary for this process (7, 20; Magdalena and Goldberg, submitted). Inert particles coated with IcsA are able to assemble actin tails in cytoplasmic extracts (Magdalena and Goldberg, submitted). Immediately after *Shigella* is released from the phagocytic vacuole, actin is recruited to *Shigella* by IcsA (7), and polar assembly of actin tails occurs shortly thereafter (21, 43). Therefore, it is likely either that IcsA is expressed on the surface of the bacterium prior to the organism's release into the host cytoplasm or that surface localization of IcsA occurs shortly thereafter.

IcsA, like most of the proteins involved in the pathogenesis of *Shigella*, is expressed on the virulence plasmid (7, 30). IcsA is unusual in that it is localized exclusively at the old pole (i.e., relative to binary fission) of the bacillus, the site where actin is assembled to form actin tails for intracellular motility (19). Therefore, IcsA belongs to a growing number of bacterial proteins that are directed to specific regions of bacteria to perform specific functions.

IcsA is a 1,102-amino acid polypeptide consisting of an amino-terminal α domain that is exposed on the bacterial surface and a 344-carboxyl-terminal β domain that anchors it in the outer membrane (19). It is localized to the old pole following direct targeting of the protein to that site during its secretion (58) (Fig. 2). Following insertion into the outer membrane, IcsA is slowly cleaved at the Arg-758—Arg-759 bond at the junction of the α and β domains. During exponential growth, approximately 20% of the IcsA is cleaved and the α domain is released from the bacterial surface as a 95-kDa polypeptide, and 80% remains located within the outer membrane as the full-length form of IcsA (21). Cleavage of IcsA is mediated by IcsP (SopA), an outer membrane serine protease (15, 55). *icsP* mutant strains show a marked reduction in the cleavage of IcsA in that less than 0.2% of the protein is released into the media (55). We have proposed that, following its insertion into the outer membrane at the pole, IcsA diffuses laterally in the outer membrane and that IcsP aids in the maintenance of a polar cap of IcsA by slowly processing surface IcsA that has diffused laterally from the pole (58). Surprisingly, our data indicate that IcsP also processes IcsA at the pole of the bacteria. We have proposed that the rate at which IcsA is directed to the pole in dividing cells is faster than the rate at which IcsP cleaves it. In addition, the rate of lateral diffusion of IcsA is approximately equivalent to the rate of IcsP-mediated cleavage, thereby maintaining the tight polar cap of IcsA (58).

A second factor that has been shown to contribute to the maintenance of IcsA at the pole is the composition of the lipopolysaccharide (LPS). Truncation of the LPS (i.e., rough mutants) leads to delocalization of IcsA from the pole, which suggests that interactions between IcsA and the LPS aid in maintaining IcsA at the pole (47, 48). We have proposed that this occurs by one or both of the following mechanisms: (i) noncovalent interactions between LPS molecules and IcsA, and/or (ii) improper insertion of IcsP into the outer membrane, causing a decrease in its ability to cleave IcsA. This would result in a pseudo-*icsP* phenotype. The more truncated the LPS, the more delocalized is the IcsA. Thus, in strains lacking the entire outer core and O-antigen of the LPS (e.g., *galU*, UDP-glucose

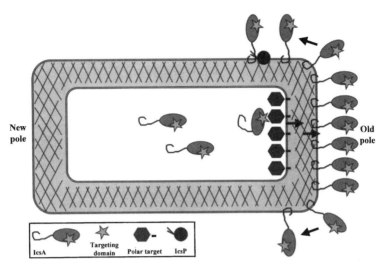

FIGURE 2 Proposed mechanism of targeting of IcsA to the bacterial old pole. Following translation, the targeting domain of IcsA (star) recognizes a polar target (hexagon) associated with the cytoplasmic membrane at the old pole. IcsA is then translocated across the cytoplasmic membrane at the pole and its carboxy-terminal β domain is inserted into the outer membrane, such that the amino-terminal α domain is exposed on the bacterial surface. Following insertion into the outer membrane, IcsA diffuses laterally in the membrane. IcsP cleaves IcsA from all surfaces of the bacteria at a rate that is slower than the rate at which IcsA is inserted at the pole, such that a polar cap of IcsA is maintained.

pyrophosphorylase mutant), IcsA is present in an almost circumferential fashion on the surface of *Shigella*. While in the *rfe* background (*N*-acetylglucosamine transferase-deficient mutant), in which only the O antigen is absent, IcsA is present both at the pole and to some extent along the lateral sides of the bacillus (48).

While it is well established that IcsA is targeted directly to the pole of the *Shigella* bacillus, the physiological significance of unipolar localization appears to be complex. It seems logical that unipolar distribution of IcsA would be important in the intra- and intercellular spread of *Shigella*; however, the tools to rigorously test this are not yet available. The unipolar localization of the actin tail as mediated by the localization of IcsA should permit the most directional movement possible. One might imagine that the recruitment of actin filaments at the pole of *Shigella* is anal-

ogous to the distribution of flagella on gram-negative bacteria, which induces motility in response to chemoattractants and repellents in the external milieu (56). For example, *E. coli* exhibits a "tumbly" behavior when the immediate environment is rich in nutrients and low in metabolic wastes; this is correlated with a peritrichous distribution of the flagella. However, when the environment is not ideal, the flagella form bundles at the pole of the bacteria, which is correlated with smooth, directional swimming to a more ideal environment.

The importance of the tight polar cap of IcsA in actin tail assembly and in spread from one cell into another has not been directly tested. However, several lines of evidence support the idea that the presence of IcsA at the pole of *Shigella* plays an important role in this process. For example, asymmetric particles that are coated with IcsA are able to form actin tails in cytoplasmic extracts, while spherical beads that are coated with IcsA are not (Magdalena and Goldberg, unpublished data). Also, mutations that affect the surface processing of IcsA also affect the localization of this protein and the intracellular movements of *Shigella*.

The SSRRASS phosphorylation consensus sequence in IcsA contains the site for processing by the IcsP (SopA) outer membrane protease. Site-directed mutagenesis of one or both of the arginines to asparagines in this domain leads to loss of IcsA cleavage and alters the surface distribution of IcsA. d'Hauteville et al. (13) demonstrated that serotype 5 *S. flexneri* strains that express the R759D or R758D/R759D mutation of IcsA spread more rapidly through HeLa cells and produce the membrane ruffling associated with intercellular dissemination. Fukuda et al. demonstrated that a serotype 2a *S. flexneri* strain that expresses the R759D mutant derivative of IcsA on an integrated single copy plasmid behaves as the wild-type strain (17). Egile et al. (15) and Shere et al. (55) have independently isolated the outer membrane protease (IcsP) that cleaves IcsA at the arginine residues. Egile et al. demonstrated that a targeted disruption of *icsP* in a serotype 5 strain leads to shorter protrusions from the surface of infected epithelial cells and to the formation of smaller plaques on Caco-2 monolayers (15). However, Shere et al. demonstrated that a disruption of *icsP* in a serotype 2a strain accelerates actin-based motility, but is similar to wild-type *Shigella* strains in terms of its ability to form protrusions from the surface of infected mammalian cell monolayers and to form plaques on epithelial monolayers (55). The reasons for these observed differences are unclear; however, in all cases the authors have demonstrated that the cleavage of IcsA is drastically reduced or abolished and that IcsA is found primarily at the pole of *Shigella* with some along the sides of the bacilli.

Lastly, mutations in *galU* and *rfe* genes both affect wild-type expression of bacterial LPS, and, in both backgrounds, IcsA is expressed and the strains are able to invade HeLa cell monolayers (48). However, these strains are highly attenuated with respect to their ability to generate membrane protrusions, "fireworks," and to form plaques on a confluent monolayer of HeLa cells—both manifestations of intercellular dissemination. In the *galU* background, IcsA is expressed in a circumferential fashion on the surface of *Shigella*. The strain is incapable of forming plaques on confluent eukaryotic cell monolayers. In the *rfe* background, IcsA is expressed to a limited extent at the pole of the bacillus with some along the sides; this strain forms small plaques (compared to wild-type strains) on monolayers. One important caveat to these data is that the *galU* mutation not only abolishes the polar localization of IcsA, but also leads to rapid death of the bacterium in the host cell cytoplasm. (The intracellular survival of the *rfe* mutant has not been tested.)

Taken together, these studies suggest that the polar distribution of IcsA is important for intercellular dissemination. The common theme of these studies is that the phenotypes that are manifestations of intra- and intercellular spread can be correlated with the polar localization of IcsA. To rigorously test this hypothesis, it will be necessary to isolate mutants that adversely affect the unipolar targeting of IcsA and lead to the circumferential distribution of IcsA without altering the LPS.

The complexity of unraveling the role of unipolar localization of IcsA in intracellular motility and intercellular dissemination of *Shigella* is illustrated by a recombinant strain of *E. coli* that expresses IcsA in a circumferential pattern. Studies utilizing this organism have demonstrated the ability of this strain to assemble actin tails in a polar fashion (20). This suggests that in this system, the asymmetry inherent in the oblong shape of the bacillus may be sufficient to enable polar assembly of the actin tail, perhaps by generating regional differences in the density of IcsA over the bacterial surface.

The factors that are responsible for targeting IcsA to the pole of *Shigella* are unknown. Sandlin and Maurelli (49) have recently demonstrated that IcsA is found at the pole in an *S. flexneri* strain cured of the virulence plasmid and in an avirulent strain of *E. coli* that has an intact LPS. As described above, we have shown that IcsA expressed by a mutant strain of *E. coli* K-12 presenting a truncated LPS is distributed in a circumferential fashion on the bacterial surface (20). This again suggests that the presence of a full-length LPS is important to maintain the tight polar cap of IcsA. More importantly, the work by Sandlin and Maurelli indicates that the unipolar targeting of IcsA is either inherent to IcsA or mediated by a chromosomally encoded factor common to both *Shigella* spp. and nonpathogenic *E. coli*.

HOST FACTORS INVOLVED IN *SHIGELLA* ACTIN ASSEMBLY

IcsA is the only *Shigella* protein required for *Shigella* actin assembly. Recent work suggests that IcsA assembles actin by subverting a host cell signal transduction pathway that normally mediates the formation of filopodia (34) (Fig. 3). In this pathway, the GTP-binding protein Cdc42 and phosphatidylinositol (4,5) biphosphate (PI(4,5)P_2) bind to the amino terminus of N-WASP, a member of the WASP (Wiskott-Aldrich syndrome protein) family of proteins (46). This interaction unmasks the carboxy terminus of N-WASP, enabling it to bind to the Arp2/3 complex and trigger a marked increase in actin polymerization (46).

Shigella-induced actin assembly has been shown to be dependent on N-WASP. A dominant negative N-WASP inhibits *Shigella* actin tail assembly in intact cells, and depletion of N-WASP inhibits *Shigella* actin tail assembly in *Xenopus* cytoplasmic extracts (60). Furthermore, N-WASP localizes to the growing end of the actin tail, and IcsA is capable of interacting with N-WASP in vitro (60). Thus, it seems likely that IcsA unmasks the carboxy terminus of N-WASP in an analogous manner to Cdc42 and PI(4,5)P_2 and thereby interdigitates into the host cell signaling pathway. The verprolin domain of N-WASP, which lies between the domain to which Cdc42 and PI(4,5)P_2 bind and the domain to which the Arp2/3 complex binds, is required for IcsA binding (60). How the barbed ends of the growing filaments are kept uncapped and how actin filaments generated in this process are bundled into a growing actin tail are unknown. Several other actin-associated proteins have been shown to colocalize with the *Shigella* actin tail, but have not been shown to have a functional role in actin-tail assembly. The requirement of vinculin for this process is controversial (18, 28).

SECRETION OF IcsA

Genetic and biochemical studies demonstrate that IcsA is an autotransported protein and therefore can be included in the family of proteins that use the Type IV secretion pathway (23, 59). IcsA bears remarkable sequence similarity to the best-studied member of this family, the IgA protease of *Neisseria gonorrhoeae* (42). Regions of similarity include an extensive (52-amino-acid long) Sec-like signal sequence and a β domain that is thought to form an outer membrane channel (β barrel pore) through which the α domain is translocated to the surface of the bacterium (59). Like many autotransporters, IcsA also contains a serine protease recognition site at which cleavage occurs. Once the α domain of IcsA reaches the external milieu, as discussed above, cleavage of it from the bacterial surface occurs slowly, such that during exponential growth only about 20% of the protein is found in the culture supernatant. Thus, the dual function of the β domain, as both a channel and a membrane anchor, distinguishes IcsA from other autotransported proteins that have been characterized. IcsA differs from those proteins in that following the passage of the α domain through the channel formed by the β domain, the α domain is efficiently cleaved and released into the external milieu on all molecules (23). Less commonly, the β domain serves as an outer membrane anchor for the α domain with no cleavage occurring (40).

Although IcsA contains a Sec-like signal sequence, there is no direct evidence that it uses

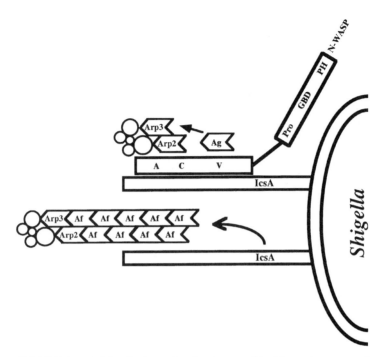

FIGURE 3 Model of IcsA-mediated actin assembly. The amino-terminal glycine-rich repeat domain of IcsA (shaded bar) binds the verprolin (V) domain of N-WASP (open bar). This binding leads to unmasking of carboxy-terminal domains of N-WASP, thereby allowing binding of the Arp2/3 complex to the N-WASP acidic domain (A) and G-actin (Ag) to the N-WASP verprolin (V) domain. In the presence of the Arp2/3 complex actin is nucleated, with the Arp2/3 complex capping the pointed end of the actin nucleus. G-actin (Ag) is rapidly polymerized onto the uncapped barbed end. Subsequently, N-WASP and the growing actin filament are released from IcsA and crosslinked into the actin tail. Af, filamentous actin.

the Sec machinery for secretion across the inner membrane, but it seems likely that leader peptidase is involved in cleavage of the signal peptide from the mature protein. Recent studies have shown that F pilin, which has an extensive Sec-like signal sequence, is secreted across the inner membrane in a Sec-independent fashion (29). Furthermore, the nature of transit of IcsA and other autotransported proteins across the periplasm is unclear (23). Recombinant forms of IcsA or IgA protease, where the β domain is fused to a surrogate protein, have periplasmic intermediates (26, 59). However, since these studies were conducted using recombinant proteins, it is uncertain whether the secretion pathway is the

same as that used by the native proteins. Studies in our laboratory indicate that the native form of IcsA does not have a soluble periplasmic intermediate (Brandon and Goldberg, submitted). This suggests that if IcsA is present in the periplasm, its presence is transient and/ or its secretion across the inner and outer bacterial membranes is temporally coupled.

LATE STEPS: INVASION OF ADJACENT EPITHELIAL CELLS

Within the host cell cytoplasm, *Shigella* uses the force generated by actin polymerization to reach the inner face of the host cell membrane, where the process of intercellular dissemination begins. In the initial stage of this process, *Shigella* induces the formation of finger-like protrusions extending from the cell surface (43). Eighty-five percent of the protrusions are formed at focal adhesion sites (43), suggesting that these protein complexes are involved in targeting the bacteria and forming the protrusions. The new pole of the bacteria is seen at the tips of these extensions, with the actin tail extending from the old pole of the bacteria toward the body of the host cell (19). Adjacent cells take up the bacterium-containing tips of the protrusions, thereby engulfing the bacteria within a double-membrane vesicle. Host cell cadherin expression is required for generation of protrusions and their engulfment by adjacent cells (54). IcsB, a 57-kDa protein whose gene is located 1.5 kilobases upstream of the *mxi-spa* cluster (4), is required for the escape of the bacteria from the double-membrane vesicles, a process that occurs quickly. *icsB* mutants are unable to lyse the double membranes of these vesicles; instead, the mutants and the protein are found trapped within them (4). Because IcsB is located in the bacterial periplasm, it seems likely that it functions prior to the molecule that lyses the membranes.

EXPRESSION OF *SHIGELLA*-DERIVED VIRULENCE FACTORS

Genes on the large virulence plasmid encode most of the factors responsible for the virulence of *Shigella*. However, expression of these genes is modulated by chromosomal gene products whose primary roles are likely to include sensing optimal environmental conditions for the expression of virulence genes (Fig. 4). The chromosomal factors include, but are not limited to, the CpxA/CpxR and EnvZ/OmpR sensor kinase/effector proteins and the H-NS DNA binding protein.

VirF (a protein expressed on the virulence plasmid) is the global regulator of the virulence genes on the large 220-kb plasmid (3). Most of the genes known to be regulated by VirF have been discussed in this chapter or in chapter 14 in this edition; however, it is likely that VirF also modulates the expression of

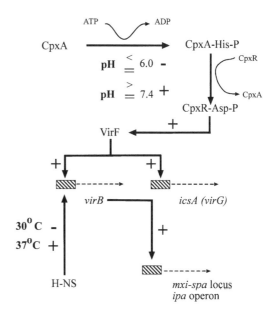

FIGURE 4 Model for modulating the expression of virulence genes. VirF is the global regulator of genes found on the virulence plasmid. The chromosomally encoded CpxA/CpxR two-component regulatory proteins regulate VirF. VirF activates the transcription of *icsA* and *virB*. The latter's gene product activates the transcription of the *mxi-spa* locus and the *ipa* operon. The chromosomally encoded H-NS protein is an antagonistic regulatory factor that represses the expression of *virB* at 30°C, but which is released from the *virB* promoter at 37°C.

other uncharacterized genes. VirF directly activates the transcription of *icsA* and *virB*, the latter of which directly activates transcription of genes in the *mxi/spa* locus (3, 40, 64).

CpxA and CpxR, which belong to the superfamily of two-component sensor kinase/effector proteins, activate the expression of VirF. The CpxA sensor kinase responds to extracellular pH. When bacteria are introduced into environments where the pH is 6.0 or less, CpxA is not phosphorylated; however, under physiological pH conditions, CpxA is phosphorylated on a histidine residue and rapidly transfers the phosphate moiety to CpxR. In its phosphorylated state, CpxR binds to the *virF* promoter, thereby activating its transcription (38, 39). These findings are corroborated by studies that show that bacterial cell envelope perturbations, such as those mediated by elevated pH, increase the levels of phosphorylated CpxR (44).

Further evidence that optimal physiological conditions are involved in the modulation of expression of virulence genes is provided by investigations of the roles that medium osmolarity and temperature play in the expression of these genes. Virulence gene expression is enhanced at 300 mOsM, the osmolarity found within epithelial cells. The EnvZ/OmpR two-component system plays an important role in sensing the osmolarity of the surrounding environment. Mutations in the corresponding genes adversely affect the modulation of expression of the virulence genes in response to medium osmolarity and the ability of the mutants to survive intracellularly (6). Furthermore, the expression of virulence factors and the invasive capabilities of *Shigella* are dependent on temperature (33). H-NS, a histonelike protein, represses the expression of the *virB* operon at 30°C, even in the presence of the VirF activator, but not at 37°C (3, 22, 33).

CONCLUSIONS

The pathogenesis of shigellosis involves a series of intimate interactions between *Shigella* and the host cell. During these interactions, the bacilli exploit host cell functions on many levels. These include (i) binding of *Shigella* proteins to host cell receptors; (ii) recruitment of components of the host cell cytoskeleton for uptake, intracellular motility, and intercellular dissemination; and (iii) induction of apoptosis of macrophages and an associated inflammatory response that promotes the intracellular survival of this organism. The interaction of the organism with the host cell triggers the release of bacterial factors that stimulate host cell signaling pathways. Such host cell signaling induces rearrangements in cytoskeletal proteins that lead to the uptake of the bacteria. The bacteria rapidly lyse the phagocytic vacuole and are released into the cell cytoplasm, where they divide rapidly and recruit host-cell actin and associated cytoskel-

etal proteins for intracellular motility and intercellular spread. The bacteria also secrete factors that induce the inflammatory response, thus destroying the integrity of the epithelial cell barrier so that the uptake of additional bacteria is enhanced. Finally, the expression of *Shigella* virulence factors involved in pathogenicity and the intracellular existence of the organism are modulated by factors that sense when the environment is suitable for their expression.

This review has addressed the detailed molecular mechanisms involved in the aforementioned processes. An understanding of the mechanisms employed by pathogens like *Shigella* to interact with host cells has provided invaluable insight into how these mechanisms can be exploited to eradicate bacterial diseases. This is especially important with the increase in antibiotic-resistant bacterial pathogens. Such knowledge can be used to generate safe and efficacious vaccines against infection and to develop alternative treatments for infections. Furthermore, an understanding of how bacterial virulence factors are secreted and how they interact with host cells and host-cell components might be used to develop bacterial-mediated delivery systems for the eradication of diseases that are not bacterial in origin. Finally, *Shigella* may be employed as a tool to better understand the molecular events involved in the organization and reorganization of cytoskeletal proteins in eukaryotic cells.

REFERENCES

1. **Adam, T., M. Arpin, M. C. Prevost, P. Gounon, and P. J. Sansonetti.** 1995. Cytoskeletal rearrangements and the functional role of T-plastin during entry of *Shigella flexneri* into HeLa cells. *J. Cell. Biol.* **129:**367–381.
2. **Adam, T., M. Giry, P. Boquet, and P. J. Sansonetti.** 1996. Rho-dependent membrane folding causes *Shigella* entry into epithelial cells. *EMBO J.* **15:**3315–3321.
3. **Adler, B., C. Sasakawa, T. Tobe, S. Makino, K. Komatsu, and M. Yoshikawa.** 1989. A dual transcriptional activation system for the 230 kb plasmid genes coding for virulence-

associated antigens of *Shigella flexneri*. *Mol. Microbiol.* **3:**627–635.

4. **Allaoui, A., J. Mounier, M.-C. Prevost, P. J. Sansonetti, and C. Parsot.** 1992. *icsB*: a *Shigella flexneri* virulence gene necessary for the lysis of protrusions during intercellular spread. *Mol. Microbiol.* **6:**1605–1616.

5. **Bahrani, F. K., P. J. Sansonetti, and C. Parsot.** 1997. Secretion of Ipa proteins by *Shigella flexneri*: inducer molecules and kinetics of activation. *Infect. Immun.* **65:**4005–4010.

6. **Bernardini, M. L., A. Fontaine, and P. J. Sansonetti.** 1990. The two-component regulatory system OmpR-EnvZ controls the virulence of *Shigella flexneri*. *J. Bacteriol.* **172:**6274–6281.

7. **Bernardini, M. L., J. Mounier, H. d'Hauteville, M. Coquis-Rondon, and P. J. Sansonetti.** 1989. Identification of *icsA*, a plasmid locus of *Shigella flexneri* that governs bacterial intra- and intercellular spread through interaction with F-actin. *Proc. Natl. Acad. Sci. USA* **86:**3867–3871.

8. **Chen, Y., M. Smith, K. Thirumalai, and A. Zychlinsky.** 1996. A bacterial invasin induces macrophage apoptosis by binding directly to ICE. *EMBO J.* **15:**3853–3860.

9. **Clerc, P., and P. J. Sansonetti.** 1987. Entry of *Shigella flexneri* into HeLa cells: evidence for directed phagocytosis involving actin polymerization and myosin accumulation. *Infect. Immun.* **55:**2681–2688.

10. **Cornelis, G. R., and H. Wolf-Watz.** 1997. The *Yersinia* Yop virulon: a bacterial system for subverting eukaryotic cells. *Mol. Microbiol.* **23:**861–867.

11. **DeGeyter, C., B. Vogt, Z. Benjelloun-Touimi, P. J. Sansonetti, J.-M. Ruysschaert, C. Parsot, and V. Cabiaux.** 1997. Purification of IpaC, a protein involved in entry of *Shigella flexneri* into epithelial cells and characterization of its interaction with lipid membranes. *FEBS Lett.* **400:**149–154.

12. **Dehio, C., M. C. Prevost, and P. J. Sansonetti.** 1995. Invasion of epithelial cells by *Shigella flexneri* induces phosphorylation of cortactin by a pp60c-src-mediated signalling pathway. *EMBO J.* **14:**2471–2482.

13. **d'Hauteville, H., R. D. Lagelouse, F. Nato, and P. J. Sansonetti.** 1996. Lack of cleavage of IcsA in *Shigella flexneri* causes aberrant movement and allows demonstration of a cross-reactive eukaryotic protein. *Infect. Immun.* **64:**511–517.

14. **DuPont, H. L.** 1995. *Shigella* species (bacillary dysentery), p. 2033–2039. *In* G. L. Mandell, J. E. Bennett, and R. Dolin (ed.), *Principles and Practice of Infectious Diseases*. Churchill Livingstone, New York, N.Y.

15. **Egile, C., H. d'Hauteville, C. Parsot, and P. J. Sansonetti.** 1997. SopA, the outer membrane protease responsible for polar localization of IcsA in *Shigella flexneri*. *Mol. Microbiol.* **23:**1063–1073.

16. **Finlay, B. B., and P. Cossart.** 1997. Exploitation of mammalian host cell functions by bacterial pathogens. *Science* **276:**718–725.

17. **Fukuda, I., T. Suzuki, H. Munakata, N. Hayashi, E. Katayama, M. Yoshizawa, and C. Sasakawa.** 1995. Cleavage of *Shigella* surface protein VirG occurs at a specific site, but the secretion is not essential for intracellular spreading. *J. Bacteriol.* **177:**1719–1726.

18. **Goldberg, M. B.** 1997. *Shigella* actin-based motility in the absence of vinculin. *Cell Motil. Cytoskeleton* **37:**44–53.

19. **Goldberg, M. B., O. Barzu, C. Parsot, and P. J. Sansonetti.** 1993. Unipolar localization and ATPase activity of IcsA, a *Shigella flexneri* protein involved in intracellular movement. *J. Bacteriol.* **175:**2189–2196.

20. **Goldberg, M. B., and J. A. Theriot.** 1995. *Shigella flexneri* surface protein IcsA is sufficient to direct actin-based motility. *Proc. Natl. Acad. Sci. USA* **92:**6572–6576.

21. **Goldberg, M. B., J. A. Theriot, and P. J. Sansonetti.** 1994. Regulation of surface presentation of IcsA, a *Shigella* protein essential to intracellular movement and spread, is growth phase-dependent. *Infect. Immun.* **62:**5664–5668.

22. **Hale, T. L.** 1991. Genetic basis of virulence of *Shigella* species. *Microbiol. Rev.* **55:**206–224.

23. **Henderson, I. R., F. Navarro-Garcia, and J. P. Nataro.** 1998. The great escape: structure and function of the autotransporter proteins. *Trends Microbiol.* **6:**370–378.

24. **High, N., J. Mounier, M. C. Prevost, and P. J. Sansonetti.** 1992. IpaB of *Shigella flexneri* causes entry into epithelial cells and escape from the phagocytic vacuole. *EMBO J.* **11:**1991–1999.

25. **Isberg, R.** 1991. Discrimination between intracellular uptake and surface adhesion of bacterial pathogens. *Science* **252:**934–938.

26. **Jose, J., J. Kramer, T. Klauser, J. Pohlner, and T. F. Meyer.** 1996. Absence of periplasmic DsbA oxidoreductase facilitates export of cysteine-containing passenger proteins to the *Escherichia coli* cell surface via the Iga beta autotransporter pathway. *Gene* **178:**107–110.

27. **Kotloff, K. L., J. P. Winickoff, B. Ivanoff, J. D. Clemens, D. L. Swerdlow, P. J. Sansonetti, G. K. Adak, and M. M. Levine.** 1999. Global burden of *Shigella* infections: implications for vaccine development and implementation of control strategies. *Bull. W. H. O.* **77:**651–666.

28. Laine, R. O., W. Zeile, F. Kang, D. L. Purich, and F. S. Southwick. 1997. Vinculin proteolysis unmasks an ActA homolog for actin-based *Shigella* motility. *J. Cell Biol.* **138:** 1255–1264.

29. Majdalani, N., and K. Ippen-Ihler. 1996. Membrane insertion of the F-pilin subunit is Sec independent but requires leader peptidase B and the proton motive force. *J. Bacteriol.* **178:**3742–3747.

30. Makino, S., C. Sasakawa, K. Kamata, T. Kurata, and M. Yoshikawa. 1986. A genetic determinant required for continuous reinfection of adjacent cells on large plasmid in S. flexneri 2a. *Cell* **46:**551–555.

31. Marquart, M. E., W. L. Picking, and W. D. Picking. 1996. Soluble invasion plasmid antigen C (IpaC) from *Shigella flexneri* elicits epithelial cell responses related to pathogen invasion. *Infect. Immun.* **64:**4182–4187.

32. Maurelli, A. T., B. Baudry, H. d'Hauteville, T. L. Hale, and P. J. Sansonetti. 1985. Cloning of virulence plasmid DNA sequences involved in invasion of HeLa cells by *Shigella flexneri*. *Infect. Immun.* **49:**164–171.

33. Maurelli, A. T., and P. J. Sansonetti. 1988. Identification of a chromosomal gene controlling temperature-regulated expression of *Shigella flexneri*. *Proc. Natl. Acad. Sci. USA* **85:**2820–2824.

34. Miki, H., T. Sasaki, Y. Takai, and T. Takenawa. 1998. Induction of filopodium formation by a WASP-related actin-depolymerizing protein NWASP. *Nature* **391:**93–96.

35. Monack, D. M., B. Raupach, A. E. Hromockyj, and S. Falkow. 1996. *Salmonella typhimurium* invasion induces apoptosis in infected macrophages. *Proc. Natl. Acad. Sci. USA* **93:** 9833–9838.

36. Mounier, J., V. Laurent, A. Hall, P. Fort, M. F. Carlier, P. J. Sansonetti, and C. Egile. 1999. Rho family GTPases control entry of *Shigella flexneri* into epithelial cells but not intracellular motility. *J. Cell. Sci.* **112:**2069–2080.

37. Mounier, J., T. Vasselon, R. Hellio, M. Lesourd, and P. Sansonetti. 1992. *Shigella flexneri* enters human colonic Caco-2 epithelial cells through the basolateral pole. *Infect. Immun.* **60:** 237–248.

38. Nakayama, S.-I., and H. Watanabe. 1995. Involvement of *cpxA*, a sensor of a two-component regulatory system, in the pH-dependent regulation of expression of *Shigella sonnei virF* gene. *J. Bacteriol.* **177:**5062–5069.

39. Nakayama, S.-I., and H. Watanabe. 1998. Identification of *cpxR* as a positive regulator essential for the expression of the *Shigella sonnei virF* gene. *J. Bacteriol.* **180:**3522–3528.

40. O'Toole, P. W., J. W. Austin, and T. J. Trust. 1994. Identification and molecular characterization of a major ring-forming surface protein from the gastric pathogen *Helicobacter mustelae*. *Mol. Microbiol.* **11:**349–361.

41. Perdomo, J. J., P. Gounon, and P. J. Sansonetti. 1994. Polymorphonuclear leukocyte transmigration promotes invasion of colonic epithelial monolayer by *Shigella flexneri*. *J. Clin. Invest.* **93:**633–643.

42. Pohlner, J., R. Halter, K. Beyreuther, and T. F. Meyer. 1987. Gene structure and extracellular secretion of *Neisseria gonorrhoeae* IgA protease. *Nature* **325:**458–462.

43. Prevost, M. C., M. Lesourd, M. Arpin, F. Vernel, J. Mounier, R. Hellio, and P. J. Sansonetti. 1992. Unipolar reorganization of F-actin layer at bacterial division and bundling of actin filaments by plastin correlate with movement of *Shigella flexneri* within HeLa cells. *Infect. Immun.* **60:**4088–4099.

44. Raivio, T. L., and T. J. Silhavy. 1997. Transduction of envelope stress in *Escherichia coli* by the Cpx two-component system. *J. Bacteriol.* **179:** 7724–7733.

45. Rajkumar, R., H. Devaraj, and S. Niranjali. 1998. Binding of *Shigella* to rat and human intestinal mucin. *Mol. Cell. Biochem.* **178:**261–268.

46. Rohatgi, R., L. Ma, H. Miki, M. Lopez, T. Kirchhausen, T. Takenawa, and M. W. Kirschner. 1999. The interaction between N-WASP and the Arp2/3 complex links Cdc42-dependent signals to actin assembly. *Cell* **97:**221–231.

47. Sandlin, R. C., M. B. Goldberg, and A. T. Maurelli. 1996. Effect of O side chain length and composition on the virulence of *Shigella flexneri* 2a. *Mol. Microbiol.* **22:**63–73.

48. Sandlin, R. C., K. A. Lampel, S. P. Keasler, M. B. Goldberg, A. L. Stolzer, and A. T. Maurelli. 1995. Avirulence of rough mutants of *Shigella flexneri*: requirement of O antigen for correct unipolar localization of IcsA in bacterial outer membrane. *Infect. Immun.* **63:**229–237.

49. Sandlin, R. C., and A. T. Maurelli. 1999. Establishment of unipolar localization of IcsA in *Shigella flexneri* 2a is not dependent on virulence plasmid determinants. *Infect. Immun.* **67:**350–356.

50. Sansonetti, P. J. 1998. Pathogenesis of shigellosis: from molecular and cellular biology of epithelial cell invasion to tissue inflammation and vaccine development. *Jpn. J. Med. Sci. Biol.* **51**(Suppl):S69–S80.

51. Sansonetti, P. J., J. Arondel, J.-M. Cavaillon, and M. Huerre. 1995. Role of interleukin-1 in the pathogenesis of experimental shigellosis. *J. Clin. Invest.* **96:**884–892.

52. **Sansonetti, P. J., J. Arondel, M. Huerre, A. Harada, and K. Matsushima.** 1999. Interleukin-8 controls bacterial transepithelial translocation at the cost of epithelial destruction in experimental shigellosis. *Infect. Immun.* **67:**1471–1480.

53. **Sansonetti, P. J., D. J. Kopecko, and S. B. Formal.** 1982. Involvement of a plasmid in the invasive ability of *Shigella flexneri. Infect. Immun.* **35:**852–860.

54. **Sansonetti, P. J., J. Mounier, M. C. Prevost, and R.-M. Mege.** 1994. Cadherin expression is required for the spread of Shigella flexneri between epithelial cells. *Cell* **76:**829–839.

55. **Shere, K. D., S. Sallustio, A. Manessis, T. G. D'Aversa, and M. B. Goldberg.** 1997. Disruption of IcsP, the major *Shigella* protease that cleaves IcsA, accelerates actin-based motility. *Mol. Microbiol.* **25:**451–462.

56. **Silversmith, R. E., and R. B. Bourret.** 1999. Throwing the switch in bacterial chemotaxis. *Trends Microbiol.* **7:**16–22.

57. **Skoudy, A., G. Tran Van Nhieu, N. Mantis, M. Arpin, J. Mounier, P. Gounon, and P. Sansonetti.** 1999. A functional role for ezrin during Shigella flexneri entry into epithelial cells. *J. Cell Sci.* **112:**2059–2068.

58. **Steinhauer, J., R. Agha, T. Pham, A. W. Varga, and M. B. Goldberg.** 1999. The unipolar *Shigella* surface protein IcsA is directly targeted to the old pole; IcsP cleavage of IcsA occurs over the entire bacterial surface. *Mol. Microbiol.* **32:**367–378.

59. **Suzuki, T., M.-C. Lett, and C. Sasakawa.** 1995. Extracellular transport of VirG protein in *Shigella. J. Biol. Chem.* **270:**30874–30880.

60. **Suzuki, T., H. Miki, T. Takenawa, and C. Sasakawa.** 1998. Neural Wiskott-Aldrich syndrome protein is implicated in the actin-based motility of *Shigella flexneri. EMBO J.* **17:**2767–2776.

61. **Tran Van Nhieu, G., A. Ben-Ze'ev, and P. J. Sansonetti.** 1997. Modulation of bacterial entry into epithelial cells by association between vinculin and the *Shigella* IpaA invasin. *EMBO J.* **16:**2717–2729.

62. **Tran Van Nhieu, G., E. Caron, A. Hall, and P. J. Sansonetti.** 1999. IpaC induces actin polymerization and filopodia formation during *Shigella* entry into epithelial cells. *EMBO J.* **18:** 3249–3262.

63. **Wachter, C., C. Beinke, M. Mattes, and M. A. Schmidt.** 1999. Insertion of EspD into epithelial target cell membranes by infecting enteropathogenic *Escherichia coli. Mol. Microbiol.* **31:** 1695–1707.

64. **Watanabe, H., E. Arakawa, K.-I. Ito, J.-I. Kato, and A. Nakamura.** 1990. Genetic analysis of an invasion region by use of a Tn3-*lac* transposon and identification of a second positive regulator gene, *invE*, for cell invasion of *Shigella sonnei*: significant homology of InvE with ParB of plasmid P1. *J. Bacteriol.* **172:**619–629.

65. **Watarai, M., S. Funato, and C. Sasakawa.** 1996. Interaction of Ipa proteins of *Shigella flexneri* with $\alpha 5 \beta 1$ integrin promotes entry of the bacteria into mammalian cells. *J. Exp. Med.* **183:** 991–999.

66. **Zychlinsky, A., J. J. Perdomo, and P. J. Sansonetti.** 1994. Molecular and cellular mechanisms of tissue invasion by *Shigella flexneri. Ann. N. Y. Acad. Sci.* **730:**197–208.

67. **Zychlinsky, A., K. Thirumalai, J. Arondel, J. Cantey, A. Aliprantis, and P. Sansonetti.** 1996. In vivo apoptosis in *Shigella flexneri* infections. *Infect. Immun.* **64:**5357–5365.

BACTERIAL INDUCTION OF CYTOKINE SECRETION IN PATHOGENESIS OF AIRWAY INFLAMMATION

Alice Prince

13

Many bacterial species colonize mucosal surfaces, but relatively few are capable of evading the numerous host defenses to gain access to epithelial cells and cause disease. The respiratory tract is continually exposed to potential pathogens, yet pulmonary infection occurs relatively infrequently. Successful mucosal pathogens must express specific virulence factors to elude innate defense mechanisms, mucociliary clearance, the activity of antimicrobial peptides, and the physical barrier provided by host mucin glycopeptides. Such pathogens can then recognize specific receptors on the apical surfaces of epithelial cells, attach, colonize, and establish infection. The clinical signs and symptoms of infection are attributed to the local accumulation of polymorphonuclear leukocytes (PMNs) and subsequent destruction of host tissue. This inflammatory response, while often signaled by "professional" immune cells such as macrophages and lymphocytes, can also be elicited by the epithelial cells that line mucosal surfaces and are most likely to encounter perceived pathogens first. For some mucosal pathogens,

such as *Helicobacter pylori, Escherichia coli, Salmonella*, and several other gastrointestinal pathogens, specific adhesins and corresponding receptors have been defined (23). These pathogens elicit a prompt and highly specific epithelial response that may include substantial cytoskeletal rearrangements, the transcription of various cytokines, or the induction of apoptosis. However, particularly for pathogens that are confined to the surface of epithelial cells, exactly how ligation of receptors signals the epithelial cytokine response has not been clearly established.

In the respiratory tract, little is known about the molecular basis for the host-pathogen interactions that are important in the pathogenesis of pneumonia, for either alveolar or airway infection. Although some bacterial ligands have been defined, most of the corresponding epithelial receptors are uncharacterized, and it is not clear how these organisms signal the epithelial cell to activate defense mechanisms. Much of the recent data establishing the molecular basis for host-pathogen interactions in the respiratory tract have been derived from studies to explain the pathogenesis of *Pseudomonas aeruginosa* infection in cystic fibrosis (CF). In this disease, commonly inhaled organisms from the environment, particularly *P. aeruginosa*, elicit a

Alice Prince   College of Physicians & Surgeons, Columbia University, 650 W. 168th St., New York, NY 10032.

Virulence Mechanisms of Bacterial Pathogens, 3rd ed., Edited by K. A. Brogden et al.
©2000 ASM Press, Washington, D.C.

florid epithelial inflammatory response that eventually results in the destruction of the lung and pulmonary failure (5, 30). It remains unclear exactly why mutations in the cystic fibrosis transmembrane conductance regulator (CFTR), a chloride channel, predispose CF patients to pulmonary infection. However, efforts to understand the pathophysiological basis for CF have provided a great deal of useful information to characterize the interactions between respiratory epithelial cells and bacteria. In studying the exaggerated host response to the pathogens associated with CF, it has been possible to define some of the general mechanisms that are important in the pulmonary epithelial response to pathogens in normal subjects as well as those with underlying immune and genetic defects.

The clinical signs and symptoms of airway infection in CF are mediated primarily by the influx of PMNs from the circulation to the site of infection in the airway (36). This signaling is accomplished by the expression of chemokines, especially interleukin-8 (IL-8), at the site of infection and the upregulation of PMN receptors on endothelial and epithelial cells to target the translocation of PMNs to the infected area (21, 43). Bacteria in the airway lumen provide a potent stimulus for the epithelial expression of the proinflammatory cytokines, TNF-α, IL-1β, and IL-8. Although the influx of PMNs from the circulating pool to the airways may be helpful in the clearance of organisms that have eluded the innate defenses of the airway, the release of their damaging products further stimulates epithelial IL-8 expression, more PMN influx, and host cell damage. Neutrophils contain a number of potent bactericidal agents that serve not only to kill ingested organisms, but to further stimulate epithelial IL-8 signaling (31). The toxic effects of neutrophil elastase on the respiratory epithelial cell are well characterized. Therapeutic strategies include administration of protease inhibitors to neutralize excessive protease activity in an attempt to reduce the pathologic damage caused by unrestricted protease activity in the airway. Other

PMN products, such as reactive oxygen intermediates and superoxide, are potent bactericidal agents that are damaging to host tissue as well as to pathogens. The histopathologic changes associated with CF are in many ways reflections of the damage caused by inflammation and fibrosis, the consequence of excessive amounts of neutrophil elastase and other membrane-damaging products (5, 33). As IL-8 is the major PMN chemokine in the airway, directing and activating the PMNs that are mobilized from the vascular compartment to the site of infection, this chemokine has been the focus of studies to understand how bacterial infection stimulates the host response in the lung (43).

BACTERIAL STIMULATION OF EPITHELIAL IL-8 PRODUCTION

The epithelial cells lining mucosal surfaces act as a peripheral component of the immune system, serving to both signal and respond to the threat of microbial pathogens (23). In addition to providing a barrier function, mucosal epithelial cells express an array of receptors that provide a sensing mechanism for perceived pathogens. As at other mucosal sites, a specific ligand-receptor interaction is required to activate the epithelial cytokine signaling response. The nature of the bacterial stimulus that activates epithelial IL-8 expression may be different at specific mucosal sites. The lung and lower airways, which are normally sterile, respond to different signals than the gastrointestinal or urogenital mucosae, which are chronically exposed to bacteria. At those sites, it appears that invasion of the epithelial cell is necessary to stimulate epithelial cytokine signaling (16). Pulmonary epithelial cells, which line a more pristine environment, respond to entirely superficial stimuli. Several discrete *P. aeruginosa* gene products elicit an IL-8 response. Secreted *P. aeruginosa* gene products isolated from cell culture supernatants (28) as well as ligands associated with the bacterial cell surface efficiently stimulate epithelial IL-8 expression in biologically relevant amounts (13). Lipopolysaccharide (LPS), a potent stimulus for macrophages and lymphocytes, is not a stimulus for epithelial IL-8 expression in airway cells, nor do pulmonary epithelial cells appear to express CD14, a component of the LPS-binding complex on immune cells (13, 25). Cellular fractionation studies have demonstrated that *P. aeruginosa* outer membrane proteins, pilin and flagellin, elicit epithelial IL-8 expression (13). In addition, the small homoserine lactone molecule associated with quorum sensing, the *P. aeruginosa* autoinducer, a *lasR*-dependent gene product, is also able to activate epithelial IL-8 expression (13). Adherent *Neisseria* (32) and *E. coli* expressing type 1 fimbriae (20) also stimulate epithelial cytokine expression, whereas the mucosal cells of the gastrointestinal tract usually respond only to internalized organisms or gene products (16). Thus, the types of ligands that activate epithelial responses may be highly tissue specific.

ADHERENT BACTERIA ACTIVATE EPITHELIAL IL-8 TRANSCRIPTION

One of the major tenets of bacterial pathogenesis is the requirement for a specific ligand receptor interaction between the host and pathogen to initiate infection. Although large numbers of *P. aeruginosa* are not usually adherent to normal respiratory epithelial cells, a reproducible fraction of a *P. aeruginosa* inoculum does bind to the epithelial cell. Approximately 50% of this binding is mediated by type IV pili of *P. aeruginosa* (38, 41, 45). These adherent *P. aeruginosa* stimulate an epithelial cytokine response. Comparison of the ability of several *P. aeruginosa* mutants to trigger epithelial IL-8 expression confirmed that surface components, but not LPS, mediated this response (13). Mutants of strains PAO, PAK, and PA1244 lacking pilin were associated with a significantly decreased IL-8 response, whereas mutants expressing neither pilin nor flagellin did not elicit IL-8 expression at all. Similarly, PAK*fliC* (Fla⁻) was able to stimulate IL-8, but PAK*fliA* (Pil⁻ Fla⁻) was not. Purified pilin isolated from strain PAO1 stimulated IL-8 expression in a dose-dependent fashion,

and piliated bacteria were able to stimulate an IL-8 response, but not organisms expressing pilin with a mutation in the carboxy-terminus adhesin domain (13). Purified flagellin from PAO1 also is a stimulus for epithelial IL-8 expression in a dose-dependent fashion, indicating that flagellin can function as an epithelial ligand (17). These observations demonstrate that adherence, the ligation of a specific epithelial receptor by a bacterial adhesin (either pilin or flagellin), is sufficient to trigger the epithelial production of IL-8.

Other bacteria that can bind directly to respiratory epithelial cells also elicit epithelial IL-8 expression. The pulmonary pathogens *Staphylococcus aureus* and *Haemophilus influenzae* are common pathogens in patients with CF. Adherent *S. aureus* strains were found to stimulate epithelial IL-8 expression and *agr* and *agr, sar* double mutants (RN6911 and ALC135, respectively), which adhere to a lesser degree, were also less stimulatory (Fig. 1) (6). Similarly, *H. influenzae* strains found to be capable of epithelial binding also activated IL-8 expression. Choline modification of *H. influenzae* lipooligosaccharide (LOS) has been reported to enhance colonization of the respiratory tract (47). However, both ChoP⁺ (H394) or ChoP⁻ (H311 and H395) variants of *H. influenzae* were able to both adhere and efficiently induce epithelial IL-8, suggesting that an LOS-independent adhesin is responsible for this interaction. Adherent *H. influenzae* and *S. aureus*, but not *agr* mutants, compete with *P. aeruginosa* for epithelial binding (Fig. 2), indicating that these respiratory pathogens all can recognize a common receptor.

IDENTIFICATION OF EPITHELIAL RECEPTORS FOR BACTERIAL BINDING TO AIRWAY MUCOSAL CELLS

There are numerous components of the airway epithelium that could act as receptors for bacterial ligands. *P. aeruginosa* recognizes several different types of glycoconjugates, including components of mucin and epithelial

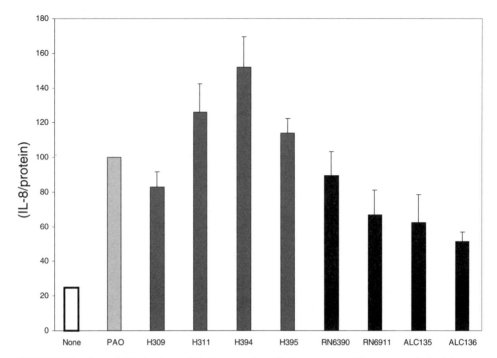

FIGURE 1 Bacterial induction of IL-8 expression. IL-8 was assayed by enzyme-linked immunosorbent assay 18 h after a 1-h exposure of 1HAEo− airway epithelial cells to the strains listed, followed by sterilization of the monolayer with gentamicin and overnight incubation in tissue culture media. *P. aeruginosa* PAO1, which elicits 35 pg of IL-8 per μg of protein, was considered 100%. The *H. influenzae* strains included H309 (strain Rd), H311 (ChoP⁻ variant), H394 (strain Eagan, ChoP⁺), and H395 (strain Eagan, ChoP⁻ variant). *S. aureus* strains tested included wild-type RN6390, RN6911 (*agr*), ALC 135 (*agr, sar*), and ALC 136 (*sar*).

glycolipids. The specific glycolipid components that function as receptors for pulmonary pathogens were defined in vitro by Krivan, who demonstrated that the GalNAcβ1-4Gal moiety exposed on asialylated glycolipids such as asialoGM1, but not available on fully sialylated glycoconjugates, function to bind *S. aureus* as well as *H. influenzae, P. aeruginosa,* and many other organisms (24). Fluorescein isothiocyanate-labeled antibody to asialoGM1 was used to screen for the presence of potential receptors for bacterial attachment on respiratory epithelial cells by flow cytometry. Binding sites were found on 2 to 3% of respiratory epithelial cells in primary culture derived from nasal polyp tissue. Cells from CF patients had significantly more receptors with superficial asialoGM1 found on 12 to 15% of

the epithelial cells studied, and almost 20% of the epithelial cells had asialoGM1 exposed following incubation with *P. aeruginosa* culture supernatants containing protease and neuraminidase activity (8, 38). The increased number of receptors on CF cells was consistent with a relative lack of sialylation on host cell glycoconjugates in CF (22). Studies of explants of human respiratory epithelium similarly have demonstrated increased numbers of asialylated receptors for *P. aeruginosa* binding around areas of epithelial damage and regeneration (12).

P. aeruginosa adhesins, ¹²⁵I-labeled pilin and flagellin, were shown to bind directly to asialoGM1 (17, 41). The biological relevance of pilin- or flagellin-induced IL-8 expression in the pathogenesis of airway inflammation

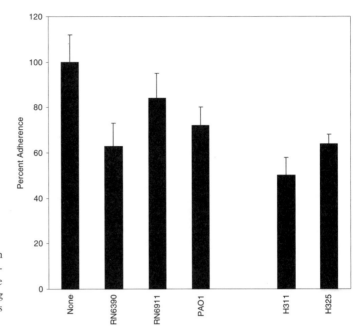

FIGURE 2 Binding competition studies. The percentage of the ^{35}S-PAO1 inoculum that bound in the presence of each of the competing strains in five- to eightfold excess is shown.

was further established in an animal model of respiratory tract infection (45, 46). The purified ligands, pilin, or flagellin, compared with a bovine serum albumin control at the same concentration, were inoculated into the nares of neonatal mice and the lungs were examined 18 h later. The histopathology elicited by the purified adhesins, airways filled with PMNs, was very similar to that stimulated by the introduction of wild-type, piliated, and flagellated bacteria by the same route of infection (45). Instillation of bovine serum albumin did not elicit a PMN response (17). These studies demonstrated that ligation of asialoGM1 receptors by *P. aeruginosa* adhesins is sufficient to cause PMN-dominated inflammation in the airways.

Pilin-mediated adherence to asialylated receptors, in addition to stimulating IL-8 expression, may also be a prerequisite for other types of *P. aeruginosa* virulence. Recent studies have demonstrated that the expression and cytotoxicity associated with the secretion of type III-dependent *P. aeruginosa* toxins, such as exoenzymes S, T, and U, require the close apposition of the host cell and bacteria, which is

dependent upon pilin-mediated adherence (10). These cytotoxic strains are often invasive, producing sepsis in addition to local pulmonary infection. The requirement for pilin-mediated adherence for the efficient secretion of *P. aeruginosa* cytotoxins into host cells further emphasizes the central role of pilin-mediated attachment to mucosal sites in the pathogenesis of *P. aeruginosa* infection.

DISTRIBUTION OF ASIALOGM1 RECEPTORS IN THE RESPIRATORY TRACT

Although many pulmonary pathogens are capable of ligating the GalNAcβ1-4Gal receptor exposed on asialylated glycolipids in vitro, it is not clear how important this receptor is in initiating epithelial IL-8 signaling because of its limited distribution on normal cells. AsialoGM1 receptors are most abundant in areas of epithelial regeneration (e.g., following trauma or viral infection with neuraminidase-producing organisms). These receptors are rarely found in normal epithelial cells in primary culture, which maintain tight junctions. However, cells with mutations in CFTR, or with abnormal CFTR function caused by constitutive overexpression of the regulatory domain of the protein, all have increased amounts of asialylated glycoconjugates and increased bacterial binding as mediated by these receptors (39). As a consequence of increased bacterial binding, comparable monolayers of epithelial cells differing only in the structure and/or function of CFTR have significantly elevated IL-8 expression (38). A cell line with normal CFTR structure, but overexpression of the CFTR regulatory domain, exhibits "CF-like" physiology and does not secrete chloridion (Cl⁻) in response to ATP. These cells have increased asialoGM1, bind increased numbers of piliated *P. aeruginosa*, and express increased amounts of proinflammatory cytokines such as IL-8 (7).

The exposure of asialylated glycolipids on the normal epithelium following a viral infection with a neuraminidase-producing organism such as influenza may also be responsible for the predisposition of these patients to significant *S. aureus* infection, as a complication of a primary viral infection. Similarly, hospitalized patients with traumatized airways following intubation or other manipulation would be expected to be at increased risk for nosocomial pneumonia, which is a frequent complication of modern intensive care. Areas of epithelial regeneration after damage are enriched in asialylated glycolipids, and hence have increased susceptibility to infection by opportunistic pathogens such as *P. aeruginosa* or *S. aureus*, which recognize these receptor sites. The lack of available receptors on the normal airway may help to explain why such infections are relatively unusual. The altered distribution of surface glycolipids after influenza infection may explain why *S. aureus* is a frequent cause of secondary bacterial infection as a complication of viral disease.

ACTIVATION OF TRANSCRIPTION FACTORS BY ADHERENT BACTERIA

Having identified the major components of the epithelial cytokine signaling pathway in airway epithelial cells, the adhesins, and their receptors, the signaling system within the epithelial cell that mediates the cytokine response remains to be identified. Portions of this signaling cascade appear to be similar to those of other cells that mediate host responses to external stimuli. The transcription of IL-8 is regulated by NF-κB, a transcription factor that is normally complexed to inhibitory proteins (IκBs) that are phosphorylated and targeted for degradation in the proteosome after activation of signaling kinases (2). Removal of these inhibitors exposes a nuclear localization sequence, facilitating translocation of the NF-κB heterodimer into the nucleus and consequent transcriptional activation (3). IL-8 transcription is also activated by AP-1, a transcriptional activator, which is often stimulated by the JNK family of mitogen-activated protein kinases (MAPKs) (40). AP-1 and NF-κB synergistically activate IL-8 expression in respiratory epithelial cells (29). Adherent *P. aeruginosa*, pilin, or antibody to asialoGM1, the pilin receptor, all stimulate NF-κB translocation to the nucleus. This has been demonstrated by gel-shift assays and by the identification of the p65 component of NF-κB in the nuclei of epithelial cells exposed to *P. aeruginosa*, but not unstimulated cells (14). Inhibition of proteosome protease activity with TPCK or TLCK (tosylamido-2 phenylalanine or tosyl-L-lysine chloromethyl ketone) inhibited the amount of NF-κB available to activate IL-8 transcription and decreased the amount of IL-8 produced by *P. aeruginosa*-stimulated epithelial cells. This appears to be the same pathway that is activated in respiratory epithelial cells by the proinflammatory cytokine IL-1β. The activation of NF-κB and AP-1 by respiratory pathogens shares several features in common with the events following infection of epithelial cells with *Neisseria* species (32). Signaling events further upstream, between the ligation of the asialoGM1 receptor and the activation of NF-κB, remain to be defined.

BACTERIAL ACTIVATION OF EPITHELIAL Ca²⁺ SIGNALING SYSTEMS

A major mechanism involved in relaying external stimuli from the environment to the cell nucleus is through the generation of Ca^{2+} signals (9). Calcium acts as an important second messenger, relaying information from specific receptors to downstream signaling elements. Although the role of Ca^{2+} in signal transduction is better established in excitatory cells and T cells (18), it is clear that other nonexcitatory cells transfer information by modulating levels of intracellular Ca^{2+} (11). Using Ca^{2+}-sensitive fluorescent markers, changes in the levels of $[Ca^{2+}]_i$ can be readily detected. In response to addition of either *P. aeruginosa*, *S. aureus*, or *H. influenzae*, fura-2AM loaded 1HAEo-epithelial cells responded within minutes with a 100 to 200 nM increase in $[Ca^{2+}]_i$ (Fig. 3). Nonadherent organisms did not stimulate this response, nor did the addition of LPS or lipoteichoic acid from staphylococci (37). The isolated ligands that stimulated IL-8 transcription, antibody to asialoGM1, purified pilin, or flagellin, also stimulated Ca^{2+} increases. To determine whether the transient increase in $[Ca^{2+}]_i$ was associated with the subsequent IL-8 transcrip-

tion, epithelial cells were treated with thapsigargin, which causes a transient increase in Ca^{2+} by inhibiting the activity of the Ca^{2+}-dependent ATPase. Thapsigargin stimulated 1HAEo- cells to produce IL-8.

The involvement of Ca^{2+}-dependent signaling cascades, and specifically the Ca^{2+}-dependent phosphatase calcineurin, was further confirmed by the inhibition of *P. aeruginosa*-induced IL-8 production by either cyclosporin or FK506. Many Ca^{2+} signals result in modulation of calcineurin activity, a phosphatase that is intimately associated with the IP_3-R Ca^{2+} channel and mediates the activation of NF-AT transcriptional activators (35). The translocation of NF-AT from the cytoplasm of 16HBE- cells into the nucleus is stimulated by exposure to *P. aeruginosa*, and nuclear NF-AT3 is detectable minutes after PAO1 addition to epithelial monolayers (Fig. 4). Indirect evidence for the role of Ca^{2+} in mediating the epithelial response to adherent bacteria is suggested by the effects of several compounds that inhibit or stimulate Ca^{2+}-

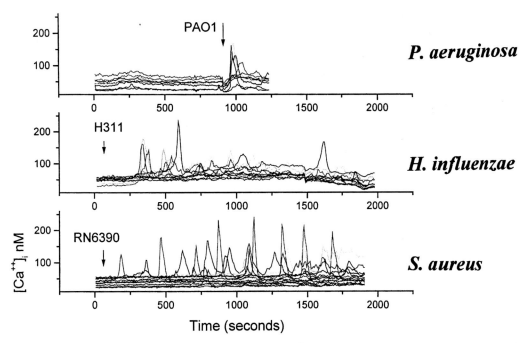

FIGURE 3 Ca^{2+} fluxes in 1HAEo− cells. Changes in Ca^{2+} following the addition of bacteria to fura-2 loaded cells (arrows) are shown. Composite images of single cell Ca^{2+} transients are depicted.

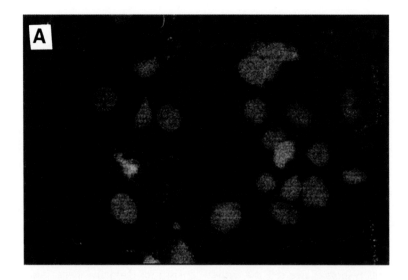

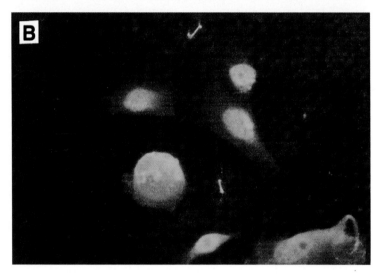

FIGURE 4   Nuclear localization of NF-AT3 in 16HBe- respiratory epithelial cells. NF-AT3 immunoflorescence   is detected using polyclonal antibody tagged with florescein   isothiocyanate-labeled goat anti-rabbit antibody. (A) Control conditions. (B) Fifteen minutes after addition of PAO1 to the epithelial cells showing nuclear localization of NF-AT3.

dependent effectors on IL-8 production (15) (Fig. 5). IL-8 production was blocked by TEMPO, an inhibitor of Fe^2 - and Zn^2 - dependent enzymes including calcineurin. These Ca^2 flxes appear to be generated by the release of Ca^2 from intracellular stores. Inhibition of external Ca^2 channels with ei-

ther verapamil or $NiCl_2$ did not diminish epithelial IL-8 signaling. Addition of EGTA to chelate external Ca^2 did not diminish the IL-8 response, whereas an intracellular Ca^2 chelator BAPTA/AM or an inhibitor of the IP_3-R Ca^2 channel, heparin, effectively blocked IL-8 signaling. This pathway appears

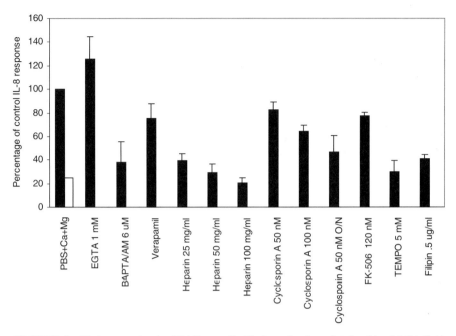

FIGURE 5   IL-8 expression in 1HAEo− cells. IL-8 production stimulated by PAO1 (5 × 10⁸CFU/ml) in the presence of the several selective kinase inhibitors and compounds expected to block MAPK signaling was measured by ELISA. Solid bars, with PAO; white bar, without PAO.

to be a general epithelial signaling response, because *S. aureus, P. aeruginosa,* and *H. influenzae* all stimulated Ca^{2+} fluxes of similar amplitude and duration and also activated IL-8 expression.

MAPK ACTIVATION BY ADHERENT BACTERIA

The likely signaling pathway that links the activation of Ca^{2+} fluxes and the NF-κB-dependent transcription of IL-8 is a kinase cascade. MAPKs often mediate such signaling pathways, although there are numerous members of this family, with different functions in specific cell types (40). The family of MAPKs consists of three modules of serine/threonine-specific protein kinases that serve to link cell-surface receptors to the transcription factors, which control resulting gene expression. Each module may be preferentially activated by specific stimuli, and the mechanisms that provide

the specificity for each signaling pathway have been a topic of active investigation. Activation of many cellular responses requires synergistic effects from more than one of the MAPK families (48). Involvement of MAPKs in mediating the epithelial IL-8 response to adherent bacteria was demonstrated in several experiments. 9HTEo- epithelial cells preincubated with the MAPK ERK 1/2 inhibitor PD98059 showed a significant inhibition in IL-8 production, as did epithelial cells incubated with the p38 inhibitor SB202190 (Fig. 6). By Western hybridization, phosphorylated ERK is detectable in 9HTEO- cells within 15 min of exposure to adherent bacteria or following treatment with thapsigargin, an inhibitor of the Ca^{2+}-dependent ATPase, to cause a transient increase in intracellular Ca^{2+} (Fig. 7). These studies provide evidence that several families of MAPKs are likely to be involved in mediating the signal of adherent organisms to result in IL-8 expression.

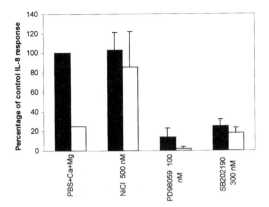

FIGURE 6   IL-8 expression induced by antibody to asialoGM1. IL-8 expression was inhibited in 1HAEo− cells pretreated with the p38 MAPK inhibitor SB202190 or with the ERK1/1 inhibitor PD98059 but was not altered by NiCl₂. Solid bars, with antibody to asialoGM1; white bars, without antibody to asialoGM1.

INVOLVEMENT OF CAVEOLI AND GLYCOLIPID COMPLEXES IN SIGNALING PATHWAYS ACTIVATED BY ADHERENT BACTERIA

The molecular events that link recognition of asialoGM1 and the release of $[Ca^{2+}]_i$ are undefined. Because the receptor for pulmonary pathogens, asialoGM1, is a ceramide conjugate, it is possible that the generation of free ceramide is important in the signaling cascade (4, 19). This has been suggested previously to act as the signaling moiety in the uroepithelium activated by ligation of the P fimbria to their glycolipid receptors (20). Activation of receptor tyrosine kinases might also be in-

volved in mediating the response to adherent *P. aeruginosa* (27). However, there is no clear association between the G proteins usually activated in these pathways and the glycolipid receptors identified as the *P. aeruginosa* ligands. It is also possible that glycolipid components of caveolae mediate the response of the respiratory epithelial cell to adherent bacteria. Caveolae are flask-like clusters of membrane structures enriched in cholesterol, often containing signaling molecules such as GPI-anchored proteins (e.g., Src, IP₃-R, the Ca^{2+}-dependent ATPase, or caveolin) (1, 27, 42). Caveolae can be isolated from a triton-insoluble fraction of lung epithelial cells (1). Recent colocalization studies of triton-insoluble fractions of 9HTEo− cells suggest that asialoGM1, caveolin, and IP₃-R colocalize (Fig. 8), and that IL-8 signaling by these cells can be disrupted in the presence of filipin, which intercalates into cholesterol rich membranes and interrupts signaling (Fig. 5). Although the inhibition of Src family kinases with PP1 did not decrease *P. aeruginosa* induction of IL-8 expression, the phosphoryl-

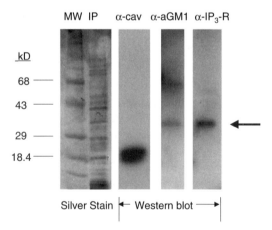

FIGURE 8   Colocalization of caveolin, IP3-R, and asialoGM1 in triton-insoluble fraction of 9HTEo− cells. Proteins in the triton insoluble fraction of 9HTEo− cell lysates were immunoprecipitated with anti-caveolin polyclonal antibody. Aliquots of the IP are shown stained with silver on a sodium dodecyl sulfate-polyacrylamide gel (left) or hybridized with anti-caveolin, anti-asialoGM1, or anti-IP3-R1.

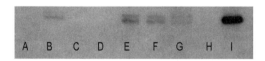

FIGURE 7   Phosphorylated ERK1/2 in response to *P. aeruginosa*. A Western hybridization using antibody to phospho-ERK1/2 is shown. Lane A, 0− unstimulated 9HTEo− cells; lanes B, C, and D, 15, 30, and 60 min after stimulation with thapsigargin (100 nM); lanes E, F, and G, 15, 30, and 60 min after stimulation with PAO1; lane H, negative control; lane I, positive control.

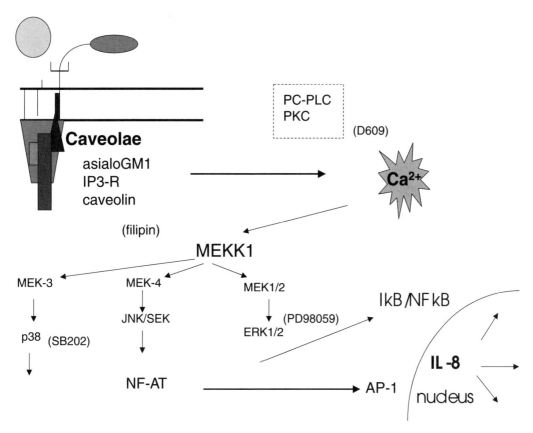

FIGURE 9 A model of epithelial activation. Adherent bacteria stimulate asialoGM1 receptors contained within caveolae and signal IL-8 expression through a Ca^{2+}-dependent MAPK cascade, resulting in translocation of AP-1 and NF-κB and gene transcription.

choline specific phospholipase C inhibitor D609 was effective in blocking IL-8 production (26). PC-PLC activity has also been found to activate IP_3-R activity in caveolae (34). The accumulated data thus far are entirely consistent with a role for caveolae in mediating the initial epithelial response to adherent bacteria that recognize asialylated glycolipid receptors. The activation step in this pathway may provide the specificity of the host response, not necessarily the MAPKs or the actual cytokine genes transcribed (44).

CONCLUSIONS

The response of respiratory epithelial cells to adherent bacteria capable of ligating an asialoGM1 receptor is schematized in Fig. 9.

Although this pathway has been studied primarily using *P. aeruginosa* as a model pathogen, it appears to be a general signaling system that responds to many organisms, both gram positive and gram negative. We postulate that this is a "fail-safe" mechanism activated promptly in response to the perceived bacterial pathogen and that it occurs entirely independently of "professional" immune cells. Few bacteria are normally capable of gaining access to the epithelial cell surface. The organisms that elude mucociliary clearance, are resistant to antimicrobial peptide activity in the airways, and are not cleared by encounters with phagocytes, may reach the epithelial surface. In the normal host, infection still does not occur because of the absence of receptors for bacterial attachment. However, in areas of damage or regeneration, or in cells with CFTR mutations, a few asialoGM1 receptors may provide a niche for the attachment of piliated organisms and initiation of infection. Binding of *P. aeruginosa* pilin to asialoGM1 receptors, perhaps as a component of caveolae, is the initial stimulus activating Ca^{2+} release and a relatively nonspecific activation of the MAPK cascade, which culminates in AP-1 and NF-κB translocation and IL-8 transcription. Thus, this is a pathological response to infection, not a normal clearance mechanism. Epithelial stimulation through asialoGM1 receptors evokes the influx of PMNs to the area of adherent bacteria, despite the consequent damage to the airways themselves from PMN products. As with other immune responses, the regulation of the pathway is critically important. While evoking PMN-dependent clearance mechanisms is clearly an important component of the host defense against pulmonary pathogens, in diseases such as CF, the excessive activation of this pathway may ultimately lead to destruction of the lung.

REFERENCES

1. **Anderson, R. G. W.** 1998. The caveolae membrane system. *Annu. Rev. Biochem.* **67**:199–225.
2. **Baeuerle, P. A., and D. S. Baltimore.** 1996. NF-$\kappa\beta$: ten years after. *Cell* **87**:13–20.
3. **Baldwin, A. S.** 1996. The NF-$\kappa\beta$ and I$\kappa\beta$ proteins: new discoveries and insights. *Annu. Rev. Immunol.* **14**:649–681.
4. **Ballou, L. R., S. J. F. Laulederkine, E. Rosloniec, and R. Raghow.** 1996. Ceramide signaling and the immune response. *Biochim. Biophys. Acta* **1301**:273–287.
5. **Bonfield T. L., J. R. Panuska, M. W. Konstan, K. A. Hilliard, J. B. Hilliard, H. Ghnaim, and M. Berger.** 1995. Inflammatory cytokines in cystic fibrosis lungs. *Amer. J. Respir. Crit. Care Med.* **152**:2111–2118.
6. **Booth, M. C., A. L. Cheung, K. L. Hatter, B. D. Jett, M. C. Callegan, and M. S. Gilmore.** 1997. Staphylococcal accessory regulator (*sar*) in conjunction with *agr* contributes to *Staphylococcus aureus* virulence in endophthalmitis. *Infect. Immun.* **65**:1550–1556.
7. **Bryan, R., A. Perez, D. Kube, P. Davis, and A. Prince.** 1998. Overproduction of the CFTR R-domain leads to increased levels of asialoGM1 and increased *Pseudomonas aeruginosa* binding by epithelial cells. *Am. J. Respir. Cell Mol. Biol.* **19**:269–277.
8. **Cacalano, G., M. Kays, L. Saiman, and A. Prince.** 1992. Production of the *Pseudomonas aeruginosa* neuraminidase is increased under hyperosmolar conditions and is regulated by genes involved in alginate expression. *J. Clin. Invest.* **89**:1866–1874.
9. **Clapham, D. E.** 1995. Calcium signaling. *Cell* **82**:259–268.
10. **Comolli, J. C., L. L. Waite, K. E. Mostov, and J. N. Engel.** 1999. Pili binding to asialoGM1 on epithelial cells can mediate cytotoxicity or bacterial internalization by *Pseudomonas aeruginosa*. *Infect. Immun.* **67**:3207–3214.
11. **Crabtree, G. R.** 1999. Generic signals and specific outcomes: signaling through Ca^{2+}, calcineurin, and NF-AT. *Cell* **96**:611–614.
12. **DeBentzmann, S., P. Roger, F. Dupuit, O. Bajolet-Laudinat, C. Fuchley, M. Plotkowski, and E. Puchelle.** 1996. AsialoGM1 is a receptor for *Pseudomonas aeruginosa* adherence to regenerating respiratory epithelial cells. *Infect. Immun.* **64**:1582–1588.
13. **DiMango, E., H. J. Zar, R. Bryan, and A. Prince.** 1995. Diverse *Pseudomonas aeruginosa* gene products stimulate respiratory epithelial cells to produce interleukin-8. *J. Clin. Invest.* **96**:2204–2210.
14. **DiMango, E., A. Ratner, R. Bryan, S. Tabibi, and A. Prince.** 1998. Activation of NF-κB in normal and cystic fibrosis respiratory epithelial cells. *J. Clin. Invest.* **101**:2598–2606.
15. **Dixon, S. J., D. Stewart, S. Grinstein, and S. Spiegel.** 1987. Transmembrane signaling by the B subunit of cholera toxin: increased cytoplasmic free calcium in rat lymphocytes. *J. Cell Biol.* **105**:1153–1161.
16. **Eckmann, L., M. F. Kagnoff, and J. Fierer.** 1993. Epithelial cells secrete the chemokine interleukin-8 in response to bacterial entry. *Infect. Immun.* **61**:4569–4574.
17. **Feldman, M., R. Bryan, S. Rajan, L. Scheffler, S. Brunnert, H. Tang, and A. Prince.** 1998. The role of flagella in pathogenesis of *Pseudomonas aeruginosa* pulmonary infection. *Infect. Immun.* **66**:43–51.
18. **Garcia-Cozar, F. J., H. Okamura, J. F. Aramburu, K. T. Y. Shaw, L. Pelletier, R. Showalter, E. Villafranca, and A. Rao.** 1998. Two-site interaction of nuclear factor of activated T cells with activated calcineurin. *J. Biol. Chem.* **273**:23877–23883.
19. **Gouy, H., P. Deterre, P. Debre, and G. Bismuth.** 1994. Cell calcium signaling via GM1 cell

surface gangliosides in the human Jurkat T cell line. *J. Immunol.* **152:**3271–3281.

20. **Hedlund, M., M. Svensson, A. Nilsson, R. Duan, and C. Svanborg.** 1996. Role of the ceramide-signaling pathway in cytokine responses to P-fimbriated *Escherichia coli. J. Exp. Med.* **183:** 1037–1044.

21. **Huber, A. R., S. J. Kunkel, R. F. Todd, and S. J. Weiss.** 1991. Regulation of transendothelial neutrophil migration by endogenous interleukin-8. *Science* **254:**99–102.

22. **Immundo, L., J. Barasch, A. Prince, and Q. Al-awqati.** 1995. CF epithelial cells have a receptor for pathogenic bacteria on their apical surface. *Proc. Natl. Acad. Sci. USA* **92:**3019–3023.

23. **Kagnoff, M. F., and L. Eckmann.** 1997. Epithelial cells as sensors for microbial infection. *J. Clin. Invest.* **100:**6–10.

24. **Krivan, H. C., D. D. Roberts, and V. Ginsburg.** 1988. Many pulmonary pathogenic bacteria bind specifically to the carbohydrate sequence GalNAcβ1-4Gal found in some glycolipids. *Proc. Natl. Acad. Sci. USA* **85:**6157–6161.

25. **Li, J.-D., W. Feng, M. Gallup, J.-H. Kim, Y. Kim, and C. Basbaum.** 1998. Activation of NF-κB via a *Src*-dependent Ras-MAPK-pp90rsk pathway is required for *Pseudomonas aeruginosa*-induced mucin overproduction in epithelial cells. *Proc. Natl. Acad. Sci. USA* **95:**5718–5723.

26. **Li, X., H. Yu, L. M. Graves, and H. S. Earp.** 1997. Protein kinase C and protein kinase A inhibit calcium dependent but not stress-dependent c-Jun N-terminal kinase activation in rat liver epithelial cells. *J. Biol. Chem.* **272:**14996–15002.

27. **Liu, J., T. Horner, R. A. Rogers, and J. Schnitzer.** 1997. Organized endothelial cell surface signal transduction in caveolae distinct from glycosylphosphatidylinositol-anchored protein microdomains. *J. Biol. Chem.* **272:**7211–7222.

28. **Massion, P., H. Inoue, J. Richman-Eisenstat, D. Grunberger, P. G. Jorens, B. Houssel, J. P. Wiener-Kronisch, and J. Nadel.** 1994. Novel Pseudomonas product stimulates interleukin-8 production in airway epithelial cells in vitro. *J. Clin. Invest.* **93:**26–32.

29. **Mastronarde, J. G., M. M. Monick, N. Mukaida, K. Matushima, and G. W. Hunninghake.** 1998. Activator protein 1 is the preferred transcription factor for cooperative interaction with nuclear factor κB in respiratory syncytical virus induced interleukin-8 gene expression in the airway epithelium. *J. Infect. Dis.* **177:**1275–1281.

30. **McElvaney, N. G., H. Nakamura, P. Birrer, C. A. Hebert, et al.** 1992. Modulation of airway inflammation in cystic fibrosis. *J. Clin. Invest.* **90:**1296–1301.

31. **Nakamura, H., K. Yoshimura, N. G. McElvaney, and R. G. Crystal.** 1992. Neutrophil elastase in respiratory epithelial lining fluid of individuals with cystic fibrosis induces IL-8 expression in a bronchial epithelial cell line. *J. Clin. Invest.* **89:**1478–1484.

32. **Naumann, M., S. Webler, C. Rartsch, B. Wieland, and T. F. Meyer.** 1997. *Neisseria gonorrhoeae* epithelial cell interaction leads to the activation of the transcription factors nuclear factor κB and activator protein 1 and the induction of inflammatory cytokines. *J. Exp. Med.* **186:**247–258.

33. **Noah, T. L, H. R. Black, P. W. Cheng, R. E. Wood, and M. W. Leigh.** 1997. Nasal and bronchoalveolar lavage fluid cytokines in early cystic fibrosis. *J. Infect. Dis.* **175:**638–647.

34. **Nofer, J.-R., M. Tepel, M. Walter, U. Seedort, G. Assmann, and W. Zidek.** 1997. Phosphatidylcholine-specific phospholipase C regulates thapsigargin-induced calcium influx in human lymphocytes. *J. Biol. Chem.* **272:**32861–32868.

35. **Okamoto, S., N. Mukaida, K. Yasumoto, N. Rice, Y. Ishikawa, H. Horiguchi, S. Murakami, and K. Matsushima.** 1994. The interleukin-8 AP-1 and κB-like sites are genetic end targets of FK506-sensitive pathway accompanied by Ca^{2+} mobilization. *J. Biol. Chem.* **269:** 8582–8589.

36. **Parsons, P. E., G. S. Worthen, and P. E. Henson.** 1991. Injury from inflammatory cells. *Lung. Sci. Found.* **2:**1981–1992.

37. **Ratner, A. J., R. Bryan, J. E. Gelber, M. Heath, M. Davis, and A. Prince.** 1998. Pseudomonas aeruginosa attachment to respiratory epithelial cells induces intracellular Ca^{2+} fluxes which are required for cytokine expression. *Am. J. Respir. Crit. Care Med.* **157:**A200.

38. **Saiman, L., and A. Prince.** 1993. *Pseudomonas aeruginosa* pili bind to asialoGM1 which is increased on the surface of cystic fibrosis epithelial cells. *J. Clin. Invest.* **92:**1875–1880.

39. **Saiman, L., G. Cacalano, D. Gruenert, and A. Prince.** 1992. Comparison of adherence of *Pseudomonas aeruginosa* to respiratory epithelial cells from cystic fibrosis patients and healthy subjects. *Infect. Immun.* **60:**2808–2814.

40. **Schaeffer, H. J., and M. J. Weber.** 1999. Mitogen-activated protein kinases: specific messages from ubiquitous messengers. *Mol. Cell. Biol.* **19:** 2435–2444.

41. **Sheth, H. B., K. K. Lee, W. Y. Wong, et al.** 1994. The pili of *Pseudomonas aeruginosa* strains PAO and PAK bind specifically to the carbohy-

drate sequence βGalNAc(1-4)β Gal found in glycolipids asialoGM1 and asialoGM2. *Mol. Microbiol.* **11:**715–723.

42. **Song, K. S., S. Li, T. Okamoto, L. A. Quillam, M. Sargiacomo, and M. P. Lisanti.** 1996. Co-purification and direct interaction of Ras with caveolin, an integral membrane protein of caveolae microdomains. *J. Biol. Chem.* **271:** 9690–9697.

43. **Standiford, T. J., S. L. Kunkel, M. A. Basha, W. Hensue, J. P. Lynch III, G. B. Toews, J. Westwick, and R. M. Strieter.** 1990. Interleukin-8 gene expression by a pulmonary epithelial cell line. A model for cytokine networks in the lung. *J. Clin. Invest.* **86:**1945–1951.

44. **Su, B., and M. Karin.** 1996. Mitogen-activated protein kinase cascades and regulation of gene expression. *Curr. Opin. Immunol.* **7:**402–411.

45. **Tang, H. B., E. DiMango, R. Bryan, M. Gambello, B. H. Iglewski, J. B. Goldberg, and A. Prince.** 1996. Contribution of specific *Pseudomonas aeruginosa* virulence factors to pathogenesis of pneumonia in a neonatal mouse model of infection. *Infect. Immun.* **64:**37–43.

46. **Tang, H. B., M. Kays, and A. Prince.** 1995. The role of pili in the development of acute *Pseudomonas aeruginosa* respiratory tract infection of the neonatal mouse. *Infect. Immun.* **63:**1278–1285.

47. **Weiser, J. N., N. Pan, K. L. McGowan, D. Musher, A. Martin, and J. Richards.** 1998. Phosphorylcholine on the lipopolysaccharide of *Haemophilus influenzae* contributes to persistence in the respiratory tract and sensitivity to serum killing mediated by C-reactive protein. *J. Exp. Med.* **187:**631–640.

48. **Wesselborg, S., M. K. A. Bauer, M. Vogt, M. L. Schmitz, and K. Schulze-Osthoff.** 1997. Activation of transcription factor NF-κB and p38 mitogen-activated protein kinase is mediated by distinct and separate stress effector pathways. *J. Biol. Chem.* **272:**12422–12429.

THE TYPE III SECRETION PATHWAY: DICTATING THE OUTCOME OF BACTERIAL-HOST INTERACTIONS

Raymond Schuch and Anthony T. Maurelli

14

The exploitation of specialized animal or plant host niches by invading microorganisms is largely governed by secreted bacterial products, including toxigenic proteins and elements of elaborate bacterial surface organelles (i.e., flagella, pili, and piluslike structures). These extracellular proteins and surface structures directly engage the eukaryotic host and can promote avoidance of defense measures and ensure proliferation and transmission at the host's expense (25, 45). A major problem for gram-negative bacterial pathogens is getting the virulence effector proteins to the host target. Gram-negative bacteria contain two membranes that act as barriers separating the cytosol from the outside environment. Between the inner (cytoplasmic) membrane (IM) and the outer membrane (OM) is a periplasmic space and a rigid peptidoglycan layer. A limited number of solutions have evolved among gram-negative pathogens to resolve the problem of delivering these proteins from their sites of synthesis in the bacterial cytosol to the extracellular environment. Given the conserved structure of the bacterial envelope of gram-negative bacterial genera, this limited repertoire may not seem unusual. However, when one considers the diversity of secreted effector molecules and the array of pathogenic lifestyles that they support, it is indeed surprising.

Four major transmembrane secretion pathways have been identified in gram-negative bacteria and are referred to as types I through IV (37, 45, 62). "Secretion" is generally defined as the active transport of proteins from the cytoplasm across the IM and OM into the bacterial supernatant or onto the surface of the bacterial cell (37, 59). By contrast, export refers to the transport of proteins from the cytoplasm into the periplasmic space. Virulence proteins specifically targeted to and secreted by each of these pathways have been identified. The striking primary protein sequence similarities between sets of secretory system components from different organisms reflect the homologous molecular structures that define each pathway. Additionally, subunits comprising each distinct system of a particular pathway are usually encoded within tightly linked chromosomal or plasmid loci, which display well-conserved gene orders. Pathway elements that are conserved at the amino acid level may share identical in vivo localizations

Raymond Schuch and Anthony T. Maurelli Department of Microbiology and Immunology, F. Edward Hébert School of Medicine, Uniformed Services University of the Health Sciences, Bethesda, MD 20814-4799.

Virulence Mechanisms of Bacterial Pathogens, 3rd ed., Edited by K. A. Brogden et al.
©2000 ASM Press, Washington, D.C.

and activities and can be used to functionally complement each other. These findings imply an evolutionary relatedness among the systems comprising each secretion pathway and suggest that the basic mechanisms of secretion among these systems will be similar.

In this chapter, we review the general structure of type III secretion systems in gram-negative bacterial pathogens and the means by which these systems may have arisen and spread. Certain salient features are focused upon, including the regulation of type III secretion induction, subcellular targeting of secreted substrates, and the surface assembly of type III system-associated appendages. Finally, we describe the present state of knowledge regarding the structural details of type III secretion in the *Shigella flexneri* system (referred to as Mxi-Spa). By presenting both the common features ascribed to the type III pathway and the specific structural details of one such system, we hope to illustrate how bacterial pathogens may exploit one basic mechanism for adaptation to a broad range of host interactions.

GENERAL FEATURES OF PROTEIN SECRETION PATHWAYS

The model bacterial secretion pathway for a gram-negative bacterium contains several elements. A set of envelope-associated proteins forms a transmembrane pathway, which recognizes target substrate macromolecules at the cytoplasmic face of the IM and delivers them to the extracellular face of the OM. Because this process is energy dependent, an element of the secretion apparatus is dedicated to providing the energy required for secretion. At the OM face, the secreted substrates may diffuse freely into the extracellular matrix or assemble surface structures that serve as secretory channel extensions and/or effectors of locomotion or adhesion. Peripheral components like cytoplasmic chaperones may also exist to stabilize target proteins in the cytoplasm and to deliver these target proteins to translocase structures. In addition, synthesis of pathway components or regulation of function may by

controlled by specific transcriptional activators or repressors. Features of each secretion pathway are discussed below, and major attributes that are either pathway-specific or conserved are highlighted.

Type I secretion is the simplest system and requires only three distinct proteins: an IM transport ATPase, an IM fusion protein, and an OM pore-forming protein. Substrates for type I secretion, like α-hemolysin of *Escherichia coli*, are recruited to an energized IM complex (translocase). This interaction likely induces formation of a transient bridge to the outer membrane element, creating a transmembrane channel for target protein secretion in a single step (i.e., involving no periplasmic intermediate) (69). Successive rounds of ATP binding and hydrolysis continue the process of substrate engagement and secretion.

The type II pathway, or general secretion pathway (GSP), requires over 20 distinct proteins, which together direct a two-step secretion mechanism involving separable IM and OM translocation events. Proteins secreted via this pathway are characterized by a short (about 20 to 30 amino acids) amino-terminal signal sequence. The signal sequence is cleaved by a periplasmic signal peptidase during translocation through the periplasm. Substrates for secretion via the GSP are delivered by the SecB cytoplasmic chaperone and the SecA ATPase to a large IM translocase structure made of additional Sec proteins, which form a ring-shaped pore (24). Cycles of ATP binding to SecA, target protein translocation, and ATP hydrolysis ensue, driving progressive protein transfer through the translocase into the periplasmic space. Periplasmic GSP substrates can then transit across the OM via several different "terminal branches," which are sets of 10 or more envelope-associated proteins dedicated to protein transfer across OM pores into the extracellular environment (59). Distinct terminal branches have been identified that are specific for the secretion of various extracellular toxins and hydrolases as well as type IV pili. Components of these branches are similar both to each other and to elements involved in other transmembrane traffic systems, including the type III pathway, and systems for filamentous phage biogenesis and extracellular DNA uptake (10, 35). Interestingly, most secretion systems (including those of the type III pathway) require a functional type II system to assemble their structural components.

The type III pathway is also a translocase structure, which consists of 20 or more envelope-associated proteins that can recognize and secrete proteins associated with cytoplasmic chaperones (37). Unlike the proteins secreted via the type II GSP, proteins secreted by the type III pathway have no cleavable signal sequence at their amino terminus. These proteins have virulence-related functions and are directed through the type III pathway to eukaryotic membrane, intracytosolic, and nuclear targets. In many cases, piluslike structures are elaborated on the bacterial surface and may serve to direct protein traffic. While detailed knowledge regarding type III mechanisms of function is lacking, secretion is believed to be a one-step process across a continuous gated transmembrane channel. Direct physical contact between a bacterium and the host cell can induce secretion in a process likely requiring ATP hydrolysis. A set of up to 10 IM-associated type III pathway components are evolutionarily related to the IM-associated components of flagellar secretory pathways from both gram-positive and gram-negative bacteria, based on highly conserved primary amino acid sequences (37, 68). This group of flagellar proteins comprises the bulk of the C- and MS-rings of the flagellum basal body. Elements of these structures recognize and translocate axial subunits (rod, hook, and filament components) across the IM, alter the direction of flagellar movement, and form a section of the rivet complex. Type III pathway elements of the OM, on the other hand, are highly related to the OM elements of type II secretion pathways. These observations are taken as evidence that a progenitor type III system arose by fusion of a set of duplicated flagellar secretory proteins with the outer membrane channel-forming proteins of type II systems (68). The success of this hybrid secretion pathway is attested to by the range of bacterial pathogens that have acquired type III secretion as a major virulence mechanism (Table 1).

The type IV pathway consists of a conserved set of at least 9 membrane-associated proteins, which together direct either the one- or two-step transfer of proteins or protein–DNA complexes out of the bacterium (45). Secretion can be cell contact-dependent and may be directed specifically into either a prokaryotic or eukaryotic target cytosol via a type IV system-associated piluslike structure. Pathogenic macromolecules secreted through this versatile pathway have been implicated in both tumorigenesis in plants (in the case of *Agrobacterium tumefaciens*) and disease in humans (*Bordetella pertussis*); a limited number of type IV pathway homologs have also been identified in secretion systems of both *Legionella pneumophila* and *Helicobacter pylori*. The type IV pathway also includes sets of proteins involved in the conjugal transfer of self-transmissible plasmids. This set of proteins is believed to be the progenitor type IV system, which was subsequently modified to serve virulence-related needs. Details regarding the actual mechanism of type IV secretion are poorly understood and appear to vary dramatically between the different systems identified.

TYPE III SECRETION SYSTEMS: A RAPIDLY EXPANDING UNDERSTANDING OF VIRULENCE PROTEIN DELIVERY

To date, partial or complete sets of type III systems have been identified in over 20 pathogenic bacterial species (Table 1). Techniques that include genetic screens for avirulent mutants or genes expressed exclusively within host cells or tissues, PCR screens for loci encoding well-conserved type III system components, and large-scale genomic sequencing projects have contributed to extending this list of organisms. The systems identified are quite

TABLE 1 Type III virulence protein secretion systems[a]

Bacterium(-a)[b]	Disease caused	Relevant aspects of pathogenic lifestyle	Activities attributed to type III system	Effector injection	Contact induced[c]	Extracellular structures[d]
EPEC	Infantile diarrhea	Predominantly extracellular during infection, colonizes intestinal mucosa	A/E lesion formation on epithelial cells (alter host signaling pathways and induce cytoskeletal changes)	Yes	Yes	Yes
Yersinia spp.	Plague and gastroenteritis	Predominantly extracellular during infection, tropism for lymphoid tissue	Inhibit macrophage phagocytic and signaling capacities (alter inflammatory response and induce apoptosis)	Yes	Yes	Yes
Salmonella enterica[e]	Typhoid fever and gastroenteritis	Extracellular or in phagosome during infection, penetrates intestinal mucosa, may cause systemic disease	SPI-1: enterocyte invasion, transepithelial signaling, macrophage apoptosis. SPI-2: intramacrophage survival, alter intracellular vacuolar traffic	Yes (SPI-1, SPI-2)	Yes (SPI-1)	Yes (SPI-1)

Plant pathogens	See below	Extracellular during infection, colonizes intercellular spaces in plants, triggers cell death	HR elicitation in nonhost plant species, parasitic growth in host species	Likely	Likely	Yes
Shigella spp.	Dysentery	Extracellular or intracytosolic during infection, penetrates colonic epithelium, induces an intense inflammatory reaction	Direct invasion of epithelial cell cytosol, cell-to-cell movement, alter macrophage signaling capacities/induce apoptosis, transepithelial signaling	No	Yes	Yes

[a] Information here was compiled predominantly from references 28 and 37.

[b] Other bacterial species that possess or are likely to possess type III systems (and the diseases caused or functions served) include the following: Enterohemorrhagic E. coli (hemorrhagic colitis and hemolytic uremic syndrome in humans) (47), diffusely adhering E. coli (diarrhea in humans) (9), Hafnia alvei (diarrhea in humans) (47), Citrobacter freundii (colonic hyperplasia in mice) (47), enteroinvasive E. coli (dysentery-like enteritis in humans) (36), rabbit-specific EPEC (diarrhea in rabbits) (47), Bordetella bronchiseptica (respiratory illness in swine, dogs, and rodents) (28), Chlamydia spp. (sexually transmitted, respiratory, and ocular disease in humans) (28), P. aeruginosa (severe opportunistic infections in humans) (37), Burkholderia cepacia (severe opportunistic infections in humans) (C. D. Mohr and D. W. Martin, Abstr. 98th Gen. Meet. Am. Soc. Microbiol. 1998, abstr. B-230, p. 94, 1998), Burkholderia pseudomallei (severe opportunistic infections in humans) (76), Erwinia amylovora (fire blight of apple or pear) (28), Erwinia chrysanthemi (soft rots and parenchymatal necroses) (28), Erwinia herbicola pv. gypsophila (gypsophila galls) (28), Erwinia stewartii (Stewart's wilt of corn) (28), P. syringae (water-soaked lesions in bean and tobacco) (28), Ralstonia solanacearum (bacterial wilt of tomato) (28), Xanthomonas spp. (foliar spots and blights) (28), and Rhizobium spp. (N_2 fixation and root nodule symbiosis) (28).

[c] Bacterial-host cell contact-dependent transfer of at least a subset of effector proteins into extracellular milieu or directly into target host cell.

[d] Elaboration of either filamentous structures (Shigella, Salmonella, EPEC, and plant pathogens) or undefined high-molecular-weight aggregates (Yersinia).

[e] S. enterica encodes two distinct type III secretion pathways, SPI-1 and SPI-2.

versatile with respect to the array of pathogenic lifestyles that they support, and these systems can direct the parasitization of either plant or animal host tissues. This diversity is particularly apparent among the animal pathogens when considering the range of infection types caused (opportunistic or frank, localized or systemic, extracellular or intracellular, self-limiting or life-threatening) and the routes by which they are acquired (oral, respiratory, sexual, injury, or insect bite).

Over the past 10 years, a tremendous amount of research on the subject of virulence protein secretion has focused on the identification and functional analysis of these type III pathways. The net result of this work is an abundance of new insights into both the evolution of pathogenic bacteria and the mechanisms by which they interact with and manipulate basic cellular processes of a host to promote their survival. These issues are discussed below.

TYPE III SYSTEM GENETIC STRUCTURE AND THE HORIZONTAL EXPANSION OF PATHOGENIC TRAITS

A hallmark of the type III pathway concerns the degree of homology between a large subset of components found in all such systems (37). This subset, which may be structurally similar to elements of the progenitor type III pathway, consists predominantly of the essential envelope-associated structural components, or core system, which are encoded within tightly linked operons. Another subset of components, including the secreted effectors and some transcriptional regulators, are poorly conserved between the different systems and are often encoded in clusters at positions adjacent to that of the core components. Proteins in this group likely evolved independently from the core secretory elements and have been gradually integrated into these systems as part of an adaptation to the environment in which each system functions.

Ample evidence indicates that the genetic blocks encoding existing type III systems were acquired through processes of horizontal gene transfer, i.e., the mobilization of DNA between bacterial species by transducing phage, conjugative plasmids, or DNA transformation (48). In most cases, type III systems are encoded within either large plasmids (some belonging to a family of potentially mobilizable plasmids) or chromosomal pathogenicity islands. Pathogenicity islands (reviewed in reference 32) are large, often unstable, DNA regions that encode virulence loci and differ from the bulk of the genome with respect to their percent G+C content and codon usage patterns (indicative of "foreign" DNA). These islands are often inserted at tRNA loci, known hot spots for insertion of extrachromosomal elements, and are flanked by inverted repeat sequences and inactive IS element, transposon, or bacteriophage genes. Thus, it is likely that pathogenicity islands were mobile at one time. Association of type III systems with such putative mobility factors and remnants thereof may explain how a common secretion pathway became so widespread.

Acquisition of type III systems should represent a major advance in the evolution of certain bacteria as pathogens, in particular those recipients in which type III secretion complements existing capacities and allows new or significantly improved mechanisms of host exploitation. Not surprisingly then, these systems are never found in nonpathogenic bacterial strains or in species that are closely related to a type III system-bearing pathogen. Once acquired, type III system functions were modified in a variety of ways, depending on the host and the demands of a pathogenic lifestyle. Such forces have likely driven divergent evolution of the ancestral type III system, yielding the systems apparent today. A major modification of each system, which is unique to each, regards its integration into existing regulatory networks that restrict virulence gene expression to appropriate conditions (37). In *Shigella*, for instance, temperature-regulated induction of the virulence plasmid-encoded type III secretion system is strictly controlled by the chromosomal product, H-NS (22). H-NS represses virulence gene expression at low temperatures by binding to specific promoter sites (see chapter 12). At higher temperatures, sensed by H-NS through alterations in DNA supercoiling, binding is relieved and virulence functions, including the type III secretion system, are induced. In other pathogens, control elements such as two-component response regulators, alternative sigma factors, and histonelike proteins (like H-NS) are employed to tailor type III secretion gene transcription to particular lifestyles (37).

Another significant modification of type III systems, and a force in their divergent evolution, was the acquisition of new secreted effector genes. A source of many type III system substrates appears to be the eukaryotic host, since many substrates demonstrate activities more suited to eukaryotic organisms. Elements that interact with eukaryotic signal-transduction pathways are one example. How such elements are acquired and integrated into type III systems is unknown, but it is clearly the means by which the repertoire of type III system functions may progressively increase and adapt to become more efficient to particular host environments.

FUNCTIONS OF TYPE III SYSTEMS

The widespread distribution of type III secretory functions among pathogenic bacterial species can be partly attributed to the fact that type III secretion efficiently executes a basic aspect of the host-pathogen interaction—delivery of pathogenic bacterial proteins to the host. Its efficiency lies in the ability to deliver virulence proteins to a variety of cellular positions and targets, including the host plasma membrane, cytoplasmic components of signal transduction pathways or the cytoskeleton, and the nucleus. General features of the process by which virulence proteins are delivered to these positions via the type III pathway will be discussed with particular attention to secretion regulation, effector protein delivery, and

the elaboration of type III-specific surface organelles.

Regulation of Effector Secretion

A major feature of type III systems is the induction or upregulation of virulence protein secretion upon intimate contact with target host cells or by conditions or molecules that mimic contact. *S. flexneri* centrifuged onto semiconfluent HeLa cell monolayers releases most of its cytoplasmic pool of Ipa invasins (IpaB, C, D, and A, the target substrates for type III secretion in *Shigella*) into the extracellular environment within 5 min (50). When centrifuged onto tissue culture dishes lacking HeLa cells, little or no secretion was observed. With *Yersinia*, only extracellular-adherent organisms were found competent to induce translocation of several Yop proteins (the type III substrates of *Yersinia*) into target HeLa cells (61). Treatments that may mimic host cell contact massively induce secretory functions, as observed upon incubation of *Shigella* with Congo red dye or extracellular matrix glycoproteins (collagen type IV, fibronectin, or laminin), or *Yersinia* spp. in the absence of calcium (14, 55, 74). The induction of translocase activity near a target eukaryotic cell may ensure high extracellular concentrations of effector proteins at the host membrane (as with the Ipas of *Shigella*) or reflect a requirement for target proteins to be directly injected from the bacterium into the host (as with some of the Yop proteins of *Yersinia*). While issues regarding type III-dependent protein injection are discussed below, it is worth noting that enteropathogenic *E. coli* (EPEC), *Pseudomonas*, *Salmonella*, and probably the plant pathogens also inject effector proteins into target host cells.

While contact-dependent secretion is clearly a requirement for virulence in some cases, secretion may also be induced by other physiologically relevant signals. Ipa secretion by *S. flexneri*, for example, can be induced by treatment with fetal bovine serum (50) or bile salts (58). Invasin secretion by *Salmonella enterica* serovar Typhimurium (16) or attaching and effacing (A/E) lesion effector secretion by EPEC (39) can be induced by a pH shift or growth in mildly alkaline conditions, respectively. The induction of secretion in response to a variety of environmental conditions could sustain multiple secretory bursts during the course of infection and allow application of type III secretory functions to interactions other than that occurring upon direct cell contact.

An unusual finding regarding the regulation of type III secretion in *S. flexneri* regards the role of the Ipa substrates in this process. Wild-type *Shigella* growing in liquid media normally secretes a small subset of its large cytoplasmic Ipa pool (referred to as leakage). *ipaB* or *ipaD* null mutants grown similarly spontaneously release all type III substrates at levels sufficient to cause their visible aggregation (50, 55). This finding, coupled with the isolation of IpaB-IpaD complexes from the bacterial envelope (likely from the OM), led to the hypothesis that IpaB and IpaD interact with each other and the envelope-associated Mxi-Spa translocase to form a secretory "plug." This structure would prevent Ipa secretion in the absence of appropriate signals (such as cell contact). Experimental evidence suggests the existence of a similar antisecretion complex in *S. enterica* (38). In *Yersinia*, three proteins (YopN, TyeA, and LcrG) have been identified that act as sensors and stop-valves to control not only Yop secretion but also synthesis (37). Together, these antisecretion proteins prevent the release of a *yop* transcriptional repressor, LcrQ, until appropriate host cell contact has been established (which in turn upregulates both Yop expression and secretion). Mutant derivatives lacking YopN, TyeA, or LcrG exhibit cell contact-independent Yop synthesis/secretion, much like the spontaneous secretion observed with the *ipaB* and *ipaD* mutants of *Shigella*. *Pseudomonas aeruginosa* encodes a type III pathway closely related to that of *Yersinia* and likely regulates secretion in a similar manner.

While *Shigella* and *Salmonella* maintain large presynthesized pools of certain effector pro-teins prior to host contact, the expression of another set of effector proteins in these systems may actually require host cell contact (in a manner similar to that observed in *Yersinia*). Firstly, homologs of the YopN contact-sensor from *Yersinia* are found in both *Shigella* (MxiC) and *Salmonella* (InvE). Secondly, a level of regulation was identified recently in *Shigella* by which transcription and presumably secretion of two type III system substrates, VirA and IpaH9.8, were upregulated by cell contact-induced Ipa secretion and epithelial cell invasion (20). The induction of VirA and IpaH9.8 synthesis/secretion upon host cell contact also suggests secretion of a repressor with a function similar to that of LcrQ of *Yersinia* described above. This proposed repressor has been tentatively identified (MxiE) and is currently under investigation in our laboratory. The presence of this type of regulation suggests that type III secretion may in some cases occur in waves. In *Shigella*, for example, secretion immediately induced by contact would mediate invasion, whereas synthesis/secretion upregulated by repressor release would mediate the subsequent process of intercellular spread.

Effector Delivery to Host Cells

Once triggered, secretion via the type III pathway can be directed to particular extra-cellular or intracellular host targets. The nature of these targets varies according to the diverse requirements of the invading pathogen. This adaptability, with respect to targets that may be exploited, is one of the major advantages afforded by the type III pathway over other pathways of virulence protein delivery.

Several type III substrates are targeted to the host plasma membrane during infection. At least two mechanisms for mediating such function exist, as exemplified by the type III secretion of IpaB and IpaC by *Shigella* and of the Tir protein by EPEC. During the process by which *Shigella* contacts and invades epithelial cells, IpaB and IpaC are secreted to the extracellular milieu, where they form a soluble complex (51). When it is used to coat inert latex beads, this complex triggers bead internalization by HeLa cells (49). Studies analyzing the interactions between purified IpaB and/or IpaC with artificial membrane bilayers and target host cells suggest that the Ipa complex self-inserts into the host membrane and forms a channel that induces extensive cytoskeletal reorganization in the host (19, 70). The Tir protein of EPEC, unlike the Ipa proteins, is directly inserted into the host membrane without a detectable extracellular intermediate (40). Insertion of Tir requires an intact type III system as well as one of its secreted substrates, EspA. EspA forms extracellular surface structures in EPEC, which are proposed to mediate this direct insertion of Tir into the host membrane (41). Once translocated, Tir is a receptor for the bacterial adhesin intimin, which is required for intimate host cell attachment and for A/E lesion formation.

In several extensively studied type III systems, secretion is coupled with effector translocation directly into the target host cell cytosol (Table 1). This direct injection of virulence proteins is one reason why type III systems are often referred to as molecular syringes. Syringe functions have been visualized and confirmed using a variety of techniques, including immunofluorescence and confocal microscopy, immunoelectron microscopy, a reporter-enzyme strategy, and direct fractionation of infected host cells. Within each system capable of such translocation, only a subset of secreted effectors may be so targeted. Within these subsets, only a portion of the total secreted pool may be detected inside of target cells. Those that are injected may stimulate or interfere with certain cellular processes, usually related to cytoskeletal dynamics. In *Yersinia,* six type III-translocated Yop proteins have been detected at different positions within target cells (37, 67): YopE, enriched in the perinuclear region, disrupts actin microfilaments (likely through interactions with host GTPases); YopH, enriched in the host cytosol and plasma membrane, is a protein tyrosine phosphatase that dephosphorylates the host proteins p130^{CAS} and focal adhesion kinase; YpkA, enriched in the inner surface of the plasma membrane, is a serine/threonine kinase that drives cytoskeletal contraction; YopT, at an unknown position in the host, alters signal transduction and disrupts actin microfilaments; YopJ, in the cytoplasm of host cells, activates the apoptotic pathway of murine macrophages; YopM, transported to the nucleus (one of only three bacterial proteins known to do so), serves an unknown but essential role in virulence. Together these proteins make up the translocated antihost arsenal of *Yersinia* and allow a coordinated attack on host cell functions, which serves to inhibit phagocytosis and promote bacterial survival in lymphoid tissues.

The translocated antihost system of *Salmonella* (called SPI-1) consists of a set of proteins (including an exchange factor for rho GTPases, an inositol phosphate phosphatase, an actin binding protein, and a tyrosine phosphatase) that together remodel the host cytoskeleton and induce membrane ruffling and macropinocytotic events, which results in bacterial uptake (27). After entry, *Salmonella* upregulates expression of another type III system (SPI-2), which translocates proteins from intravacuolar bacteria into the host cytosol to alter vacuolar maturation and promote intramacrophage survival. The *P. aeruginosa* translocated antihost system includes an ADP-ribosylating toxin that targets GTP-binding proteins of the Ras superfamily to disrupt the cytoskeletal structure of macrophages and inhibit phagocytosis (26). Additional injected type III substrates of *P. aeruginosa* include an adenylate cyclase and a potent cytotoxin. In plant pathogens, several different translocated proteins have been tentatively identified, including an activator of a serine/threonine kinase signaling pathway as well as a probable transcription factor (which encodes a nuclear localization sequence). These proteins likely induce the hypersensitive response in nonpermissive plants and promote growth and eventual disease in permissive plants (34). Clearly, the injector function of the type III pathway can be manipulated to support a broad range of bacterial parasitizations.

Type III System-Specific Surface Appendages

An emerging theme in type III secretion concerns the assembly of large filamentous surface structures, which likely act as extracellular extensions of type III systems to deliver effector proteins (Table 1). Scanning electron microscopic analysis of interactions between serovar Typhimurium and cultured epithelial cells first revealed the presence of such appendages. These invasomes (as they were referred to) are induced by cell contact, are dependent on a functional type III system for assembly, are ~60 nm in diameter and 0.3-1 μm in length, and are shed upon initiation of the membrane ruffling events triggered by type III effector proteins (31). Using an array of microscopic techniques, type III system-dependent structures similar in size to invasomes were also observed mediating direct physical contact between the surfaces of EPEC and its target host cell (41). The EPEC structures are hollow tubes comprising EspA (a type III substrate), which are required for protein translocation functions and are lost after A/E lesion formation. Fine-detail analysis revealed that the EspA structures are constructed from aggregates of smaller piluslike filaments, 7 to 8 nm in diameter. These filaments, rather than the larger element, are more similar to the type III system pili elaborated by the plant pathogen, *Pseudomonas syringae*. The *P. syringae* pilus consists of at least one type III substrate (HrpA), has a diameter of 8 nm and a length of ~2 μm, and is proposed to pierce the plant cell wall to allow protein delivery (60). The type III system of *S. flexneri* also produces a pilus (5 nm in diameter and 120 nm in length). This structure bridges intravacuolar shigellae with the host membrane during the process of intercellular spread and is proposed to act as an Ipa delivery structure (65). Intravacuolar chlamydiae elaborate similar pili (10 to 13 nm in diameter), possibly constructed by a type III system for the purpose of nutrient acquisition

(8). *Yersinia*, while not known to elaborate structures via its type III system, does produce Yop aggregates in culture, which may represent appendages formed as part of the Yop injection system (52). The widespread nature of the type III system-dependent pili clearly suggests an important role for these structures in the secretion process. Because evidence for these pili as secretory extensions is indirect at best, other roles, such as in the process of adhesion, may also exist.

TYPE III SECRETION MACHINES: EXAMPLE FROM THE *SHIGELLA* SYSTEM

In 1992, homologies between virulence-associated secretory genes in *S. flexneri* and *Y. pestis* were first reported (7). Subsequently, additional homologies were also reported to putative secretory genes in serovar Typhimurium and several plant pathogenic bacteria (73). Thus, the type III pathway was recognized. Since then, a great deal of work has focused on the homologies between the components of each system and the functional characterization of secreted effector proteins that distinguish each system. It has become clear that these studies will reveal more detail about mechanisms of bacterial pathogenesis and that they will uncover novel targets for antimicrobial development with broad spectrum potential. Little work has been done, however, on the structure and function of the envelope-associated type III secretion machine, i.e., how substrates are transported across the bacterial membranes. This gap is surprising, considering that the bulk of genetic information specifying the type III pathway concerns this transmembrane structure and that the components of this structure are the most broadly conserved elements. This dearth of knowledge certainly hampers detailed understanding of what has already been learned, and it will impede attempts to develop broad-spectrum therapies targeting the type III system-bearing pathogens as a group. For these reasons, characterization of the type III transmembrane elements should be important in future research undertakings. In this section, we discuss what is known regarding the inventory of type III secretion genes, using the type III system of *Shigella* (called Mxi-Spa) as a model, and describe how these proteins may contribute to a supramolecular secretion complex.

Assignment of type III secretion components (Mxi-Spa or otherwise) to the category of structural subunit is based on features such as predicted secondary structure (including membrane-spanning hydrophobic domains), lipoprotein modification sites, signal sequences recognized by the type II (Sec-dependent) secretion system, homologies with characterized proteins, and direct localization studies. Five classes of envelope-associated type III components have been discerned, including proteins that are (i) peripherally IM-associated, (ii) IM-anchored, (iii) OM-anchored, (iv) IM-OM spanning proteins, and (v) mobile secretory elements. Elements encoded in the *mxi-spa* gene cluster that fall into these categories (and others) are discussed below and/or in Table 2.

The IM-anchored Mxi-Spa elements include a set of at least five proteins (MxiA, Spa24, Spa9, Spa29, and Spa40) that are strongly predicted to possess multiple membrane-spanning domains. These proteins are required for Ipa secretion and *Shigella* invasiveness and are the most highly conserved type III system components (6, 37, 63). Like the flagellar homologs, the IM *Shigella* proteins presumably form a protein-conducting pore, possibly defined by their transmembrane domains. Large cytoplasmic domains are also predicted for MxiA, Spa24, and Spa40 (366aa, 90aa, and 135aa, respectively) and could form an Ipa substrate docking structure. The cytoplasmic domains of MxiA and Spa24 are predicted to adopt coiled-coil structures (sites of homo- or heteropolymeric interactions between different polypeptides common among large multiprotein assemblies [54]) which could mediate the protein-protein interactions necessary within a docking complex. That this complex could receive Ipa substrates is supported by results of domain swapping experiments among the MxiA homologs in *Salmonella* and *Yersinia* (InvA and LcrD, respectively) (30). The transmembrane domains of InvA and LcrD (and by extension, MxiA) are functionally interchangeable, whereas the cytoplasmic extensions (which are the least-conserved region of these proteins) are not. The cytoplasmic domains may therefore impose specificity on each system, possibly through interactions with system-specific proteins like the secreted substrates. Spa40 also likely participates in docking processes, based on findings that the flagellar homolog, FlhB, controls the substrate specificity of the flagellar export apparatus (75).

Ipa delivery to the putative cytoplasmic docking structure of Mxi-Spa likely occurs in a manner similar to that of the type II pathway, in which the SecB chaperone and SecA ATPase bind to target proteins and insert them into the IM translocase. While type III secretory components similar in sequence to SecB do not exist, cytoplasmic chaperones that bind to and stabilize proteins targeted for secretion are common in type III systems (37). Additionally, all type III systems encode a well-conserved cytoplasmic ATPase (37), which could be a functional homolog of the energy producing SecA protein (though no sequence similarities exist). The putative ATPase in the *Shigella* type III system, Spa47, is required for Ipa secretion and invasion (63). As a family, the Spa47-like proteins of type III systems are highly homologous over a large central region to the catalytic domains of the water-soluble F_1 component of F_0F_1 ATPase and are therefore presumed to energize type III secretion or the process of translocon synthesis. The structure of Spa47 and its homologs in type III secretion consists of a moderately divergent N-terminal region, a conserved central domain (containing ATP/GTP binding sites), and a conserved C-terminal domain, which is strongly predicted to form a coiled-coil structure. Based on this structure, the Spa47 family could serve a SecA-like function in type III secretion as follows: the N-terminal domain binds target sub-

TABLE 2 Products of the Mxi-Spa locus of *S. flexneri*

Protein	Size (kDa)	Location[a]	Type III system homologs and relevent information[b]	Function in *Shigella* (demonstrated or inferred)[c]	Reference(s)
IpgD	59	S	SopB (44%), inositol phosphate phosphatase injected into host	Not required for invasion or intercellular spread (may affect intestinal secretory/inflammatory response)	4
IpgE	13	C*	PipC (28%), proposed SopB chaperone	Unknown role (may stabilize cytoplasmic IpgD)	29
IpgF	17	P	IagB (44%) and a family of lytic transglycosylases involved in secretory pathway assembly	Not required for invasion or intercellular spread; may help to degrade cell wall during Mxi-Spa assembly	4, 44
MxiG	41	IM/OM	PrgH (23%) is the only homolog	Required for invasion and Ipa secretion; internal RGD sequence is required for interaction with host membrane during intercellular spread	3
MxiH	9	C*	PrgI (68%); *Shigella* *mxiI* is a paralog (18%); homologs in animal systems	Unknown role; coiled-coil domain suggests participation in Mxi-Spa complex	
MxiI	11	C*	PrgJ (37%); homologs in animal and plant systems	Unknown role; coiled-coil domain suggests participation in Mxi-Spa complex	

MxiJ	27	IM/OM*	PrgK (46%); homologs in animal, plant, and flagellar systems; conserved structure: lipidated N terminus in OM and C-terminal TM domain in IM	Lipoprotein required for Ipa secretion and invasion; formation of bridge spanning the IM and OM	1
MxiK	47	C*	OrgA (24%) is the only homolog	Unknown role	
MxiL	15	C*	Weak identity only to *Yersinia* chaperone YerA (18%)	Unknown role; size, acidity, and similarity to YerA suggest it is a chaperone; has coiled-coil domain	37
MxiM	16	OM (inner face)	No homolog in any bacterial secretion system	Lipoprotein required for Ipa secretion, invasion, and intercellular spread; stabilizes MxiD expression and formation of high Mr MxiD complexes in OM	64, 65, Schuch and Maurelli, unpublished data
MxiE	23	C*	InvF (37%); homologs in animal and plant systems; AraC family of transcriptional regulators	No role in invasion but likely postinvasion functions; postinvasion regulation of *mxi-spa* expression	37, Schuch and Maurelli, unpublished data
MxiD	64	OM	InvG (46%); homologs in animal and plant systems and in type II pathways; multimerizes in OM to form a gated ring structure	Required for Ipa secretion and invasion; requires MxiM for stabilization; amphipathic β sheet structures at C terminus; predicted OM pore	10, Schuch and Maurelli, unpublished data
MxiC	39	OM*	InvE (31%); homologs in animal and plant systems; *Yersinia* homolog controls secretion induction	Unknown role; coiled-coil domains suggest interaction with Mxi-Spa complex; may regulate secretion	37

Continued on following page

TABLE 2 (*Continued*)

Protein	Size (kDa)	Location[a]	Type III system homologs and relevent information[b]	Function in *Shigella* (demonstrated or inferred)[c]	Reference(s)
MxiA	73	IM	InvA (64%); homologs in animal, plant, and flagellar systems; conserved structure: 8 TM domains and a large C-terminal cytoplasmic domain	Required for Ipa secretion and invasion; coiled-coil region in cytoplasmic domain; may form part of IM pore and the cytoplasmic Ipa docking structure	6
Spa15	15	C*	InvB (33%) is the only homolog; weak similarity to flagellar export pathway component FliH	Not required for invasion; soluble, highly acidic protein may be peripherally associated with Mxi-Spa	63
Spa47	47	C*	InvC (58%); homologs in animal, plant, and flagellar systems (also F_0F_1 proton-translocating ATPases); conserved structure: ATP/GTP binding motifs	Required for Ipa secretion and invasion; coiled-coil domain at C terminus suggests interactions within Mxi-Spa; may energize the process of Ipa secretion	37, 63
Spa13	13	S*	SpaM (23%); homologs in animal and plant systems; a "mobile secretory element" in *Yersinia*	Required for Ipa secretion and invasion; coiled-coil domains suggest interactions within Mxi-Spa; may be mobile secretory protein	56, 63
Spa32	32	OM	SpaN (19%); homologs in animal and plant systems; a "mobile secretory element" in *Yersinia*	Required for Ipa secretion and invasion; controls Ipa surface release; may be mobile secretory protein	57, 63

Spa33	33	S	SpaO (24%); homologs in animal, plant, and flagellar systems (N-terminal regions are highly divergent); a "mobile secretory element" in *Salmonella*	Required for Ipa secretion, invasion, and intercellular spread; overexpression blocks Ipa secretion; may be mobile secretory protein	13, 63, Schuch and Maurelli, unpublished data
Spa24	24	IM*	SpaP (63%); homologs in animal, plant, and flagellar systems; conserved structure: 4 TM domains, large central cytoplasmic domain	Required for Ipa secretion and invasion; coiled-coil domain in central cytoplasmic domain; may form part of IM pore and the cytoplasmic Ipa docking structure	63
Spa9	9	IM*	SpaQ (64%); homologs in animal, plant, and flagellar systems; conserved structure: 2 TM domains and well-conserved periplasmic domain	Required for Ipa secretion and invasion; may form part of IM pore and periplasmic bridge	63
Spa29	29	IM*	SpaR (45%); homologs in animal, plant, and flagellar systems; conserved structure: 6 TM domains	Required for Ipa secretion and invasion; may form part of IM pore	63
Spa40	40	IM*	SpaS (49%); homologs in animal, plant, and flagellar systems; conserved structure: 4-6 TM domains and C-terminal cytoplasmic domain	Required for Ipa secretion and invasion; may form part of IM pore and the cytoplasmic Ipa docking structure	63

[a] Indicates the subcellular location of each protein. Proteins for which no localization data exist were analyzed using the PSORT computer prediction and are marked with an asterisk (*). C, cytoplasmic, but peripherally associated with the IM translocase; IM, inner membrane; OM, outer membrane; P, periplasmic; S, secreted; and IM/OM, spanning the two membranes.

[b] Homologies and % identities were identified by BLAST (5) and ALIGN, respectively. The *Salmonella* SPI-1 system homolog (and % identity to the respective Mxi-Spa protein) is listed except where indicated and is the most closely related system to Mxi-Spa based on protein similarities; these SPI-1 homologs are listed with the % identities in parentheses. Transmembrane (TM) domains in the *Shigella* proteins were identified using SOSUI and TopPred 2 computer analyses.

[c] Coiled-coil domains were identified using the COILS program (46).

strates (explaining the divergence of this region among different type III systems), while the C-terminal domain interacts with the Ipa docking complex/translocase at the IM (via interactions between coiled-coil domains of Spa47, MxiA, and Spa24), and ATP binding/hydrolysis in the central domain drives Ipa insertion into an IM pore structure. Interestingly, interactions between the SecA ATPase and the IM translocase occur over a C-terminal region of SecA that contains a coiled-coil domain (23, 72).

Since periplasmic intermediates of proteins secreted via the type III pathway are not normally detectable, a one-step translocation mechanism has been proposed in which secretion flows across the periplasm in a channel linking the IM and OM. The likely candidate for this transmembrane channel element is MxiJ, a member of a highly conserved family of type III components. MxiJ is absolutely required for Ipa secretion and invasiveness (1). It is a lipidated membrane protein presumably anchored at its N terminus to the OM via lipid moieties and at its C terminus to the IM via a transmembrane domain. This structure is highly conserved among the MxiJ family members, suggesting a pathway-wide role as the transperiplasmic bridge. Experimental support for this function is derived primarily from work on the plant system homolog, HrcJ, which is detectable in both IM- and OM-associated proteins fractions, and is required for efficient type III protein transfer across the IM (12, 21). Additionally, the MxiJ family proteins are homologous over their N-terminal half with the flagellar basal body component, FliF. The N-terminal region of FliF forms the inner core of a ring structure through which flagellar subunits are transferred (71).

The sequences of both MxiD and MxiM suggest that these proteins form the outer membrane structure of Mxi-Spa. Indeed, both proteins are detected in the OM and are absolutely required for Ipa secretion and *Shigella* invasiveness (2, 64). MxiD is highly homologous to a large family of proteins called secre-

tins, which form gated OM channels involved in secretory systems such as type IV pilus biogenesis, filamentous phage extrusion, and type II and type III secretions (10). Secretins form multimeric complexes in the outer membrane, which appear as ringlike structures in electron micrographs. The N-terminal half of secretins has been shown in one case to bind secretion intermediaries (66) and likely forms the inner face of the OM ring. The C-terminal half of secretins forms amphipathic β strands, defining the outer edges of the OM ring (11), and serves as the binding site for another OM protein—the secretin pilot (17, 18, 33). Pilots are secretin-specific, display no sequence homology, and may serve up to three functions, including protection of secretins from proteolysis (15, 33), OM insertion of secretins (15, 18, 42), and as a structural subunit of the OM pore (53). We propose that MxiM is the MxiD pilot based on findings that, like other pilots, it is an OM-linked lipoprotein, which stabilizes the expression of MxiD and promotes the assembly of MxiD into OM-linked high-molecular-weight, partially heat-resistant complexes (R. Schuch and A. T. Maurelli, unpublished data). Additionally, the OM insertion of MxiM, while not required to stabilize MxiD, is required for formation of the MxiD complexes. The interaction between MxiM and MxiD likely results in the formation of a cylindrical OM pore for Ipa traffic.

Among the Spa proteins required for Ipa secretion and invasiveness, Spa13, Spa32, and Spa33 are unusual in that they display only weak homologies to other type III system elements (37). Also notable are the facts that Spa32 is secreted to an attachment site at the OM surface (74) and Spa33 is secreted to the supernatant (R. Schuch and A. T. Maurelli, unpublished data). The *Yersinia* and *Salmonella* homologs of these proteins, including Spa13, also display the unusual feature of being secreted proteins required for secretion or mobile secretory elements (13, 56, 57). In *Yersinia*, these findings have been taken to suggest a dynamic moving core for the type

III system, which serves as part of the mechanism pumping out the target Yop proteins. The weak sequence conservation among this group of mobile secretory elements is believed to reflect interactions with the system-specific secreted substrates.

Recently, a largely intact supramolecular type III secretion complex was recovered from the envelope of serovar Typhimurium and visualized by electron microscopy (43). A hollow needlelike structure was observed connected at its base to a cylindrical set of at least four rings. Considering its physical similarity with the flagellar basal body structure,

the ringed base is likely the transmembrane structure, while the needle element extends from the secretory channel into the extracellular environment. Interestingly, this "needle complex" appears to be similar to the baseplate structure of the bacteriophage T4. Because the *Salmonella* type III pathway components are most similar to that of *Shigella* (at the level of amino acid identity), a similar secretory structure can be envisioned in *Shigella*. Based on this, and the information presented here regarding Mxi-Spa, a model for the *Shigella* type III system structure and function is presented (Fig. 1). In response to con-

A.

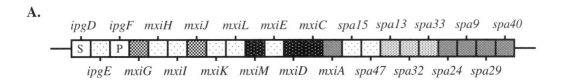

B.

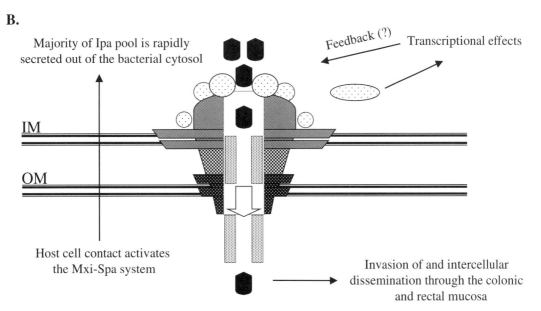

FIGURE 1 (A) The ~20 kb Mxi-Spa locus of the *S. flexneri* virulence plasmid. With the exception of at least IpgD and IpgE (which are predicted to be a secreted effector protein and its chaperone, respectively), the products of these loci are predicted to assemble as the transmembrane Mxi-Spa type III apparatus. The known or putative location of each of these proteins is denoted within the indicated box: ⬚, cytoplasmic or peripherally associated with the IM; ▨, mobile secretory elements; ▪, IM-anchored; ▩, spanning the IM and OM; ▪, OM-anchored; S, secreted; P, periplasmic. (B) Model for Ipa secretion through Mxi-Spa. Subunit shading corresponds to the gene shading above. The Ipa proteins are shown as black bullets.

tact-induction (removal of the IpaB–IpaD plug), cytoplasmic Ipas are delivered to an inner membrane translocase consisting of five proteins (MxiA, Spa24, Spa9, Spa29, and Spa40), which together form an Ipa docking structure and a trans-IM channel. These proteins may form the lower set of rings observed in the purified *Salmonella* structure. ATP hydrolysis by Spa47 likely energizes Ipa transfer across the translocase rings into the axial channel elements, which include a MxiJ transperiplasmic channel connected to a MxiM–MxiD OM pore. These axial channel elements probably form the outer rings of the type III system base structure and serve to direct Ipa flow into the needlelike portion of the pathway. Putative needle elements may include the mobile secretory proteins described (secreted proteins that serve extracellular roles in secretion) and would further direct Ipa traffic into either the extracellular environment or extended surface structures. Based on the systemwide similarities concerning the structural components of type III secretion, the mechanics of secretion described here for *Shigella* are predicted to be basically the same among type III systems from other pathogens described. Most differences lie in the array of secreted effectors released and cellular targets that are attacked.

CONCLUSIONS

We have presented an abridged description of the type III secretion pathway, including details of its evolution and distinguishing characteristics. The array of functions served, including the highly regulated delivery of sets of dissimilar virulence proteins to multiple extracellular and/or intracellular targets, is a striking example of how bacterial pathogens can tailor a core virulence mechanism to multiple pathogenic lifestyles. The present widespread distribution of the type III pathway is a testament to the plasticity of type III systems to modifications and improvements, and to its role as a constantly evolving virulence mechanism. Using *Shigella* as a model, a predicted pathway for type III secretion has been presented that can likely be extended to most

type III systems. Future research should be directed toward defining the true architecture of this structure and the means by which it physically transfers virulence proteins. Undoubtedly, this will involve unraveling the network of protein-protein interactions among both the structural and secreted substrate elements of type III systems that allow this virulence strategy to function.

REFERENCES

1. **Allaoui, A., P. J. Sansonetti, and C. Parsot.** 1992. MxiJ, a lipoprotein involved in secretion of *Shigella* Ipa invasins, is homologous to YscJ, a secretion factor of the *Yersinia* Yop proteins. *J. Bacteriol.* **174:**7661–7669.
2. **Allaoui, A., P. J. Sansonetti, and C. Parsot.** 1993. MxiD: an outer membrane protein necessary for the secretion of the *Shigella flexneri* Ipa invasins. *Mol. Microbiol.* **7:**59–68.
3. **Allaoui, A., P. J. Sansonetti, R. Ménard, S. Barzu, J. Mounier, A. Phalipon, and C. Parsot.** 1995. MxiG, a membrane protein required for secretion of *Shigella* spp. Ipa invasins: involvement in entry into epithelial cells and in intercellular dissemination. *Mol. Microbiol.* **17:**461–470.
4. **Allaoui, A., R. Ménard, P. J. Sansonetti, and C. Parsot.** 1993. Characterization of the *Shigella flexneri ipgD* and *ipgF* genes, which are located in the proximal part of the *mxi* locus. *Infect. Immun.* **61:**1707–1714.
5. **Altschul, S. F., W. Gish, W. Miller, E. W. Myers, and D. J. Lipman.** 1990. Basic local alignment search tool. *J. Mol. Biol.* **215:**403–410.
6. **Andrews, G. P., A. E. Hromockyj, C. Coker, and A. T. Maurelli.** 1991. Two novel virulence loci, *mxiA* and *mxiB*, in *Shigella flexneri* 2a facilitate excretion of invasion plasmid antigens. *Infect. Immun.* **59:**1997–2005.
7. **Andrews, G. P., and A. T. Maurelli.** 1992. *mxiA* of *Shigella flexneri* 2a, which facilitates export of invasion plasmid antigens, encodes a homolog of the low-calcium response protein, LcrD, of *Yersinia pestis*. *Infect. Immun.* **60:**3287–3295.
8. **Bavoil, P. M., and R.-C. Hsia.** 1998. Type III secretion in *Chlamydia*: a case of déjà vu? *Mol. Microbiol.* **28:**859–862.
9. **Beinke, C., S. Laarmann, C. Wachter, H. Karch, L. Greune, and M. A. Schmidt.** 1998. Diffusely adhering *Escherichia coli* strains induce attaching and effacing phenotypes and secrete homologs of Esp proteins. *Infect. Immun.* **66:**528–539.

10. **Bitter, W., and J. Tommassen.** 1999. Ushers and other doorkeepers. *Trends Microbiol.* **7:**4–6.

11. **Bitter, W., M. Koster, M. Latijnhouwers, H. de Cock, and J. Tommassen.** 1998. Formation of oligomeric rings by XcpQ and PilQ, which are involved in protein transport across the outer membrane of *Pseudomonas aeruginosa. Mol. Microbiol.* **27:**209–219.

12. **Charkowski, A. O., H.-C. Huang, and A. Collmer.** 1997. Altered localization of HrpZ in *Pseudomonas syringae* pv. syringae *hrp* mutants suggest that different components of the type III secretion pathway control protein translocation across the inner and outer membranes of gram-negative bacteria. *J. Bacteriol.* **179:**3866–3874.

13. **Collazo, C. M., and J. E. Galan.** 1996. Requirement for exported proteins in secretion through the invasion-associated type III system of *Salmonella typhimurium. Infect. Immun.* **64:**3524–3531.

14. **Cornelis, G. R.** 1998. The *Yersinia* deadly kiss. *J. Bacteriol.* **180:**5495–5504.

15. **Crago, A. M., and V. Koronakis.** 1998. *Salmonella* InvG forms a ring-like multimer that requires the InvH lipoprotein for outer membrane localization. *Mol. Microbiol.* **30:**47–56.

16. **Daefler, S.** 1999. Type III secretion by *Salmonella typhimurium* does not require contact with a eukaryotic host. *Mol. Microbiol.* **31:**45–51.

17. **Daefler, S., I. Guilvout, K. R. Hardie, A. P. Pugsley, and M. Russel.** 1997. The C-terminal domain of the secretin PulD contains the binding site for its cognate chaperone, PulS, and confers PulS dependence on pIV^{f1} function. *Mol. Microbiol.* **24:**465–475.

18. **Daefler, S., and M. Russel.** 1998. The *Salmonella typhimurium* InvH protein is an outer membrane lipoprotein required for the proper localization of InvG. *Mol. Microbiol.* **28:**1367–1380.

19. **De Geyter, C., B. Vogt, Z. Benjelloun-Touimi, P. J. Sansonetti, J.-M. Ruysschaert, C. Parsot, and V. Cabiaux.** 1997. Purification of IpaC, a protein involved in entry of *Shigella flexneri* into epithelial cells and characterization of its interaction with lipid membranes. *FEBS Lett.* **400:**149–154.

20. **Demers, B., P. J. Sansonetti, and C. Parsot.** 1998. Induction of type III secretion in *Shigella flexneri* is associated with differential control of transcription of genes encoding secreted proteins. *EMBO J.* **17:**2894–2903.

21. **Deng, W.-L., and H.-C. Huang.** 1999. Cellular locations of *Pseudomonas syringae* pv. syringae HrcC and HrcJ proteins, required for harpin secretion via the type III pathway. *J. Bacteriol.* **181:**2298–2301.

22. **Dorman, C. J., and M. E. Porter.** 1998. The *Shigella* virulence gene regulatory cascade: a paradigm of bacterial gene control mechanisms. *Mol. Microbiol.* **29:**677–684.

23. **Economou, A.** 1999. Following the leader: bacterial protein secretion through the Sec pathway. *Trends Microbiol.* **7:**315–320.

24. **Fekkes, P., and A. J. M. Driessen.** 1999. Protein targeting to the bacterial cytoplasmic membrane. *Microbiol. Mol. Biol. Rev.* **63:**161–173.

25. **Finlay, B. B., and S. Falkow.** 1997. Common themes in microbial pathogenicity revisited. *Microbiol. Mol. Biol. Rev.* **61:**136–169.

26. **Frithz-Lindsten, E., Y. Du, R. Rosqvist, and Å. Forsberg.** 1997. Intracellular targeting of exoenzyme S of *Pseudomonas aeruginosa* via type III-dependent translocation induces phagocytosis resistance, cytotoxicity and disruption of actin microfilaments. *Mol. Microbiol.* **25:**1125–1139.

27. **Galan, J. E.** 1998. Interactions of *Salmonella* with host cells: encounters of the closest kind. *Proc. Natl. Acad. Sci. USA* **95:**14006–14008.

28. **Galan, J. E., and A. Collmer.** 1999. Type III secretion machines: bacterial devices for protein delivery into host cells. *Science* **284:**1322–1328.

29. **Galyov, E. E., M. W. Wood, R. Rosqvist, P. B. Mullan, P. R. Watson, S. Hedges, and T. S. Wallis.** 1997. A secreted effector protein of *Salmonella dublin* is translocated into eukaryotic cells and mediates inflammation and fluid secretion in infected ileal mucosa. *Mol. Microbiol.* **25:**903–912.

30. **Ginocchio, C. C., and J. E. Galan.** 1995. Functional conservation among members of the *Salmonella typhimurium* InvA family of proteins. *Infect. Immun.* **63:**729–732.

31. **Ginocchio, C. C, S. B. Olmsted, C. L. Wells, and J. E. Galan.** 1994. Contact with epithelial cells induces the formation of surface appendages on *Salmonella typhimurium. Cell* **76:**717–724.

32. **Hacker, J., G. Blum-Oehler, I. Muhldorfer, and H. Tschape.** 1997. Pathogenicity islands of virulent bacteria: structure, function and impact on microbial evolution. *Mol. Microbiol.* **23:**1089–1097.

33. **Hardie, K. R., A. Seydel, I. Guilvout, and A. P. Pugsley.** The secretin-specific, chaperone-like protein of the general secretory pathway: separation of proteolytic protection and piloting functions. *Mol. Microbiol.* **22:**967–976.

34. **He, S. Y.** 1997. Hrp-controlled interkingdom protein transport: learning from flagellar assembly? *Trends Microbiol.* **5:**489–495.

35. **Hobbs, M., and J. S. Mattick.** 1993. Common components in the assembly of type 4 fimbriae,

DNA transfer systems, filamentous phage, and protein-secretion: a general system for the formation of surface-associated protein complexes. *Mol. Microbiol.* **10:**233–243.

36. **Hsia, R.-C., P. L. C. Small, and P. M. Bavoil.** 1993. Characterization of virulence genes of enteroinvasive *Escherichia coli* by Tn*phoA* mutagenesis: identification of *invX*, a gene required for entry into HEp-2 cells. *J. Bacteriol.* **175:**4817–4823.

37. **Hueck, C. J.** 1998. Type III protein secretion systems in bacterial pathogens of animals and plants. *Microbiol. Mol. Biol. Rev.* **62:**379–433.

38. **Kaniga, K., D. Trollinger, and J. E. Galan.** Identification of two targets of the type III protein secretion system encoded by the *inv* and *spa* loci of *Salmonella typhimurium* that have homology to the *Shigella* IpaD and IpaA proteins. *J. Bacteriol.* **177:**7078–7085.

39. **Kenny, B., and B. B. Finlay.** 1995. Protein secretion by enteropathogenic *Escherichia coli* is essential for transducing signals to epithelial cells. *Proc. Natl. Acad. Sci. USA* **92:**7991–7995.

40. **Kenny, B., R. DeVinney, M. Stein, D. J. Reinscheid, E. A. Frey, and B. B. Finlay.** 1997. Enteropathogenic *E. coli* (EPEC) transfers its receptor for intimate adherence into mammalian cells. *Cell* **91:**511–520.

41. **Knutton, S., I. Rosenshine, M. J. Pallen, I. Nisan, B. C. Neves, C. Bain, C. Wolff, G. Dougan, and G. Frankel.** 1998. A novel EspA-associated surface organelle of enteropathogenic *Escherichia coli* involved in protein translocation into epithelial cells. *EMBO J.* **17:**2166–2176.

42. **Koster, M., W. Bitter, H. de Cock, A. Allaoui, G. R. Cornelis, and J. Tommassen.** 1997. The outer membrane component, YscC, of the Yop secretion machinery of *Yersinia enterocolitica* forms a ring-shaped multimeric complex. *Mol. Microbiol.* **26:**789–797.

43. **Kubori, T., Y. Matsushima, D. Nakamura, J. Uralil, M. Lara-Tejero, A. Sukhan, J. E. Galan, and S.-I. Aizawa.** 1998. Supramolecular structure of the *Salmonella typhimurium* type III protein secretion system. *Science* **280:**602–605.

44. **Lehnherr, H., A.-M. Hansen, and T. Ilyina.** 1998. Penetration of the bacterial cell wall: a family of lytic transglycosylases in bacteriophages and conjugative plasmids. *Mol. Microbiol.* **30:**453–457.

45. **Lory, S.** 1998. Secretion of proteins and assembly of bacterial surface organelles: shared pathways of extracellular protein targeting. *Curr. Opin. Microbiol.* **1:**27–35.

46. **Lupas, A., M. Van Dyke, and J. Stock.** 1991. Predicting coiled coils from protein sequences. *Science* **252:**1162–1164.

47. **McDaniel, T. K., K. G. Jarvis, M. S. Donnenberg, and J. B. Kaper.** 1995. A genetic locus of enterocyte effacement conserved among diverse enterobacterial pathogens. *Proc. Natl. Acad. Sci. USA* **92:**1664–1668.

48. **Mecsas, J., and E. J. Strauss.** 1996. Molecular mechanisms of bacterial virulence: type III secretion and pathogenicity islands. *Emerg. Infect. Dis.* **2:**271–288.

49. **Ménard, R., M.-C. Prévost, P. Gounon, P. J. Sansonetti, and C. Dehio.** 1996. The secreted Ipa complex of *Shigella flexneri* promotes entry into mammalian cells. *Proc. Natl. Acad. Sci. USA* **93:**1254–1258.

50. **Ménard, R., P. J. Sansonetti, and C. Parsot.** 1994. The secretion of the *Shigella flexneri* Ipa invasins is induced by the epithelial cell and controlled by IpaB and IpaD. *EMBO J.* **13:**5293–5302.

51. **Ménard, R., P. J. Sansonetti, C. Parsot, and T. Vasselon.** 1994. Extracellular association and cytoplasmic partitioning of the IpaB and IpaC invasins of *S. flexneri*. *Cell* **79:**515–525.

52. **Michiels, T., P. Wattiau, R. Brasseur, J. M. Ruysschaert, and G. Cornelis.** 1990. Secretion of Yop proteins by yersiniae. *Infect. Immun.* **58:**2840–2849.

53. **Nouwen, N., N. Ranson, H. Saibil, B. Wolpensinger, A. Engel, A. Ghazе, and A. P. Pugsley.** 1999. Secretin PulD: association with pilot PulS, structure, and ion-conducting channel formation. *Proc. Natl. Acad. Sci. USA* **96:**8173–8177.

54. **Pallen, M. J., G. Dougan, and G. Frankel.** Coiled-coil domains in proteins secreted by type III secretion systems. *Mol. Microbiol.* **25:**423–425.

55. **Parsot, C., R. Ménard, P. Gounon, and P. J. Sansonetti.** 1995. Enhanced secretion through the *Shigella flexneri* Mxi-Spa translocon leads to assembly of extracellular proteins into macromolecular structures. *Mol. Microbiol.* **16:**291–300.

56. **Payne, P. L., and S. C. Straley.** 1998. YscO of *Yersinia pestis* is a mobile core component of the Yop secretion system. *J. Bacteriol.* **180:**3882–3890.

57. **Payne, P. L., and S. C. Straley.** 1999. YscP of *Yersinia pestis* is a secreted component of the Yop secretion system. *J. Bacteriol.* **181:**2852–2862.

58. **Pope, L. M., K. E. Reed, and S. M. Payne.** 1995. Increased protein secretion and adherence to HeLa cells by *Shigella* spp. following growth in the presence of bile salts. *Infect. Immun.* **63:**3642–3648.

59. **Pugsley, A. P.** 1993. The complete general secretory pathway in gram-negative bacteria. *Microbiol. Rev.* **57:**50–108.

60. **Roine, E., W. Wei, J. Yuan, E.-L. Nurmiaho-Lassila, N. Kalkkinen, M. Romantschuk, and S.-Y. He.** 1997. Hrp pilus: an

hrp-dependent bacterial surface appendage produced by *Pseudomonas syringae* pv. *tomato* DC3000. *Proc. Natl. Acad. Sci. USA* **94**:3459–3464.

61. **Rosqvist, R., K.-E. Magnusson, and H. Wolf-Watz.** 1994. Target cell contact triggers expression and polarized transfer of *Yersinia* YopE cytotoxin into mammalian cells. *EMBO J.* **13**:964–972.

62. **Salmond, G. P., and P. J. Reeves.** 1993. Membrane traffic wardens and protein secretion in Gram-negative bacteria. *Trends Biochem. Sci.* **18**:7–12.

63. **Sasakawa, C., K. Komatsu, T. Tobe, T. Suzuki, and M. Yoshikawa.** 1993. Eight genes in region 5 that form an operon are essential for invasion of epithelial cells by *Shigella flexneri* 2a. *J. Bacteriol.* **175**:2334–2346.

64. **Schuch, R., and A. T. Maurelli.** 1999. The Mxi-Spa type III secretory pathway of *Shigella flexneri* requires an outer membrane lipoprotein, MxiM, for invasin translocation. *Infect. Immun.* **67**:1982–1991.

65. **Schuch, R., R. C. Sandlin, and A. T. Maurelli.** 1999. A system for identifying post-invasion functions of invasion genes: requirements for the Mxi-Spa type III secretion pathway of *Shigella flexneri* in intercellular dissemination. *Mol. Microbiol.* **34**:675–689.

66. **Shevchik, V. E., J. Robert-Baudouy, and G. Condemine.** 1997. Specific interaction between OutD, an *Erwinia chrysanthemi* outer membrane protein of the general secretory pathway, and secreted proteins. *EMBO J.* **16**:3007–3016.

67. **Skrzypek, E., C. Cowan, and S. C. Straley.** Targeting of the *Yersinia pestis* YopM protein into HeLa cells and intracellular trafficking to the nucleus. *Mol. Microbiol.* **30**:1051–1065.

68. **Stephens, C., and L. Shapiro.** 1996. Bacterial pathogenesis: delivering the payload. *Curr. Biol.* **6**:927–930.

69. **Thanabalu, T., E. Koronakis, C. Hughes, and V. Koronakis.** 1998. Substrate-induced assembly of a contiguous channel for protein export from *E. coli*: reversible bridging of an inner-membrane translocase to an outer membrane pore. *EMBO J.* **17**:6487–6496.

70. **Tran Van Nhieu, G., E. Caron, A. Hall, and P. J. Sansonetti.** 1999. IpaC induces actin polymerization and filopodia formation during *Shigella* entry into epithelial cells. *EMBO J.* **18**:3249–3262.

71. **Ueno, T., K. Oosawa, and S.-I. Aizawa.** 1994. Domain structures of the MS ring component protein (FliF) of the flagellar basal body of *Salmonella typhimurium*. *J. Mol. Biol.* **236**:546–555.

72. **Van der Does, C., T. den Blaauwen, J. G. de Wit, E. H. Manting, N. A. Groot, P. Fekkes, and A. J. M. Driessen.** 1996. SecA is an intrinsic subunit of the *Escherichia coli* preprotein translocase and exposes its carboxyl terminus to the periplasm. *Mol. Microbiol.* **22**:619–629.

73. **Van Gijsegem, F., S. Genin, and C. Boucher.** 1993. Conservation of secretion pathways for pathogenicity determinants of plant and animal bacteria. *Trends Microbiol.* **1**:175–180.

74. **Watarai, M., T. Tobe, M. Yoshikawa, and C. Sasakawa.** 1995. Contact of *Shigella* with host cells triggers release of Ipa invasins and is an essential function of invasiveness. *EMBO J.* **14**:2461–2470.

75. **Williams, A. W., S. Yamaguchi, F. Togashi, S.-I. Aizawa, I. Kawagishi, and R. M. Macnab.** 1996. Mutations in *fliK* and *flhB* affecting flagellar hook and filament assembly in *Salmonella typhimurium*. *J. Bacteriol.* **178**:2960–2970.

76. **Winstanley, C., B. A. Hales, and C. A. Hart.** 1999. Evidence for the presence in *Burkholderia pseudomallei* of a type III secretion system-associated gene cluster. *J. Med. Microbiol.* **48**:649–656.

IDENTIFICATION, REGULATION, AND TRANSFER OF VIRULENCE GENES

IV

IMPACT OF HORIZONTAL GENE TRANSFER ON THE EVOLUTION OF *SALMONELLA* PATHOGENESIS

Robert A. Kingsley, Renée M. Tsolis, Stacy M. Townsend, Tracy L. Norris, Thomas A. Ficht, L. Garry Adams, and Andreas J. Bäumler

15

MOLECULAR ARCHAEOLOGY OF *SALMONELLA* INFECTIONS

What are the origins of infectious diseases such as typhoid fever or enteritis? The answer to this question lies hidden in the genomes of the pathogens causing these illnesses. Approximately 20% of the *Salmonella* genome consists of genetic material, which was introduced by phage or plasmid-mediated horizontal transfer since its divergence from the *Escherichia coli* lineage (45, 70). Some of the genes acquired in this way encode important virulence determinants of *Salmonella* serotypes (7). Identification of horizontally acquired virulence genes is therefore an important first step in reconstructing how new pathovars have emerged within the genus *Salmonella*. To understand how acquisition of this genetic material has altered host-pathogen interactions, it is necessary to synthesize data from a number of research areas, including the characterization of the virulence determinants, the signs of disease caused by the pathogen, and the defense mechanisms of the host.

The establishment of a phylogenetic tree is a prerequisite for studying the evolution of virulence (35). Multilocus enzyme electrophoresis and DNA-DNA hybridization studies showed that after its divergence from a lineage ancestral to *Shigella* and *E. coli*, the *Salmonella* lineage branched into several phylogenetic groups (Fig. 1). The current nomenclature of the genus *Salmonella* is based on this phylogenetic tree and distinguishes only two species: *Salmonella enterica* and *Salmonella bongori* (46, 67). *S. enterica* is further divided into seven subspecies designated by roman numerals (Fig. 1). Comparative sequence analysis of housekeeping genes has subsequently confirmed the clonal nature of individual *Salmonella* species or subspecies (11). Most importantly, this phylogenetic tree provides the framework necessary to infer in which lineage a particular virulence trait first emerged by accepting the hypothesis that the least number of events are most likely correct (principle of parsimony) (35). By determining the phylogenetic distribution of virulence determinants, it is then possible to establish whether acquisition of one or several genes correlates with the appearance of a new vir-

Robert A. Kingsley, Stacy M. Townsend, Tracy L. Norris, and Andreas J. Bäumler   Department of Medical Microbiology and Immunology, College of Medicine, Texas A&M University Health Science Center, 407 Reynolds Medical Building, College Station, TX 77843-1114.   *Renée M. Tsolis, Thomas A. Ficht, and L. Garry Adams*   Department of Veterinary Pathobiology, College of Veterinary Medicine, Texas A&M University, College Station, TX 77843-4467.

Virulence Mechanisms of Bacterial Pathogens, 3rd ed., Edited by K. A. Brogden et al.
©2000 ASM Press, Washington, D.C.

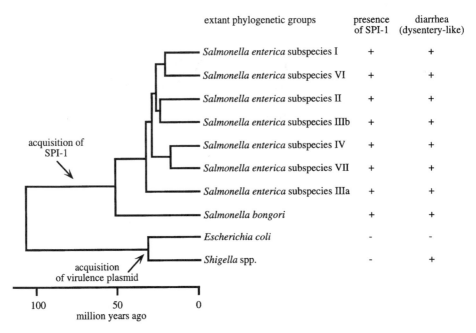

FIGURE 1 The evolution of dysenterylike disease. The branching structure of the phylogenetic tree shown on the left is based on comparative sequence analysis of housekeeping genes (reported previously) (11). Calibration of the phylogenetic tree of *E. coli* and the genus *Salmonella* using a molecular clock has been performed recently (17, 61). The phylogenetic distribution of SPI-1 genes among *S. bongori* and *S. enterica* subspecies has been described by Selander and coworkers (48). Aleksic et al. reported on the ability of *S. bongori* and *S. enterica* serotypes to cause a dysenterylike disease in humans (2). The lineages in which SPI-1 and the *Shigella* virulence plasmid were acquired have been postulated by Ochman and Groisman and are indicated by arrows (59).

ulence trait (7). Finally, the question of whether the DNA region under investigation is indeed required for a particular virulence trait can be tested using in vitro and in vivo models of infection. This approach facilitated reconstruction of the genomic archaeology of *Salmonella* serotypes and identified key events, which led to the emergence of new pathovars within the genus *Salmonella*.

EVOLUTION OF DIARRHEAL DISEASE

In 1929, P. Bruce White wrote, "All the known species [this is the equivalent of serotypes by current nomenclature] of *Salmonella* are pathogenic for man, animals or both" (87). Although fewer than 30 of the more than 2,500 *Salmonella* serotypes had been described by 1929 (40), this statement still appropriately

depicts our current knowledge of this group of pathogens. In humans, members of all phylogenetic lineages within the genus *Salmonella* can cause acute diarrhea progressing toward dysentery (2) (Fig. 1). This finding suggests that virulence factors essential for causing a dysenterylike disease were already present in a common ancestor of the genus *Salmonella* 50 million years ago. In addition to *Salmonella* serotypes, *Shigella* spp. also produce dysentery in humans. The ability of *Shigella* spp. to cause disease is dependent on the presence of a large virulence plasmid, which has been acquired recently by horizontal transfer (Fig. 1) (60). Thus, it is likely that the ability to cause enteritis was obtained independently by the lineages of *Salmonella* and *Shigella* after their divergence from a common ancestor 100 million years ago.

Salmonella pathogenicity island 1 (SPI-1), a 40-kb DNA region encoding the invasion-associated type III secretion system of *Salmonella* serotypes, is present in *S. bongori* and in all subspecies of *S. enterica*, but is absent from the *E. coli* chromosome (24, 48, 54, 66). The phylogenetic distribution of SPI-1 suggests that this pathogenicity island was acquired horizontally by the genus *Salmonella* subsequent to its divergence from the *E. coli* lineage 100 million years ago but prior to branching into extant lineages 50 million years ago (Fig. 1). Hence, acquisition of SPI-1 coincided with the appearance of diarrheal pathogens in the *Salmonella* lineage. In other words, the phylogenetic distribution of SPI-1 raises the question as to whether its acquisition contributed to the evolution of diarrheal disease or merely accompanied this event.

If acquisition of SPI-1 indeed introduced a virulence factor required for the pathogenesis of diarrheal disease, then mutational inactivation of this determinant should attenuate *Salmonella* serotypes in animal models of gastroenteritis. Enteritis can be studied during experimental *S. enterica* serotype Typhimurium infection of calves (83, 85). Serotype Typhimurium causes an infection in calves that is primarily enteric and is characterized by diarrhea and marked intestinal lesions. This disease closely resembles illness produced in humans (90). The contribution of the invasion-associated type III secretion system to serotype Typhimurium pathogenesis in this animal model has recently been investigated using strains carrying mutations in *hilA* and *prgH*. HilA is a transcriptional activator required for expression of SPI-1 genes, including the *prgHJIK* operon (6, 38). PrgH is a component of the needle complex formed by the type III export apparatus, which is required for the secretion of effector proteins and for the entry of serotype Typhimurium into epithelial cells in vitro (10, 44). Mutations in *hilA* and *prgH* markedly reduce the severity of diarrhea and result in avirulence of serotype Typhimurium during oral infection of calves (83). Furthermore, a mutation in *hilA* reduces the ability

of serotype Typhimurium to elicit fluid accumulation and polymorphonuclear leukocyte influx in bovine ligated intestinal loops (1). These data indicate that the invasion-associated type III secretion system is indeed a virulence factor, which is essential for the pathogenesis of bovine enteritis.

The phylogenetic distribution and functional analysis of SPI-1 genes suggest that acquisition of the encoded virulence determinant was a key event during the evolution of diarrheal disease within the genus *Salmonella*. It can therefore be speculated that the horizontal transfer event introducing SPI-1 into the *Salmonella* lineage led to the emergence of a new pathovar, which was able to invade the intestinal epithelium and cause diarrhea. This ancestral organism gave rise to all extant *Salmonella* serotypes, as shown by the presence of diarrheal pathogens carrying SPI-1 in all phylogenetic lineages of the genus (Fig. 1).

EVOLUTION OF SYSTEMIC DISEASE

Several serotypes are capable of causing infections that proceed beyond the mesenteric lymph nodes and become systemic. Serotypes isolated from humans with systemic disease belong to *S. enterica* subspecies I, IIIa, and IIIb (2, 82, 86) (Fig. 2). Some *S. enterica* subspecies I serotypes are highly host adapted, have a restricted host range, and are frequently associated with systemic infections characterized by high mortality rates. For example, *S. enterica* serotype Typhi causes typhoid fever in man and higher primates but is avirulent for other animal species. Nontyphoidal serotypes, on the other hand, cause clinical infections with systemic involvement at much lower frequency but exhibit a broader host range than serotype Typhi. In addition, there are differences between nontyphoidal serotypes regarding their capability to cause systemic infections in humans. For instance, in a case study of bacteremia in England and Wales, 25.2% of *S. enterica* serotype Dublin and 74.1% of *S. enterica* serotype Choleraesuis isolates were from blood, compared with only 1.1% of serotype

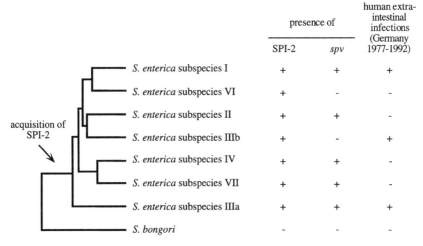

	presence of		human extra-intestinal infections (Germany 1977-1992)
	SPI-2	*spv*	
S. enterica subspecies I	+	+	+
S. enterica subspecies VI	+	-	-
S. enterica subspecies II	+	+	-
S. enterica subspecies IIIb	+	-	+
S. enterica subspecies IV	+	+	-
S. enterica subspecies VII	+	+	-
S. enterica subspecies IIIa	+	+	+
S. bongori	-	-	-

FIGURE 2 Evolution of systemic disease caused by *Salmonella* serotypes. The branching structure of the phylogenetic tree shown on the left is based on comparative sequence analysis of housekeeping genes (reported previously) (11). The phylogenetic distribution of SPI-2 has been described recently (31, 58), suggesting its acquisition by a lineage ancestral to *S. enterica* (arrow). Boyd and coworkers determined the scattered phylogenetic distribution of the *spv* gene cluster (12). Aleksic et al. reported on the ability of *S. bongori* and *S. enterica* serotypes to cause extraintestinal infections in humans (2).

Typhimurium isolates (82). These examples illustrate that the frequency at which a particular pathogen is associated with extraintestinal infection differs greatly among *Salmonella* serotypes and in some cases (e.g., serotype Typhi) reflects their degree of host adaptation. Hence, the ability to produce systemic disease appears to be a complex phylogenetic trait, which has been repeatedly modified to create various serotypes causing extraintestinal infection at different frequencies.

Most attempts to identify virulence factors required for systemic disease currently rely on the murine typhoid model of serotype Typhimurium infection. Murine typhoid is a systemic infection characterized by bacterial growth in the liver and spleen, which occurs primarily in macrophages (15, 69). Two virulence gene clusters, SPI-2 and the *Salmonella* plasmid virulence (*spv*) locus, have been found to be specifically required for systemic infection of mice. SPI-2 encodes a type III secretion system required for survival within macrophages and inhibition of fusion of se-

rotype Typhimurium-containing phagosomes with lysosomes in vitro (32, 60, 84). The first gene of the *spv* locus (*spvR*) encodes a transcriptional activator for the following four genes, *spvABCD*, whose functions are currently unknown (14). Mutations in SPI-2 or the *spv* locus result in a marked attenuation of serotype Typhimurium for mice (27, 28, 76) and in reduced bacterial multiplication in murine liver and spleen (29, 75). While SPI-2 and the *spv* locus are essential for systemic disease, these virulence determinants are dispensable during a localized infection, such as the enteritis produced by serotype Typhimurium in calves (83).

Functional analysis in the murine typhoid model suggests that the *spv* locus and SPI-2 are required for serotype Typhimurium extraintestinal infections. However, there is no perfect correlation between the presence of these virulence determinants and the ability of *Salmonella* serotypes to cause systemic disease (Fig. 2). SPI-2 has been detected in all *S. enterica* serotypes analyzed to date, but it is absent from the *S. bongori* lineage (31, 58). Thus, SPI-2 is present in all lineages commonly associated with extraintestinal infection in humans. However, *S. enterica* subspecies II, IV, VI, and VII, which are rarely or never associated with systemic disease, also contain this pathogenicity island (Fig. 2). In contrast to SPI-2, the phylogenetic distribution of the *spv* locus within the genus *Salmonella* is sporadic. Serotypes from *S. enterica* subspecies I, II, IIIa, IV, and VII hybridize with *spv*-specific DNA probes (12). Within *S. enterica* subspecies I, the *spv* locus is plasmid encoded and restricted to a small subset of serotypes, including serotypes Typhimurium, Choleraesuis, and Dublin, *S. enterica* serotype Enteritidis, *S. enterica* serotype Gallinarum, *S. enterica* serotype Abortusovis, *S. enterica* serotype Paratyphi C, and *S. enterica* serotype Heidelberg (89). The presence of the *spv* locus in nontyphoidal *S. enterica* subspecies I serotypes correlates with their ability to cause lethal infections in mice (72) and with an increased frequency of isolations from human extraintestinal infections (23, 47). However, typhoidal serotypes, such as serotype Typhi, *S. enterica* serotype Paratyphi A, or *S. enterica* serotype Paratyphi B, produce systemic infections by an *spv*-independent mechanism. Thus, *Salmonella* serotypes causing extraintestinal infections do not possess identical sets of virulence genes. Although horizontal transfer of SPI-2 and the *spv* locus has introduced virulence factors required for systemic disease, it is likely that the ability to cause extraintestinal infections has in some cases involved acquisition of other determinants that remain to be identified.

ADAPTATION TO WARM-BLOODED VERTEBRATES

S. bongori and *S. enterica* subspecies II through VII are mainly associated with cold-blooded vertebrates. In contrast, serotypes of *S. enterica* subspecies I are frequently isolated from both cold- and warm-blooded host species (65). The ability of *S. enterica* subspecies I serotypes to circulate in populations of warm-blooded vertebrates, including humans, has obvious implications for public health. *S. enterica* subspecies I contains human-adapted pathogens, such as serotype Typhi. Also, the presence of *S. enterica* subspecies I serotypes in the animal reservoir from which we draw our food supply greatly increases the risk of human infections produced by these pathogens (25, 78, 79). As a consequence, members of *S. enterica* subspecies I account for 99.55% of the 90,201 human clinical *Salmonella* isolates collected by the German National Reference Centre for Enteric Pathogens between 1977 and 1992 (2) (Fig. 1).

These data demonstrate that the expansion in host range of *S. enterica* subspecies I to include warm-blooded animals was an important step during the evolution of virulence in the genus *Salmonella*. Furthermore, an understanding of the events that enabled *S. enterica* subspecies I to circulate in populations of homoiothermic vertebrates will likely provide new insights into general mechanisms by which new pathogens emerge. However, despite its relevance for human health, there is little discussion on this topic in the literature. Below we will try to identify virulence mechanisms whose acquisition may have allowed *Salmonella* serotypes to expand their host range to include warm-blooded animals.

Immune Memory, O-Antigen Polymorphism, and Phase Variation

The identification of host defense mechanisms present in homoiothermic animals but absent from poikilothermic vertebrates is an important basis to understand the challenges *Salmonella* serotypes encountered during expansion in host range (9). One obstacle to colonizing homoiothermic animals may have been the enhanced immune memory of birds and mammals. For instance, in response to a second or subsequent administration of antigen, homoiothermic animals produce antibodies that are of higher affinity, heterogeneity, and titer than primary antibodies. In contrast, the antibody repertoire of poikilothermic vertebrates is highly restricted (18), and antibody affinity and titers do not increase even after repeated immunization (55). In mammals and birds, memory B cells regenerate in the germinal centers, where they undergo isotype switching and somatic mutation in their immunoglobulin genes. Germinal centers are histologically defined areas in lymphoid organs (e.g., lymph nodes or spleen), which develop after antigenic stimulation (43, 51, 81). These structures are recent phylogenetic acquisitions of birds and mammals and are absent from lymphoid organs of lower vertebrates, such as fish, amphibians, and reptiles (91). Because memory B cells develop in germinal centers, it has been speculated that the absence of these structures in poikilothermic animals may be responsible for the poor anamnestic qualities of the antibody response in lower vertebrates (19, 34, 55, 88).

How does the enhanced B cell memory of mammals and birds affect the host pathogen interaction following infection with *Salmonella* serotypes? The B cell memory generated by vaccination is important for protection against disease caused by *Salmonella* serotypes (F. Cabello, Letter, *Trends Microbiol.* **6**:470–472, 1998; 20, 64). The importance of antibodies for immunity is illustrated by the finding that protection of mice against serotype Typhimurium infection can be achieved by (i) immunization with killed serotype Typhimurium (33); (ii) immunization with purified serotype Typhimurium lipopolysaccharide (LPS) (73); (iii) passive transfer of immune serum (63); (iv) passive transfer of monoclonal IgG and IgM antibodies directed against the O-antigen of serotype Typhimurium LPS (16); and (v) transfer of hybridoma tumors, which secrete a monoclonal sIgA directed against the O4-antigen of the serotype Typhimurium LPS (53). Furthermore, a recent evaluation of serotype Typhi vaccine efficacy trials involving 1,866,951 human subjects concluded that the whole-cell-killed vaccine, eliciting only a humoral response, is more effective in protecting against typhoid fever than the live-attenuated vaccine (Ty21a), which elicits both humoral and cellular immunity (21). The majority of antibodies elicited by immunization with heat-killed serotype Typhimurium or with a live-attenuated serotype Typhimurium *aroA* vaccine is directed against the immunodominant O-antigen (5, 49, 50). The O-antigen of serotype Typhimurium (O-antigen formula O4,5,12) consists of three epitopes: the O12-antigen (a trisaccharide backbone consisting of mannose → rhamnose → galactose →), the O4-antigen (an abequose branch), and the O5-antigen (acetylation of the abequose branch). The O4-antigen is the dominant determinant, and immunization of mice with a serotype Typhimurium *aroA* vaccine results in anti-O4 titers that are 10-fold higher than antibody titers directed against other O-antigen epitopes (49). Furthermore, the anti-O4 titers generated during immunization with an *aroA* vaccine confer immunity to subsequent challenge with serotype Typhimurium (36, 49). These data show that in warm-blooded animals, *Salmonella* serotypes are not able to evade a secondary antibody response directed against their immunodominant O-antigen and therefore cannot cause recurrent infections.

In addition to preventing recurrent infections, the enhanced immunological memory of homoiothermic animals can generate cross-immunity between *Salmonella* serotypes sharing an immunodominant O-antigen. For instance, the O-antigen of serotype Enteritidis and serotype Gallinarum consists of the O12-antigen and the O9-antigen (a tyvelose branch), the latter of which is the immunodominant determinant. Vaccination with live serotype Gallinarum (O-antigen formula O9,12) can protect mice against subsequent challenge with serotype Enteritidis (O9,12) but not against a challenge with virulent serotype Typhimurium (O4,5,12) (15). Furthermore, vaccination of mice with a serotype Enteritidis *aroA* mutant elicits protection against subsequent challenge with a virulent serotype Typhimurium strain genetically engineered to express the O9,12-antigen but not against the serotype Typhimurium wild type (O4,5,12) (36).

If two *Salmonella* serotypes that share an immunodominant O-antigen coexist in a warm-blooded animal reservoir, cross-immunity can induce between-serotype competition by lowering the density of susceptible hosts, thereby reducing the transmissibility of both pathogens (3, 30). Mathematical models predict that in this between-serotype competition, the serotype with higher transmissibility will dominate and eventually eliminate its competitor. On the other hand, coexistence of related pathogens is possible if both serotypes do not share immunodominant surface antigens (3). We therefore propose that with the encounter of the enhanced immune memory exhibited by mammals and birds, evasion of cross-immunity provided a selective advantage for *Salmonella* serotypes, which coexist in a host population. This hypothesis predicts that the selection imposed by the immune memory of warm-blooded vertebrates will organize the pathogen population within an animal reservoir into strains, which express different immunodominant O-antigens. Indeed, *Salmonella* serotypes, which coexist and circulate in a homoiothermic animal reservoir, differ with regard to their immunodominant O-antigen, thereby evading between-serotype competition. For instance, more than 90% of salmonellosis cases in cattle are caused by two serotypes, serotype Typhimurium and serotype Dublin (71, 77 79). Both serotypes evade cross-immunity because the immunodominant O-antigen of serotype Typhimurium (O4-antigen) is different from that of serotype Dublin (O9-antigen). Similarly, serotype Choleraesuis (O7-antigen), serotype Typhimurium (O4-antigen), and serotype Dublin (O9-antigen) are associated with 85% of cases of disease in pigs (79). Since these three serotypes each express a different immunodominant O-antigen, they evade the between-serotype competition elicited by an antibody response. In humans, nontyphoidal *Salmonella* serotypes, which are frequently associated with disease, may share an immunodominant O-antigen (41). However, these nontyphoidal *Salmonella* serotypes do not circulate in the human population (person to person transmission is rare), but are rather constantly reintroduced from the animal reservoirs from which we draw our food supply.

Thus, the spectrum of nontyphoidal human isolates is a reflection of the pathogens that currently circulate in populations of livestock and domestic fowl (25). Human-adapted typhoidal *Salmonella* serotypes, on the other hand, do not possess an animal reservoir but rather circulate within the human host population (person to person transmission). The ability to cause typhoid fever evolved independently four times within *S. enterica* subspecies I (74). These four clonal lineages, represented by serotypes Typhi (O9-antigen), Paratyphi A (O2-antigen), Paratyphi B (O4-antigen), and Paratyphi C (O7-antigen), express different immunodominant O-antigens. In other words, typhoidal *Salmonella* serotypes evade between-serotype competition, which may facilitate their coexistence in the human population. Epidemiological surveys thus support the proposal that populations of pathogens, which coexist in warm-blooded animal reservoirs, are organized into discrete clones (represented by *Salmonella* serotypes) that evade cross-immunity by expressing different immunodominant O-antigens. These data suggest that the selection imposed by the enhanced immune memory of warm-blooded animals had a significant impact on the evolution of *Salmonella* serotypes and left its mark on the pathogen population structure of these organisms.

Which genetic mechanisms allowed *Salmonella* serotypes to evade cross-immunity encountered in homoiothermic vertebrates? Comparative analysis of the rough B (*rfb*) gene cluster (a DNA region involved in O-antigen biosynthesis) of several *Salmonella* serotypes revealed that the O-antigen polymorphism was generated by horizontal gene transfer (68). The *rfb* gene cluster of serotype Typhimurium is a composite of DNA regions with atypical G+C content, indicative of its complex phylogenetic history (37). All genes present in the *rfb* gene cluster have likely been acquired from distantly related bacteria after the lineages of *E. coli* and the genus *Salmonella* diverged (68). Analysis of *gnd*, a gene closely linked to (and thus frequently cotransferred with) *rfb*, provided evidence that after its acquisition, frequent transfer of the *rfb* region has occurred both between and within *S. enterica* subspecies (57, 80). The generation of O-antigen polymorphism through horizontal gene transfer was therefore a likely mechanism that allowed *Salmonella* serotypes to adapt to the enhanced immune memory encountered in warm-blooded hosts.

While *Salmonella* serotypes evade cross-immunity against LPS by O-antigen polymorphism, this mechanism does not protect against antibodies that recognize other surface structures, such as fimbriae and flagella. We have recently shown that *Salmonella* serotypes evade cross-immunity against long polar fimbrial proteins by phase variation (57a). Thus, phase variation may be another way to avoid cross-immunity, thereby facilitating the expansion in host range to include animals with an enhanced immune memory. Within the genus *Salmonella*, phase variation between the two flagellar filament structural genes *fliC* (encoding H1 phase flagellin) and *fljB* (encoding H2 phase flagellin) is only observed in biphasic *S. enterica* serotypes, belonging to a monophyletic group formed by subspecies I, II, IIIb, and VI (Fig. 3). The *fljB* gene is present in biphasic *S. enterica* subspecies but absent from monophasic *S. enterica* subspecies and *E. coli*, suggesting its acquisition by horizontal gene transfer (8). Genes required for phase variation of flagellar genes, including *hin* (encoding a site-specific invertase) and *fljA* (encoding a repressor of *fliC*), are located adjacent to and were likely cotransferred with *fljB* (62). Thus, it is probable that the mechanism of flagellar phase variation was obtained horizontally by a common ancestor of *S. enterica* subspecies I, II, IIIb, and VI. Acquisition of the flagellar phase variation mechanism did not directly precede (and thus did not trigger) the radiation of *S. enterica* subspecies I among warm-blooded animals. However, the observation that 87% of human isolates that do not belong to *S. enterica* subspecies I are biphasic (Fig. 3) (2) suggests that some adaptations to homoiothermic vertebrates may have been obtained by biphasic *Salmonella* serotypes prior to for-

mation of *S. enterica* subspecies I. Thus, the expansion in host range to include warm-blooded animals may have been a multistep process initiated before *S. enterica* subspecies I became a separate lineage.

Intestinal Colonization and Transmission

A primary pathogen can be defined as an organism capable of entering a host and finding a unique niche to multiply and avoid or subvert the host defenses, the outcome of which may be clinical disease manifestations. To ensure circulation within a population, the pathogen must be on average transmitted to at least one susceptible host (22). Members of all *S. enterica* subspecies and *S. bongori* are able to cause disease in humans (2), illustrating that they are capable of entrance, multiplication, and evasion of defense mechanisms during an encounter with a naïve host. However, serotypes of *S. enterica* subspecies I differ from *S. bongori* and *S. enterica* subspecies II through VII serotypes in their ability to circulate within populations of warm-blooded animals. Thus, the development of mechanisms ensuring effective transmission within populations of homoiothermic vertebrates was likely a prerequisite for the expansion in host range to include mammals and birds.

To circulate in a population of animals, a pathogen needs to cause at least one secondary case of infection from a primary case, the number of which is called the case reproductive number (4). The case reproductive number of *S. enterica* subspecies I serotypes for higher vertebrates must therefore be one or above, since these pathogens circulate in warm-blooded host populations. Conversely, absence of *S. bongori* and *S. enterica* subspecies II through VII serotypes from populations of livestock or domestic fowl suggests that their case reproductive number for higher vertebrates is below one, a property apparently independent of their ability to cause illness in these hosts (2). The case reproductive number is directly proportional to the transmissibility of a pathogen (4). Hence, the infrequent iso-

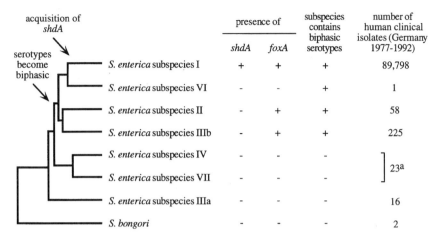

	presence of		subspecies contains biphasic serotypes	number of human clinical isolates (Germany 1977-1992)
	shdA	*foxA*		
S. enterica subspecies I	+	+	+	89,798
S. enterica subspecies VI	-	-	+	1
S. enterica subspecies II	-	+	+	58
S. enterica subspecies IIIb	-	+	+	225
S. enterica subspecies IV	-	-	-	
S. enterica subspecies VII	-	-	-	23[a]
S. enterica subspecies IIIa	-	-	-	16
S. bongori	-	-	-	2

FIGURE 3   Adaptations of *Salmonella* serotypes to circulate in populations of warm-blooded vertebrates. The branching structure of the phylogenetic tree shown on the left is based on comparative sequence analysis of housekeeping genes (reported previously) (11). The phylogenetic distributions of *shdA* and *foxA* have been determined recently (42; Kingsley et al., submitted). The H-antigens of monophasic and biphasic *Salmonella* serotypes are reviewed by Kelterborn (39). Aleksic et al. have reported on the frequency of *S. bongori* and *S. enterica* serotype isolation from clinical infections (2). *a*, The article by Aleksic et al. does not distinguish between *S. enterica* subspecies IV and VII.

lation of *S. bongori* or *S. enterica* subspecies II through VII serotypes from domesticated animals may reflect, at least in part, their ineffective transmission between warm-blooded hosts. This hypothesis implies that serotypes of *S. enterica* subspecies I may possess one or more genetic determinants, which increase their transmissibility (but not necessarily their lethality) for warm-blooded vertebrates. Transmission is a multifactorial process, which in the case of *Salmonella* serotypes probably involves factors ensuring (i) fecal shedding from a host for a prolonged period of time, (ii) growth and survival in the external environment, and (iii) implantation in the intestine of a susceptible host. As outlined above, evasion of cross-immunity is a mechanism that facilitates implantation by preventing a decline of the density of susceptible hosts, thereby increasing the probability of transmission. The rest of this section will focus on determinants involved in fecal shedding and growth in the environment.

Recently, we described a genetic determinant designated *shdA*, which is required for

efficient and prolonged shedding of serotype Typhimurium from mice (R. K. Kingsley, K. van Amsterdam, and A. J. Bäumler, submitted for publication). The deduced amino acid sequence of *shdA* has homology to members of the autotransporter family of outer membrane proteins, including AIDA-I from diarrheagenic *E. coli*, MisL from serotype Typhimurium, and IcsA from *Shigella flexneri*. Significantly, the *shdA* gene is present in 96% of *S. enterica* subspecies I isolates but is absent from serotypes of *S. bongori* and *S. enterica* subspecies II through VII (Fig. 3). The phylogenetic distribution of *shdA* is therefore consistent with its acquisition by a common ancestor of *S. enterica* subspecies I. To rationalize the selective advantage *S. enterica* subspecies I serotypes may have had when colonizing the intestine of warm-blooded animals through acquisition of *shdA*, it is important to identify differences between the alimentary tracts of higher and lower vertebrates. One component, which increases in complexity during the phylogeny of vertebrates, is the gut-associated lymphoid tissue (GALT). The

GALT of reptiles consists of solitary lymphoid aggregates located in the gut lamina propria (55). In contrast, in mammals and birds the GALT is organized in complex organs, such as Peyer's patches, tonsils, appendix, or the avian bursa of Fabricius (26, 51). Mutational inactivation of *shdA* reduces the ability of serotype Typhimurium to colonize murine Peyer's patches but does not affect colonization of the mesenteric lymph node, liver, and spleen (Kingsley et al., submitted). Although these data suggest that *shdA* may confer an adaptation to colonization of the complex GALT encountered in warm-blooded animals, it is currently not clear whether growth or survival in this tissue has an effect on bacterial shedding.

No other determinants have been studied in the context of transmissibility. Therefore, the involvement of growth in the environment and ability to become seeded in warm-blooded hosts in transmission remains unclear. However, we should note that some determinants, which likely enhance the ability of *Salmonella* serotypes to grow in the environment, have been obtained by horizontal transfer. For instance, ferrioxamine mediated iron (III) transport mediated by the outer membrane receptor FoxA is a trait acquired by serotypes of *S. enterica* subspecies I, II, and IIIb (42) (Fig. 3). Although ferrioxamines are not synthesized by *Salmonella* serotypes, they are produced by a number of microorganisms that are potentially present in the environment in coculture; therefore, the ability to utilize iron chelated in this form may increase persistence in the environment.

CONCLUSIONS

The genus *Salmonella* is a paradigm that illustrates how new pathogens may have emerged and expanded their host range. A number of features make this organism an ideal candidate for such analysis, including a wealth of information on pathogenesis, epidemiology, and genetics. In addition, recent data have elucidated the phylogenetic relationship of *Salmonella* species and subspecies and established

their clonal descent. By synthesizing this information, we have attempted to reconstruct events that contributed to the evolution of virulence in the genus *Salmonella*. An important conclusion from this work is that horizontal gene transfer was involved in a number of key steps leading to the emergence of new pathovars. Furthermore, reconstruction of these events is significant because the knowledge generated may be useful to predict how other pathogens may emerge in the future.

ACKNOWLEDGMENTS

Work in the laboratory of A. J. B. is supported by Public Health Service grants AI40124 and AI44170 and by Formula Animal Health Funding (USDA) to T. A. F. and A. J. B. R. M. T. is supported by USDA/NRICGP fellowship #9702568.

REFERENCES

1. **Ahmer, B., J. Vanreeuwijk, C. D. Timmers, P. J. Valentine, and F. Heffron.** 1998. *Salmonella typhimurium* encodes an *sdiA* homolog, a putative quorum sensor of the *luxR* family, that regulates genes on the virulence plasmid. *J. Bacteriol.* **180:**1185–1193.
2. **Aleksic, S., F. Heinzerling, and J. Bockemühl.** 1996. Human infection caused by salmonellae of subspecies II to VI in Germany, 1977–1992. *Zentralbl. Bakteriol.* **283:**391–398.
3. **Anderson, R. M.** 1995. Evolutionary pressures in the spread and persistence of infectious agents in vertebrate populations. *Parasitology* **111:**S15–S31.
4. **Anderson, R. M., and R. M. May.** 1982. Coevolution of host and parasites. *Parasitology* **85:**411–426.
5. **Angerman, C. R., and T. K. Eisenstein.** 1980. Correlation of the duration and magnitude of protection against *Salmonella* infection afforded by various vaccines with antibody titers. *Infect. Immun.* **27:**435–443.
6. **Bajaj, V., C. Hwang, and C. A. Lee.** 1995. HilA is a novel OmpR/ToxR family member that activates the expression of *Salmonella typhimurium* invasion genes. *Mol. Microbiol.* **18:**715–727.
7. **Bäumler, A. J.** 1997. The record of horizontal gene transfer in *Salmonella. Trends Microbiol.* **5:**318–322.
8. **Bäumler, A. J., and F. Heffron.** 1998. Mosaic structure of the *smpB-nrdE* intergenic region of *Salmonella enterica. J. Bacteriol.* **180:**2220–2223.

9. **Bäumler, A. J., R. M. Tsolis, T. A. Ficht, and L. G. Adams.** 1998. Evolution of host adaptation in *Salmonella enterica. Infect. Immun.* **66:** 4579–4587.

10. **Behlau, I., and S. J. Miller.** 1993. A PhoP repressed gene promotes *Salmonella typhimurium* invasion of epithelial cells. *J. Bacteriol.* **175:**4475–4484.

11. **Boyd, E. F., F.-S. Wang, T. S. Whittam, and R. K. Selander.** 1996. Molecular genetic relationship of the *Salmonellae. Appl. Environ. Microbiol.* **62:**804–808.

12. **Boyd, F. E., and D. L. Hartl.** 1998. Salmonella virulence plasmid: modular acquisition of the spv virulence region by an F-plasmid in *Salmonella enterica* subspecies I and insertion into the chromosome in subspecies II, IIIa, IV, and VII isolates. *Genetics* **149:**1183–1190.

13. **Caldwell, A. L., and P. A. Gulig.** 1991. The *Salmonella typhimurium* virulence plasmid encodes a positive regulator of a plasmid-encoded virulence gene. *J. Bacteriol.* **173:**7176–7185.

14. **Carter, P. B., and F. M. Collins.** 1974. The route of enteric infection in normal mice. *J. Exp. Med.* **139:**1189–1203.

15. **Collins, F. M., G. B. Mackaness, and R. V. Blanden.** 1966. Infection-immunity in experimental salmonellosis. *J. Exp. Med.* **124:**601–619.

16. **Colwell, D. E., S. M. Michalek, D. E. Briles, E. Jirillo, and J. R. McGhee.** 1984. Monoclonal antibodies to *Salmonella* lipopolysacchride: anti-O-polysaccharide antibodies protect C3H mice against challenge with virulent *Salmonella typhimurium. J. Immunol.* **133:**950–957.

17. **Doolittle, R. F., D. Feng, S. Tsang, G. Cho, and E. Little.** 1996. Determining divergence times of the major kingdoms of living organisms with a protein clock. *Science* **171:**470–477.

18. **Du Pasquier, L.** 1982. Antibody diversity in lower vertebrates—why is it so restricted? *Nature* **290:**311–313.

19. **Du Pasquier, L.** 1993. Phylogeny of B-cell development. *Curr. Opin. Immunol.* **5:**185–193.

20. **Eisenstein, T. K.** 1998. Intracellular pathogens: the role of antibody-mediated protection in Salmonella infection. *Trends Microbiol.* **6:**135–136.

21. **Engels, E. A., M. E. Falagas, J. Lau, and M. L. Bennish.** 1998. Typhoid fever vaccines: a meta-analysis of studies on efficacy and toxicity. *Br. Med. J.* **316:**110–116.

22. **Falkow, S.** 1997. What is a pathogen? *ASM News* **63:**359–365.

23. **Fierer, J., M. Krause, R. Tauxe, and D. Guiney.** 1992. *Salmonella typhimurium* bacteremia: association with the virulence plasmid. *J. Infect. Dis.* **166:**639–642.

24. **Galán, J. E., and R. Curtiss III.** 1991. Distribution of the *invA, -B, -C*, and *-D* genes of *Salmonella typhimurium* among other *Salmonella* serovars: *invA* mutants of *Salmonella typhi* are deficient for entry into mammalian cells. *Infect. Immun.* **59:**2901–2908.

25. **Galbraith, N. S.** 1961. Studies of human salmonellosis in relation to infection in animals. *Vet. Rec.* **73:**1296–1303.

26. **Glick, B.** 1982. RES structure and function in aves, p. 509–540. *In* N. Cohen and M. M. Sigel (ed.), *The Reticuloendothelial System,* vol. 3. Plenum, New York, N.Y.

27. **Gulig, P. A., A. L. Caldwell, and V. A. Chiodo.** 1992. Identification, genetic analysis and DNA sequence of a 7.8-kb virulence region of the *Salmonella typhimurium* virulence plasmid. *Mol. Microbiol.* **6:**1395–1411.

28. **Gulig, P. A., and R. Curtiss.** 1987. Plasmid-associated virulence of *Salmonella typhimurium. Infect. Immun.* **55:**2891 2901.

29. **Gulig, P. A., and T. J. Doyle.** 1993. The *Salmonella typhimurium* virulence plasmid increases the growth rate of salmonellae in mice. *Infect. Immun.* **61:**504–511.

30. **Gupta, S., M. C. Maiden, I. M. Feavers, S. Nee, R. M. May, and R. M. Anderson.** 1996. The maintenance of strain structure in populations of recombining infectious agents. *Nat. Med.* **2:**437–442.

31. **Hensel, M., J. E. Shea, A. J. Bäumler, C. Gleeson, F. Blattner, and D. W. Holden.** 1997. Analysis of the boundaries of *Salmonella* pathogenicity island 2 and the corresponding chromosomal region of *Escherichia coli* K-12. *J. Bacteriol.* **179:**1105–1111.

32. **Hensel, M., J. E. Shea, S. R. Waterman, R. Mundy, T. Nikolaus, G. Banks, A. Vazquez-Torres, C. Gleeson, F. C. Fang, and D. W. Holden.** 1998. Genes encoding putative effector proteins of the type III secretion system of *Salmonella* pathogenicity island 2 are required for bacterial virulence and proliferation in macrophages. *Mol. Microbiol.* **30:**163–174.

33. **Herzberg, M., P. Nash, and S. Hino.** 1972. Degree of immunity induced by killed vaccines to experimental salmonellosis in mice. *Infect. Immun.* **5:**83–90.

34. **Hinds-Frey, K. R., H. Nishikata, R. T. Litman, and G. W. Litman.** 1993. Somatic variation precedes extensive diversification of germline sequences and combinatorial joining in the evolution of immunoglobulin heavy chain diversity. *J. Exp. Med.* **178:**815–824.

35. **Holmes, E. C.** 1998. Molecular epidemiology and evolution of emerging infectious diseases. *Br. Med. Bull.* **54:**533–543.

36. **Hormaeche, C. E., P. Mastroeni, J. A. Harrison, R. Demarco de Hormaeche, S. Svenson, and B. A. Stocker.** 1996. Protection against oral challenge three months after i.v. immunization of BALB/c mice with live Aro *Salmonella typhimurium* and *Salmonella enteritidis* vaccines is serotype (species)-dependent and only partially determined by the main LPS O antigen. *Vaccine* **14**:251–259.

37. **Jiang, X. M., B. Neal, F. Santiago, S. J. Lee, L. K. Romana, and P. R. Reeves.** 1991. Structure and sequence of the *rfb* (O antigen) gene cluster of *Salmonella* serovar typhimurium (strain LT2). *Mol. Microbiol.* **5**:695–713.

38. **Johnston, C., D. A. Pegues, C. J. Hueck, A. Lee, and S. I. Miller.** 1996. Transcriptional activation of *Salmonella typhimurium* invasion genes by a member of the phosphorylated response-regulator superfamily. *Mol. Microbiol.* **22**:715–727.

39. **Kelterborn, E.** 1992. *Kauffmann-White-Schema (1989)*. Bundesgesundheitsamt, Berlin, Germany.

40. **Kelterborn, E.** 1967. *Salmonella-Species. First Isolations, Names and Occurrence.* S. Hirzel Verlag Leipzig 1967, Karl-Marx-Stadt.

41. **Khakhria, R., D. Woodward, W. M. Johnson, and C. Poppe.** 1997. *Salmonella* isolated from humans, animals and other sources in Canada, 1983–92. *Epidemiol. Infect.* **119**:15–23.

42. **Kingsley, R. A., R. Reissbrodt, W. Rabsch, J. M. Ketley, R. M. Tsolis, P. Everest, G. Dougan, A. J. Baumler, M. Roberts, and P. H. Williams.** 1999. Ferrioxamine-mediated Iron(III) utilization by *Salmonella enterica*. *Appl. Environ. Microbiol.* **65**:1610–1618.

43. **Kroese, F. G., W. Timens, and P. Nieuwenhuis.** 1990. Germinal center reaction and B lymphocytes: morphology and function. *Curr. Top. Pathol.* **84**(Pt 1):103–148.

44. **Kubori, T., Y. Matsushima, D. Nakamura, J. Uralil, M. Lara-Tejero, A. Sukhan, J. E. Galan, and S. I. Aizawa.** 1998. Supramolecular structure of the *Salmonella typhimurium* type III protein secretion system. *Science* **280**:602–605.

45. **Lan, R. T., and P. R. Reeves.** 1996. Gene transfer is a major factor in bacterial evolution. *Mol. Biol. Evol.* **13**:47–55.

46. **Le Minor, L., and M. Y. Popoff.** 1987. Designation of *Salmonella enterica* sp. nov., nom. rev., as the type and only species of the genus *Salmonella*. *Int. J. Sys. Bacteriol.* **37**:465–468.

47. **Levine, W. C., J. W. Buehler, N. H. Bean, and R. V. Tauxe.** 1991. Epidemiology of nontyphoidal *Salmonella* bacteremia during the human immunodeficiency virus epidemic. *J. Infect. Dis.* **164**:81–87.

48. **Li, J., H. Ochman, E. A. Groisman, E. F. Boyd, F. Solomon, K. Nelson, and R. K. Selander.** 1995. Relationship between evolutionary rate and cellular location among the Inv /Spa invasion proteins of *Salmonella enterica*. *Proc. Natl. Acad. Sci. USA* **92**:7252–7256.

49. **Lindberg, A. A., T. Segall, A. Weintraub, and B. A. Stocker.** 1993. Antibody response and protection against challenge in mice vaccinated intraperitoneally with a live *aroA* O4-O9 hybrid *Salmonella dublin* strain. *Infect. Immun.* **61**: 1211–1221.

50. **Lyman, M. B., B. A. Stocker, and R. J. Roantree.** 1979. Evaluation of the immune response directed against the *Salmonella* antigenic factors O4,5 and O9. *Infect. Immun.* **26**:956–965.

51. **MacLennan, I. C., Y. J. Liu, S. Oldfield, J. Zhang, and P. J. Lane.** 1990. The evolution of B-cell clones. *Curr. Top. Microbiol. Immunol.* **159**:37–63.

52. **Manning, M. J.** 1979. Evolution of the vertebrate immune system. *J. R. Soc. Med.* **72**:683–688.

53. **Michetti, P., M. J. Mahan, J. M. Slauch, J. J. Mekalanos, and M. R. Neutra.** 1992. Monoclonal secretory immunglobulin A protects mice against oral challenge with the invasive pathogen *Salmonella typhimurium*. *Infect. Immun.* **60**:1786–1792.

54. **Mills, D. M., V. Bajaj, and C. A. Lee.** 1995. A 40kb chromosomal fragment encoding *Salmonella typhimurium* invasion genes is absent from the corresponding region of the *Escherichia coli* K-12 chromosome. *Mol. Microbiol.* **15**:749–759.

55. **Muthukkaruppan, V. R., M. Borysenko, and R. El Ridi.** 1982. RES structure and function in reptilia, p. 461–508. *In* N. Cohen and M. M. Sigel (ed.), *The Reticuloendothelial System*, vol. 3. Plenum, New York, N.Y.

56. **Nahm, M. H., F. G. Kroese, and J. W. Hoffmann.** 1992. The evolution of immune memory and germinal centers. *Immunol. Today* **13**:438–441.

57. **Nelson, K., and R. K. Selander.** 1994. Intergeneric transfer and recombination of the 6-phosphogluconate dehydrogenase gene (*gnd*) in enteric bacteria. *Proc. Natl. Acad. Sci. USA* **91**: 10227–10231.

57a. **Norris, T. L., and A. J. Bäumler.** 1999. Phase variation of the lpf fimbrial operon is a mechanism to evade cross immunity between Salmonella serotypes. *Proc. Natl. Acad. Sci. USA* **96**: 13393–13398.

58. **Ochman, H., and E. A. Groisman.** 1996. Distribution of pathogenicity islands in *Salmonella* spp. *Infect. Immun.* **64**:5410–5412.

59. **Ochman, H., and E. A. Groisman.** 1995. The evolution of invasion by enteric bacteria. *Can. J. Microbiol.* **41:**555–561.

60. **Ochman, H., F. C. Soncini, F. Solomon, and E. A. Groisman.** 1996. Identification of a pathogenicity island for *Salmonella* survival in host cells. *Proc. Natl. Acad. Sci. USA* **93:**7800–7804.

61. **Ochman, H., and A. C. Wilson.** 1987. Evolution in bacteria: evidence for a universal substitution rate in cellular genomes. *J. Mol. Evol.* **26:**74–86.

62. **Okazaki, N., S. Matsuo, K. Saito, A. Tominaga, and M. Enomoto.** 1993. Conversion of the *Salmonella* phase 1 flagellin gene *fliC* to the phase 2 gene *fljB* on the *Escherichia coli* K-12 chromosome. *J. Bacteriol.* **175:**758–766.

63. **Ornellas, E. P., R. J. Roantree, and J. P. Steward.** 1970. The specificity and importance of humoral antibody in the protection of mice against intraperitoneal challenge with complement-sensitive and complement-resistant *Salmonella*. *J. Infect. Dis.* **121:**113–123.

64. **Pang, T.** 1998. Vaccination against intracellular bacterial pathogens. *Trends Microbiol.* **6:**433.

65. **Popoff, M. Y., and L. Le Minor.** 1992. *Antigenic Formulas of the Salmonella Serovars,* 5th ed. W. H. O. Collaborating Center for Reference and Research on *Salmonella,* Institute Pasteur, Paris, France.

66. **Rahn, K., S. A. De Grandis, R. C. Clarke, S. A. McEwen, J. E. Galán, C. Ginocchio, R. Curtiss III, and C. L. Gyles.** 1992. Amplification of an *invA* gene sequence of *Salmonella typhimurium* by polymerase chain reaction as a specific method of detection of *Salmonella. Mol. Cell. Probes* **6:**271–279.

67. **Reeves, M. W., G. M. Evins, A. A. Heiba, B. D. Plikaytis, and J. J. Farmer III.** 1989. Clonal nature of *Salmonella typhi* and its genetic relatedness to other salmonellae as shown by multilocus enzyme electrophoresis, and proposal of *Salmonella bongori* comb. nov. *J. Clin. Microbiol.* **27:**313–320.

68. **Reeves, P.** 1993. Evolution of *Salmonella* O antigen variation by interspecific gene transfer on a large scale. *Trends Genet.* **9:**17–22.

69. **Richter-Dahlfors, A., A. M. J. Buchan, and B. B. Finlay.** 1997. Murine salmonellosis studied by confocal microscopy: *Salmonella typhimurium* resides intracellularly inside macrophages and exerts a cytotoxic effect on phagocytes in vivo. *J. Exp. Med.* **186:**569–580.

70. **Riley, M., and A. Anilionis.** 1976. Evolution of the bacterial genome. *Annu. Rev. Microbiol.* **32:**519–560.

71. **Rothenbacher, H.** 1965. Mortality and morbidity in calves with salmonellosis. *J. Am. Vet. Med. Assoc.* **147:**1211–1214.

72. **Roudier, C., M. Krause, J. Fierer, and D. G. Guiney.** 1990. Correlation between the presence of sequences homologous to the vir region of *Salmonella dublin* plasmid pSDL2 and the virulence of twenty-two *Salmonella* serotypes in mice. *Infect. Immun.* **58:**1180–1185.

73. **Schütze, H.** 1930. The importance of somatic antigen in the production of aertrycke and gärtner immunity in mice. *J. Exp. Pathol.* **11:**34–42.

74. **Selander, R. K., P. Beltran, N. H. Smith, R. Helmuth, F. A. Rubin, D. J. Kopecko, K. Ferris, B. D. Tall, A. Cravioto, and J. M. Musser.** 1990. Evolutionary genetic relationships of clones of *Salmonella* serovars that cause human typhoid and other enteric fevers. *Infect. Immun.* **58:**2262–2275.

75. **Shea, J. E., C. R. Beuzon, C. Gleeson, R. Mundy, and D. W. Holden.** 1999. Influence of the *Salmonella typhimurium* pathogenicity island 2 type III secretion system on bacterial growth in the mouse. *Infect. Immun.* **67:**213–219.

76. **Shea, J. E., M. Hensel, C. Gleeson, and D. W. Holden.** 1996. Identification of a virulence locus encoding a second type III secretion system in *Salmonella typhimurium. Proc. Natl. Acad. Sci. USA* **93:**2593–2597.

77. **Smith, B. P., L. DaRoden, M. C. Thurmond, G. W. Dilling, H. Konrad, J. A. Pelton, and J. P. Picanso.** 1994. Prevalence of salmonellae in cattle and in the environment of California dairies. *J. Am. Vet. Med. Assoc.* **205:**467–471.

78. **Sojka, W. J., and H. I. Field.** 1970. Salmonellosis in England and Wales 1958–1967. *Vet. Bull.* **40:**515–531.

79. **Sojka, W. J., and C. Wray.** 1975. Incidence of *Salmonella* infection in animals in England and Wales, 1968–73. *Vet. Rec.* **96:**280–284.

80. **Thampapillai, G., R. Lan, and P. R. Reeves.** 1994. Molecular evolution in the *gnd* locus of *Salmonella enterica. Mol. Biol. Evol.* **11:**813–828.

81. **Thorbecke, G. J., A. R. Amin, and V. K. Tsiagbe.** 1994. Biology of germinal centers in lymphoid tissue. *FASEB J.* **8:**832–840.

82. **Threlfall, E. J., M. L. Hall, and B. Rowe.** 1992. *Salmonella* bacteraemia in England and Wales, 1981–1990. *J. Clin. Pathol.* **45:**34–36.

83. **Tsolis, R. M., T. A. Ficht, and A. J. Bäumler.** 1999. Contribution of *Salmonella typhimurium* virulence factors to diarrheal disease in calves. *Infect. Immun.* **67:**4879–4885.

84. **Uchiya, K., M. A. Barbieri, K. Funato, A. H. Shah, P. D. Stahl, and E. A. Groisman.** 1999. A Salmonella virulence protein that inhibits cellular trafficking. *EMBO J.* **18:**3924–3933.

85. **Watson, P. R., E. E. Galyov, S. M. Paulin, P. W. Jones, and T. S. Wallis.** 1998. Mutation

of *invH*, but not *stn*, reduces salmonella-induced enteritis in cattle. *Infect. Immun.* **66:**1432–1438.

86. **Weiss, S. H., M. J. Blase, F. P. Paleologo, R. E. Black, A. C. McWorther, M. A. Asbury, G. P. Carter, R. A. Feldman, and D. J. Brenner.** 1986. Occurrence and distribution of serotypes of the arizona subgroup of *Salmonella* strains in the United States from 1967–1976. *J. Clin. Microbiol.* **23:**1056–1064.

87. **White, P. B.** 1929. The *Salmonella* group, p. 86–158. *In A System of Bacteriology in Relation to Medicine*, vol. 4. His Majesty's Stationery Office, London, England.

88. **Wilson, M., E. Hsu, A. Marcuz, M. Courtet, L. Du Pasquier, and C. Steinberg.** 1992. What limits affinity maturation of antibodies in Xenopus—the rate of somatic mutation or the ability to select mutants? *EMBO J.* **11:**4337–4347.

89. **Woodward, M. J., I. McLaren, and C. Wray.** 1989. Distribution of virulence plasmids within salmonellae. *J. Gen. Microbiol.* **135:**503–511.

90. **Wray, C., and W. J. Sojka.** 1978. Experimental *Salmonella typhimurium* infection in calves. *Res. Vet. Sci.* **25:**139–143.

91. **Zapata, A. G., M. Torroba, A. Vicente, A. Varas, R. Sacedon, and E. Jimenez.** 1995. The relevance of cell microenvironments for the appearance of lymphohaemopoietic tissues in primitive vertebrates. *Histol. Histopathol.* **10:**761–778.

REGULATION OF VIRULENCE GENE EXPRESSION IN VIVO

James M. Slauch

16

Free-living bacteria have evolved sophisticated regulatory mechanisms that allow them to adapt to their environment. Pathogenic bacteria must adapt and grow in the host environment. For complex pathogens such as *Salmonella enterica* serovar Typhimurium, the host represents a series of microenvironments to which the bacteria must adjust as they progress through the infection. In this chapter, the invasion locus of serovar Typhimurium is used to exemplify several general points about the regulation of virulence gene expression in vivo. The chapter also includes a brief description of in vivo expression technology (IVET), designed to select for bacterial genes that are transcriptionally induced in the host.

REGULATION OF THE SEROVAR TYPHIMURIUM INVASION APPARATUS

As outlined elsewhere in this volume (chapter 14), type III secretion systems encode molecular machines that allow the injection of bacterial proteins into eukaryotic cells. In the case of the type III secretion system encoded on *Salmonella* pathogen island 1 (SPI1), the targets of the machine are the epithelial cells in the small intestine. Injection of the bacterial proteins leads to invasion, characterized by an actin rearrangement in the epithelial cell and engulfment of the bacteria (see reference 6 for a review).

Expression of the genes encoding the type III secretion machinery is highly regulated. Figure 1 shows a model for this regulation based on recent data (1, 5, 7, 8, 14, 22). Work from a number of laboratories has shown that invasion is optimal under conditions of low oxygen, high osmolarity, exponential growth phase, and slightly alkaline pH (1). Expression is also controlled by the PhoPQ two-component regulatory system (2), which represses invasion gene expression presumably in response to low magnesium concentration (10). Lee and colleagues have shown that many, if not all, of these environmental signals are integrated at the level of expression of HilA, a transcriptional regulator found in SPI1 (1). In other words, expression of the invasion loci is dependent on the level of HilA, which is controlled at the transcriptional level in response to these various environmental parameters. HilA directly controls the expression of some of the invasion loci and also activates the AraC-like protein InvF, which is responsible

James M. Slauch Department of Microbiology, University of Illinois, B103 Chemistry and Life Sciences, 601 S. Goodwin Ave., Urbana, IL 61801.

Virulence Mechanisms of Bacterial Pathogens, 3rd ed., Edited by K. A. Brogden et al.
©2000 ASM Press, Washington, D.C.

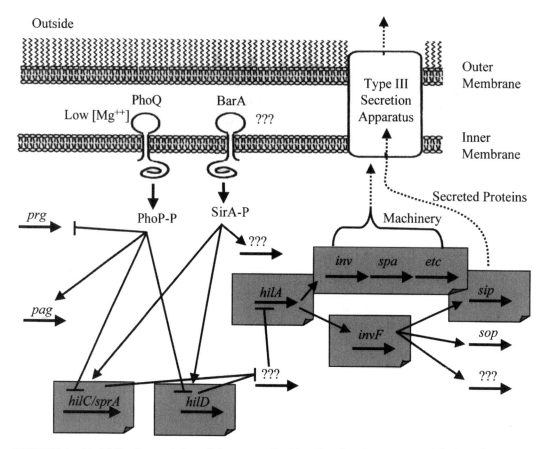

FIGURE 1 Model for the regulation of the serovar Typhimurium invasion apparatus. The boxed genes are located in SPI1. Arrows from regulatory genes indicate activation, whereas lines ending in short lines indicate repression. It is not clear in most cases whether the action is direct or indirect.

for the direct activation of many of the invasion loci, including secreted proteins that are not located on SPI1 (5, 7).

The regulation of invasion gene expression involves an increasingly large number of regulatory loci in addition to HilA and InvF. These include *hilC* (22), also called *sprA* (8) or *sirC* (14), and *hilD* (22). Both *hilC* and *hilD* are carried on SPI1. In addition to PhoPQ mentioned above, transcription of *hilA* is affected by *sirA* (14) and *barA*, which apparently encode the cognate response regulator and sensor kinase, respectively, of a two-component regulatory system (22). The genes encoding both PhoPQ and SirA/BarA are located elsewhere on the bacterial chromosome.

As mentioned, PhoPQ apparently senses the magnesium concentration (10). How the other environmental signals feed into this complex regulatory network is not clear.

REGULATION OF VIRULENCE GENES
This system illustrates several important points that apply generally to in vivo gene expression and the regulation of virulence. Point 1: Regulation is in response to normal environmental parameters. The bacteria are not sensing some specific mammalian molecule, but are apparently integrating a relatively complex combination of environmental parameters found only at the appropriate anatomical location. In the case of the invasion genes, this apparently

defines the environment that the bacteria encounter in the host at the time of invasion or just before invasion. This brings up point 2: It is usually presumed that the same environmental factors that induce expression in the laboratory also control expression of the gene in vivo. Although this seems to be true for SPI1, it is not so simple in other cases. For example, cholera toxin, encoded on a lysogenic bacteriophage in *Vibrio cholerae*, is also regulated by unlinked loci carried on other elements (for a review, see reference 24). In the laboratory, maximum toxin production is obtained under a combination of specific environmental conditions, one of which is 30°C. The toxin is not significantly expressed at 37°C, the temperature encountered in the human intestine, which is the normal site of cholera toxin production. Thus, although regulation of cholera toxin production is fairly well understood at the molecular level in vitro, we do not completely understand the regulation of these genes in vivo.

SPI1 serves as a paradigm pathogenicity island. This 40-Kb locus was apparently acquired as a unit during the evolution of serovar Typhimurium (21). Point 3: Although some of the regulatory components of the system are carried on the island, expression of the system is controlled by preexisting regulatory components found in the cell (Fig. 1). Acquisition of this island has been termed a quantum leap in the evolution of *Salmonella*, allowing the organism to gain access to a new niche (12). However, it must have been a two-step process: (i) acquisition of the genes and (ii) adaptation to preexisting regulatory systems. This point is also exemplified by another locus carried on SPI1, the *sitABCD* operon, which is apparently required for iron uptake during systemic stages of the disease (13, 31). This transport system is regulated in response to iron levels by the preexisting global iron regulator, Fur (13, 31).

It is not clear that these preexisting regulatory networks evolved to regulate virulence. For example, the PhoPQ two-component system is clearly important for regulating a number of virulence functions required for survival in macrophages (9). However, this system is found in *Escherchia coli* (11), which is not known as an intracellular pathogen. One example of a "virulence" function regulated by PhoPQ is modification of lipopolysaccharide (9). Although this confers resistance to cationic peptides, it is not clear that this is why the function evolved. Point 4: It is often difficult to distinguish between "virulence factors" and basic cellular adaptation to the environment.

IVET

Can we use these general principles to identify genes that are transcriptionally induced in the host and increase our understanding of host-pathogen interaction? We have based our studies on the following premises: (i) genes whose products are specifically required for some aspect of the infection will be regulated such that they are only expressed at the appropriate time and place in the host, and (ii) we do not understand and therefore cannot mimic the particular environment within any host tissue. Thus, we developed IVET, a genetic system designed to select for bacterial genes that are transcriptionally induced in vivo (18, 25).

The original IVET system (18, 25) was based on the fact that *purA* mutants of serovar Typhimurium are completely incapable of surviving within the animal host. We complemented this defect by providing transcriptional gene fusions to a promoterless *purA* gene. Only those bacteria containing a *purA* fusion to a gene that is transcriptionally active in the host are able to survive in the animal. This selects for fusions to genes that are expressed, or "on," in vivo. These selected bacteria are then screened for those containing a fusion that is not significantly expressed or "off" on laboratory media. Thus, we identify fusions to genes that are transcriptionally induced in the host.

The *purA* fusions are constructed using the scheme shown in Fig. 2. The vector, pIVET1 (25), encodes an artificial operon comprised of

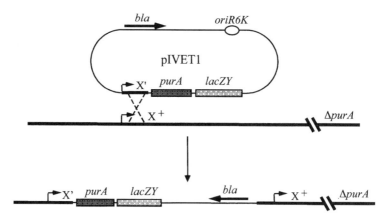

FIGURE 2 Construction of pIVET1 fusions. Random chromosomal fragments were cloned 5′ to the promoterless *purA* gene in pIVET1. The plasmid is based on oriR6K. Therefore, replication is dependent on the Pi protein, which must be supplied in trans. When the plasmids are introduced into a Pi⁻ strain of serovar Typhimurium, they must integrate into the chromosome by homologous recombination with the cloned fragment to be stably maintained. This generates a tandem duplication where one promoter drives the fusion, while the other promoter drives the wild-type copy of the gene of interest.

promoterless *purA, lacZ,* and *lacY* genes. Random fusions are generated by cloning chromosomal DNA fragments into the unique cloning site 5′ to *purA*. At some frequency, a promoter will be positioned correctly to control the expression of the *purA-lacZY* operon. The vector contains the R6K origin, such that replication is dependent on the Pi protein, which must be supplied in trans. When random fusion plasmids are introduced into a *purA* deletion strain of *Salmonella* that does not produce Pi, the plasmids must integrate by homologous recombination to be stably maintained. This creates single-copy fusions under the control of the normal chromosomal promoter. At the same time, at least for the case shown in Fig. 2, the wild-type copy of the gene of interest is maintained. In order for the *purA* deletion strain to survive in the animal, the *purA* fusion will have to be expressed. And if the gene of interest encodes a factor required in the host, the wild-type copy of the gene will also be required.

PurA is required at all stages of *Salmonella* infection in all host tissues. Thus, the selection is maintained throughout the infection and the level of "on" expression required to survive in the animal is constant. This selection can also be used in tissue-culture models of infection (13). We designed an analogous system using the chloramphenicol acetyl-transferase or *cat* gene as a reporter in place of *purA* (20). This *cat* selection can be used in bacteria that lack genetic systems that allow one to knock out *purA*, required for the pIVET1 selection. In theory, one can also control the level of expression required to answer the selection (the "on" level) by adjusting the concentration of chloramphenicol. This is certainly true in tissue-culture systems. It is more difficult to control the concentration of chloramphenicol in an animal model (20). Thus, the *purA* and *cat* systems each have advantages and disadvantages.

VARIATIONS ON A THEME

A number of researchers have subsequently developed IVET systems that are variations on a theme and have used these systems to identify in vivo-induced genes in a variety of both prokaryotic and eukaryotic pathogens (Table

1). These IVET systems fall into three categories: (i) selection systems based on metabolic (e.g., *purA*) or antibiotic (e.g., *cat*) reporters, (ii) recombination-based systems, and (iii) green fluourescent protein (GFP)-based systems. The prototypes for the latter two systems are described below.

Camilli et al. (4) developed an IVET system based on the γδ resolvase, a site-specific recombination protein. Random fusions are made to a promoterless resolvase gene. Elsewhere in the chromosome is a construct containing a tetracycline resistance (Tet^r) determinant surrounded by *res* sites, which are the targets of the resolvase protein. If the resolvase protein is produced from the fusion, it will act at the *res* sites and the Tet^r marker will be lost. A population of strains containing random fusions is first selected for Tet^r. This selects for fusions that are transcriptionally inactive on laboratory medium ("off" state). This population can then be introduced into an animal. The bacteria recovered from the host are screened for those that have become Tet^s, indicating that the fusion was expressed in vivo ("on" state).

This recombination-based system has several advantages. First, there is no selection in the animal. Thus, unlike the *purA* system, which demands that the in vivo induced gene be transcriptionally active throughout the selection, the recombination-based system can potentially identify genes that are only transiently induced in the animal. Second, the resolution of the Tet marker is a heritable change. Thus, gene expression (or more accurately, the history of gene expression) can be monitored in a single bacterium. The system indicates whether the gene was transcriptionally active at any time in the past. This is useful for identifying exactly when a gene isolated with this system or any of the IVET systems is induced in the animal.

Valdivia and Falkow (28) have used the GFP to identify bacterial genes that are induced in eukaryotic cells. Random transcriptional fusions are made using a GFP plasmid and transformed into the bacterium of interest. The original report used serovar Typhimurium. A fluorescence-activated cell sorter is used to identify bacterial genes that are induced when the bacteria are infecting macrophages, but are not transcriptionally active when the bacteria are grown in normal laboratory media. This technique has the advantage that the absolute level of expression that defines both the "on" and "off" state can be designated by the investigator. It is also simple to quantitate the level of expression in the eukaryotic cells. This technique works well with bacteria that can infect tissue culture systems. It would be more difficult, but possible, to perform experiments on bacteria recovered from whole animals.

TABLE 1 IVET systems and selections

Organism	Selection/screen	Reference(s)
Serovar Typhimurium	*purA-lacZY* (pIVET1)	18, 25
Serovar Typhimurium	*cat-lacZY* (pIVET8)	20
V. cholerae	γδ resolvase	4
Pseudomonas aeruginosa	*purEK*	29
Yersinia enterocolitica	*cat*	30
Serovar Typhimurium	GFP	28
Staphylococcus aureus	γδ resolvase	17
E. coli	*cat*	15
Streptococcus gordonii	amylase-*cat*	16
Candida albicans	Flp recombinase	26

RESULTS OF IVET SELECTIONS

We have used the *purA*-based pIVET1 system to select for serovar Typhimurium genes that are induced when the bacteria are infecting animals. PurA is required for all aspects of the infection, so we can perform selections for genes that are induced in any tissue of the animal. We have characterized 76 genes that have answered various IVET selections, which can be classified as follows: Approximately 40% of the fusions are to genes that have known metabolic and cellular functions, including genes whose products are directly involved in intermediary metabolism as well as genes involved in the synthesis of the outer surface of the bacteria, protein synthesis, and other functions. This is an important class of genes in that their identification provides information about the internal host environment in which serovar Typhimurium must adapt and grow. We must ultimately understand the metabolism of the pathogen to gain a complete view of pathogenesis.

Approximately 10% of the genes that answered these selections encode products that are known or thought to be involved in serovar Typhimurium pathogenesis. Indeed, the IVET selection has identified fusions in all of the major virulence regulons known in serovar Typhimurium. These include genes in the PhoPQ regulon (9); the SPI2 type III secretion system involved in systemic growth (23); two different iron uptake systems (13); *sodA*, encoding superoxide dismutase (27); fimbrial biosynthetic genes; and *ompR*, a regulator of virulence genes (19).

Because *E. coli* and serovar Typhimurium are evolutionarily related, we can use the known genome sequence of *E. coli* (3) to further classify the in vivo-induced fusions. Approximately 30% of fusions represent genes that are also present in *E. coli*, but whose function is unknown. Finally, 20% of genes are those that have unknown function and do not have homologs in *E. coli*. We have termed these genes "*Salmonella* specific." Thus, many of the genes identified by IVET are previously unknown genes whose products are potentially important for virulence.

We anticipated that IVET would identify two broad classes of genes. The first class includes those whose products are required for adaptation to the host environment. The second class are those whose products are specifically involved in the interaction with the host and host immune system (true virulence factors). Our results to date are consistent with this premise. However, as stated above, it is often difficult to distinguish between metabolism and virulence. One cannot simply conclude that a previously unidentified gene must be a true virulence factor.

SCREEN FOR TISSUE-SPECIFIC GENE EXPRESSION

We must prove that genes that answer an IVET selection are indeed induced in the animal. This is most easily accomplished using a competition assay (Fig. 3). A bacterial strain containing an IVET fusion to an in vivo-induced gene of interest is competed against a strain containing a fusion to a promoter that is sufficiently active to allow for survival of the

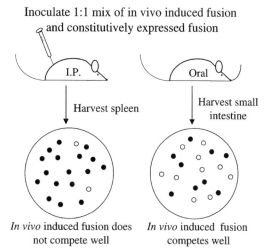

FIGURE 3 Screen for tissue-specific gene expression using IVET fusions. The figure shows the predicted outcome for a fusion that is specifically induced in the small intestine. The in vivo induced fusion strain is phenotypically Lac⁻ (open circles) on lactose MacConkey agar. The constitutive fusion strain is Lac⁺ (solid circles).

strain in any host tissue. These two strains have very different phenotypes on lactose MacConkey agar. The fusion to the potential in vivo induced gene is phenotypically Lac⁻, whereas the constitutive fusion strain is Lac⁺. Mice are inoculated either orally or intraperitoneally (i.p.) with an equal mixture of the two strains. After 3 to 5 days, the mice are sacrificed and the appropriate tissues are removed, homogenized, and plated on lactose MacConkey agar. If the gene of interest is induced, then the fusion strain will survive and compete with the constitutive fusion strain. Thus, an approximately equal number of Lac⁻ and Lac⁺ colonies will be recovered. In contrast, if the gene of interest is not sufficiently induced in the animal, it will not survive well and the vast majority of recovered bacteria will be Lac⁺. This is a critical experiment because it is possible for weakly expressed fusions to artifactually survive an IVET selection.

We have taken this competition assay one step further to identify genes that are induced in specific tissues in the animal (Fig. 3). For example, we have isolated a number of in vivo-induced fusions from the small intestine after oral inoculation. Thus, these are fusions to genes that are induced in the small intestine. We reasoned that a subset of these genes would include those whose products are specifically required for some aspect of the early infection process. This subset of genes would be expressed during the early stages but not during the systemic stages of the disease. Ac-

cordingly, we screened for fusions that were induced only in the small intestine, and not in the spleen, by performing competition assays. Mice were inoculated either orally or i.p. with an equal mixture of the two strains. After 3 to 5 days, the mice were sacrificed and the small intestine was removed from the orally infected animals, while the spleen was removed from the i.p.-infected animals. In the case of a gene that is specifically induced in the small intestine, the fusion would be induced in the oral infection, the strain would be able to compete with the constitutive fusion strain, and we would recover both fusion strains from the small intestine. In contrast, this fusion would not be induced during the systemic stages of the disease, would not survive, and would not be significantly recovered from the spleen after i.p. infection. A strain containing a fusion to the gene that we now call *gipA* (growth in Peyer's patches) clearly answered this selection (Table 2).

As mentioned above, the *sitABCD* operon encodes a putative iron transport system in SPI1 (13, 31). An IVET fusion to this operon has an expression profile opposite to that of *gipA*. Because iron is apparently available in the small intestine, the Fur-regulated *sitABCD* fusion is only induced after invasion of the intestinal epithelium (13). Thus, the fusion is out-competed after oral infection, but competes well with the constitutive fusion in an i.p. infection (Table 2). The virulence defects conferred by mutations in either *sitABCD* or

TABLE 2 Competition between in vivo-induced IVET fusions and a constitutive IVET fusion

In vivo-induced gene	Route of inoculation[a]	Organ	Median competitive index[b]
gipA	Oral	Small intestine	3.0[c]
	i.p.	Spleen/liver	0.08[c]
sitABCD	Oral	Small intestine	0.27[c]
	i.p.	Spleen/liver	1.3

[a] Groups of >5 mice each were inoculated with a 1:1 mixture of the in vivo-induced fusion strain and the constitutive-fusion strain. The dose was approximately 10⁷ oral and 10² i.p. Approximately 5 days postinoculation, the appropriate organs were removed, homogenized, and plated on lactose MacConkey agar to distinguish the in vivo induced fusion strain (Lac⁻) and the constitutive fusion strain (Lac⁺).
[b] Competitive index, output (in vivo-induced fusion strain/constitutive fusion strain)/input (in vivo-induced fusion strain/constitutive fusion strain).
[c] Results are significantly different from inoculum. $P < 0.05$ by Student's *t*-test.

gipA are perfectly consistent with these expression profiles. Strains containing mutations in *gipA* are specifically defective in growth or survival in Peyer's patches (data not shown). The mutants can initially colonize the small intestine and invade. They are also phenotypically wild type with respect to their ability to grow in systemic tissues. Strains containing mutations in *sitABCD* have the opposite phenotype in that they are primarily defective in systemic stages of growth (13).

CONCLUSIONS

Our understanding of in vivo gene expression has increased significantly in the last several years. Work concerning a variety of pathogens indicates that bacteria tightly control expression of their virulence factors. Understanding the environmental signals and regulatory proteins responsible for this control is vital to our knowledge of the disease process. Techniques such as IVET will significantly facilitate these studies. We must learn how the pathogen adapts to its niche, the host.

ACKNOWLEDGMENTS

I thank my graduate students, particularly Theresa Stanley and Anu Janakiraman, whose data are included in this chapter, and Myung Kim, for critically reading the manuscript.

This work was supported by NIH grant AI37530 and ACS Junior Faculty Research Award JFRA-633.

REFERENCES

1. **Bajaj, V., R. L. Lucas, C. Hwang, and C. A. Lee.** 1996. Co-ordinate regulation of *Salmonella typhimurium* invasion genes by environmental and regulatory factors is mediated by control of *hilA* expression. *Mol. Microbiol.* **22**:703–714.
2. **Behlau, I., and S. I. Miller.** 1993. A PhoP-repressed gene promotes *Salmonella typhimurium* invasion of epithelial cells. *J. Bacteriol.* **175**:4475–4484.
3. **Blattner, F. R., G. Plunkett III, C. A. Bloch, N. T. Perna, V. Burland, M. Riley, J. Collado-Vides, J. D. Glasner, C. K. Rode, G. F. Mayhew, J. Gregor, N. W. Davis, H. A. Kirkpatrick, M. A. Goeden, D. J. Rose, B. Mau, and Y. Shao.** 1997. The complete genome sequence of *Escherichia coli* K-12. *Science* **277**:1453–1474.
4. **Camilli, A., D. T. Beattie, and J. J. Mekalanos.** 1994. Use of genetic recombination as a reporter of gene expression. *Proc. Natl. Acad. Sci. USA* **91**:2634–2638.
5. **Darwin, K. H., and V. L. Miller.** 1999. InvF is required for expression of genes encoding proteins secreted by the SPI1 type III secretion apparatus in *Salmonella typhimurium*. *J. Bacteriol.* **181**:4949–4954.
6. **Darwin, K. H., and V. L. Miller.** 1999. Molecular basis of the interaction of *Salmonella* with the intestinal mucosa. *Clin. Microbiol. Rev.* **12**:405–428.
7. **Eichelberg, K., and J. E. Galan.** 1999. Differential regulation of *Salmonella typhimurium* type III secreted proteins by pathogenicity island 1 (SPI-1)-encoded transcriptional activators InvF and HilA. *Infect. Immun.* **67**:4099–4105.
8. **Eichelberg, K., W. D. Hardt, and J. E. Galan.** 1999. Characterization of SprA, an AraC-like transcriptional regulator encoded within the *Salmonella typhimurium* pathogenicity island 1. *Mol. Microbiol.* **33**:139–152.
9. **Ernst, R. K., T. Guina, and S. I. Miller.** 1999. How intracellular bacteria survive: surface modifications that promote resistance to host innate immune responses. *J. Infect. Dis.* **179**(Suppl. 2):S326–S330.
10. **Garcia, V. E., F. C. Soncini, and E. A. Groisman.** 1996. Mg^{2+} as an extracellular signal: environmental regulation of *Salmonella* virulence. *Cell* **84**:165–174.
11. **Groisman, E. A., F. Heffron, and F. Solomon.** 1992. Molecular genetic analysis of the *Escherichia coli phoP* locus. *J. Bacteriol.* **174**:486–491.
12. **Groisman, E. A., and H. Ochman.** 1996. Pathogenicity islands: bacterial evolution in quantum leaps. *Cell* **87**:791–794.
13. **Janakiraman, A., and J. M. Slauch.** The putative iron transport system SitABCD encoded on SPI1 is required for full virulence of *Salmonella typhimurium*. *Mol. Microbiol.*, in press.
14. **Johnston, C., D. A. Pegues, C. J. Hueck, C. A. Lee, and S. I. Miller.** 1996. Transcriptional activation of *Salmonella typhimurium* invasion genes by a member of the phosphorylated response-regulator superfamily. *Mol. Microbiol* **22**:715–727.
15. **Khan, M. A., and R. E. Isaacson.** 1998. In vivo expression of the beta-glucoside (*bgl*) operon of *Escherichia coli* occurs in mouse liver. *J. Bacteriol.* **180**:4746–4749.
16. **Kili, A. O., M. C. Herzberg, M. W. Meyer, X. Zhao, and L. Tao.** 1999. Streptococcal reporter gene-fusion vector for identification of *in vivo* expressed genes. *Plasmid* **42**:67–72.

17. **Lowe, A. M., D. T. Beattie, and R. L. Deresiewicz.** 1998. Identification of novel staphylococcal virulence genes by *in vivo* expression technology. *Mol. Microbiol.* **27:**967–976.

18. **Mahan, M. J., J. M. Slauch, and J. J. Mekalanos.** 1993. Selection of bacterial virulence genes that are specifically induced in host tissues. *Science* **259:**686–688.

19. **Mahan, M. J., J. M. Slauch, and J. J. Mekalanos.** 1996. Environmental regulation of virulence gene expression in *Escherichia, Salmonella,* and *Shigella,* p. 2803–2815. *In* F. C. Neidhardt (ed.), *Escherichia coli and Salmonella: Cellular and Molecular Biology,* 2nd ed. American Society for Microbiology, Washington, D. C.

20. **Mahan, M. J., J. W. Tobias, J. M. Slauch, P. C. Hanna, R. J. Collier, and J. J. Mekalanos.** 1995. Antibiotic-based selection for bacterial genes that are specifically induced during infection of a host. *Proc. Natl. Acad. Sci. USA* **92:**669–673.

21. **Mills, D. M., V. Bajaj, and C. A. Lee.** 1995. A 40 kb chromosomal fragment encoding *Salmonella typhimurium* invasion genes is absent from the corresponding region of the *Escherichia coli* K-12 chromosome. *Mol. Microbiol.* **15:**749–759.

22. **Schechter, L. M., S. M. Damrauer, and C. A. Lee.** 1999. Two AraC/XylS family members can independently counteract the effect of repressing sequences upstream of the *hilA* promoter. *Mol. Microbiol.* **32:**629–642.

23. **Shea, J. E., C. R. Beuzon, C. Gleeson, R. Mundy, and D. W. Holden.** 1999. Influence of the *Salmonella typhimurium* pathogenicity island 2 type III secretion system on bacterial growth in the mouse. *Infect. Immun.* **67:**213–219.

24. **Skorupski, K., and R. K. Taylor.** 1997. Control of the ToxR virulence regulon in *Vibrio cholerae* by environmental stimuli. *Mol. Microbiol.* **25:**1003–1009.

25. **Slauch, J. M., M. J. Mahan, and J. J. Mekalanos.** 1994. *In vivo* expression technology for selection of bacterial genes specifically induced in host tissues. *Methods Enzymol.* **235:**481–492.

26. **Staib, P., M. Kretschmar, T. Nichterlein, G. Kohler, S. Michel, H. Hof, J. Hacker, and J. Morschhauser.** 1999. Host-induced, stage-specific virulence gene activation in *Candida albicans* during infection. *Mol. Microbiol.* **32:**533–546.

27. **Tsolis, R. M., A. J. Baumler, and F. Heffron.** 1995. Role of *Salmonella typhimurium* Mn-superoxide dismutase (SodA) in protection against early killing by J774 macrophages. *Infect. Immun.* **63:**1739–1744.

28. **Valdivia, R. H., and S. Falkow.** 1997. Fluorescence-based isolation of bacterial genes expressed within host cells. *Science* **277:**2007–2011.

29. **Wang, J., A. Mushegian, S. Lory, and S. Jin.** 1996. Large-scale isolation of candidate virulence genes of *Pseudomonas aeruginosa* by *in vivo* selection. *Proc. Natl. Acad. Sci. USA* **93:**10434–10439.

30. **Young, G. M., and V. L. Miller.** 1997. Identification of novel chromosomal loci affecting *Yersinia enterocolitica* pathogenesis. *Mol. Microbiol.* **25:**319–328.

31. **Zhou, D., W. D. Hardt, and J. E. Galan.** 1999. *Salmonella typhimurium* encodes a putative iron transport system within the centisome 63 pathogenicity island. *Infect. Immun.* **67:**1974–1981.

IDENTIFICATION OF VIRULENCE GENES IN SILICO: INFECTIOUS DISEASE GENOMICS

George M. Weinstock, Steven J. Norris, Erica J. Sodergren, and David Smajs

17

Since the first successful application of shotgun DNA sequencing to microorganisms (7), enormous progress has been made in sequencing of whole microbial genomes. In August 1999, nearly 100 genome projects were funded and underway (Fig. 1). Most of these are aimed at microbial genomes, paving the way for a revolution in microbiology. An estimate of the number of genome projects over the last few years (Fig. 2) shows the rapid growth in this area of research. New genome projects continue to be funded; there will likely be hundreds of microbial genomes completed in the next decade. Because a typical microbial genome encodes a few thousand genes, this amounts to the discovery and description of several hundred thousand new genes, all of which will be entered into databases and made available in the public domain. This technology is being applied to a range of microbes (Fig. 1), both the well-studied organisms that are readily amenable to laboratory manipulation and the more challenging experimental systems, where progress has been hampered by experimental limitations and the research communities are smaller. Microbial genomics is a great equalizer, which promises to stimulate research and advance knowledge of these more obscure, but not necessarily unimportant, microbes to a greater extent than the better-studied systems.

How will these genome sequences affect infectious disease research? The direct, short-term goals of genome projects of pathogenic microbes are to (i) stimulate research, (ii) develop diagnostics, (iii) develop vaccines, and (iv) develop therapeutics. Stimulating research includes not only opening up new research areas but also attracting researchers to work on the sequenced organism. This is especially true for researchers who work in a specific area, e.g., transport, who may now be attracted to a new organism because many of its transport systems can be identified from the genome sequence. Development of diagnostics and vaccines requires identification of suitable sequences and antigens, particularly those of surface-localized or secreted proteins, which is aided by the sequence. Therapeutic development can take advantage of many other aspects

George M. Weinstock, Erica J. Sodergren, and David Smajs Department of Microbiology and Molecular Genetics, Center for the Study of Emerging and Re-emerging Pathogens, University of Texas—Houston Medical School, 6431 Fannin St., Houston, TX 77030. *Steven J. Norris* Department of Pathology and Laboratory Medicine, Center for the Study of Emerging and Re-emerging Pathogens, University of Texas—Houston Medical School, 6431 Fannin St., Houston, TX 77030.

Virulence Mechanisms of Bacterial Pathogens, 3rd ed., Edited by K. A. Brogden et al.
©2000 ASM Press, Washington, D.C.

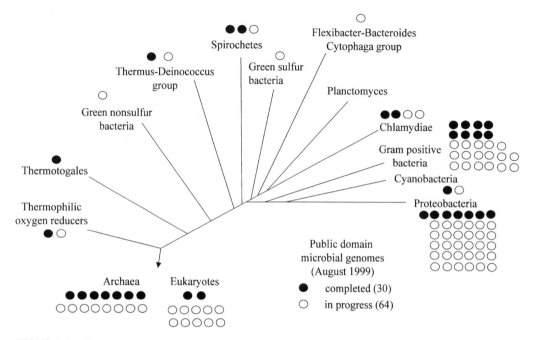

FIGURE 1 Current genome projects. The biological world, represented by the phylogeny deduced from 16S rRNA, with the eubacteria (home to microbial pathogens) emphasized. The circles represent public genome projects, i.e., those that release their data and have been funded, started, and/or completed. The data are taken from a number of tallies present on the Internet, such as those at www.tigr.org/tdb/mdb/mdb.html and www.ncbi.nlm.nih.gov/Entrez/Genome/org.html.

of the cell physiology, from metabolism to information flow, as lethal targets for antimicrobials. This process is aided by the enumeration of these functions from the genome sequence. As noted above, pathogens that have been refractory to study will benefit enormously from a whole genome sequence. For the first time, the entire collection of genes of these organisms will be described, allowing research projects to be selected from a bird's eye view of the whole genetic content. While this is not a panacea for these difficult systems, it will nevertheless have a major impact.

SEQUENCING MICROBIAL GENOMES
Figure 3 illustrates the steps in obtaining a microbial genome sequence and helps to appreciate the product of such a project. Initially the genome is sheared to fragments of about 2 kb in size, which are then cloned into standard filamentous phage (M13) or plasmid (the pUC family) vectors. Since the shearing and cloning are random, the library represents an unbiased sampling of the genome. Template DNA is made from a large number of these clones, and a sequencing reaction is performed from one or both sides of the insert. In this

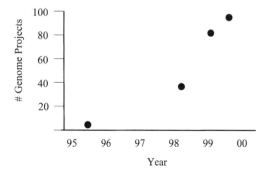

FIGURE 2 Increase in genome projects. Data such as those in Fig. 1 that were collected at several times are summarized.

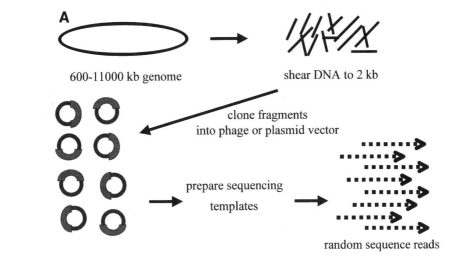

600-11000 kb genome

shear DNA to 2 kb

clone fragments
into phage or plasmid vector

prepare sequencing
templates

random sequence reads

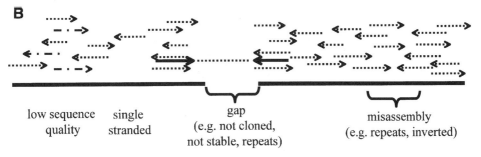

low sequence   single   gap   misassembly
quality   stranded   (e.g. not cloned,   (e.g. repeats, inverted)
                                 not stable, repeats)

Consensus sequence following initial assembly

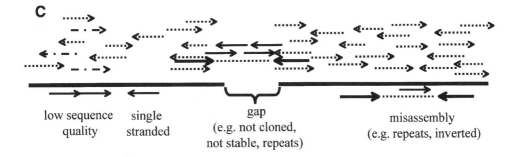

low sequence   single   gap   misassembly
quality   stranded   (e.g. not cloned,   (e.g. repeats, inverted)
                                 not stable, repeats)

high quality sequence: < 1 error/ 10,000 bases

FIGURE 3 Stages in sequencing a bacterial genome. In the initial shotgun sequencing phase (A), a random library is prepared and each clone is sequenced from one or both ends, producing a collection of random sequences (dotted arrows). These are assembled into a provisional consensus sequence (B). In addition, a few large insert clones (such as clones in phage lambda) are prepared and the ends of their inserts are sequenced (solid arrows) and assembled with the short insert clones. The inner section of these inserts is not sequenced. Resequencing of selected regions is performed in the finishing stage (C), including the inner regions of large insert clones.

way, a random set of genomic sequences is obtained. Typically, this continues until there is about seven- to eightfold coverage of the genome. This amounts to about 15,000 sequences per each megabase of the genome. The collection of random sequences is assembled by computer, using programs (4, 5, 9) that look for overlaps between the random sequences to identify how they should go together. At the end of the assembly stage, a provisional consensus sequence is obtained that has a number of imperfections (Fig. 3B). These include regions of low sequence quality (for one reason or another the sequencing reactions were not good), regions that have only been sequenced on one strand (statistically, at this depth of coverage there will always be some regions that have only single-stranded coverage), regions that have not been sequenced at all (either the region was absent due to statistical considerations or else there were biological selections against the clones and they were lost), and regions that have been misassembled (this can happen, for instance, if there are repeated sequences that can go together in more than one way). Each problem is dealt with in the finishing stage, where directed sequencing is performed to resolve these problems. In addition, a number of large insert clones (such as clones in a phage lambda vector) are prepared, and the insert ends are sequenced and assembled with the larger collection of small insert sequences. These clones provide a backbone to aid in assembly. This is particularly true for unclonable regions, where it is frequently possible to obtain a clone in a lambda vector but not a small insert vector.

Following the finishing stage, a contiguous sequence is obtained that is analyzed for coding sequences. A number of programs (14, 17, 21) are available for this purpose, and all rely on statistics of expressed sequences (e.g., codon or hexanucleotide frequencies) to distinguish open reading frames that are coding sequences from those that are not. These programs are in general about 90 to 95% accurate,

and thus most (but not all) coding sequences are identified.

Next, the predicted coding sequences are compared to sequences in databases (Gen-Bank, EMBL; found at www.ncbi.nlm .nih.gov) and "hits" are inspected to identify frameshift errors in the sequence. Often protein sequences are sufficiently highly conserved so that a good match to a protein can be used to identify such sequence errors. These appear as highly similar protein sequences where the coding sequence from the genome project is much shorter than the matched sequences or highly divergent in a region but similar elsewhere. Both of these are hallmarks of a frameshift error in the genomic sequence, which can be identified and corrected by careful inspection of the raw sequence data from this region of the sequence. Of course, this applies only to coding sequences that match something in the databases. Those sequences that are novel, i.e., do not match known sequences, could still retain their errors. At the end of this process the sequence is finished. The result of this analysis for the genome of *Treponema pallidum*, the causative agent of syphilis, is shown in Fig. 4.

FISHING FOR VIRULENCE FACTORS
Sequence gazing is no different than other genetic "fishing" methods. Since the advent of genomic sequencing and genomic expression array analysis, much has been written about the role of research that is not hypothesis driven (1, 8). However, most genetic screens are no more than fishing without a hypothesis just like sequence gazing. These include many classic experimental screening approaches such as those for antigens, genes with common regulation, genes encoding secreted proteins, genes expressed in vivo (15, 16, 23), and genes required in vivo (10, 11). They also include the many traditional mutant hunts, e.g., for temperature-sensitive or cell-cycle mutants, or genetic footprinting with a more genomic view (24, 25). Thus, the modern approaches to mining whole genome sequences have much in common with these classic ap-

- 1,138,006 - 1,138,011 bp (± a few)
- 1040 protein-coding regions predicted (94% of genome)
  - 9 with frameshifts → 1031 coding regions
  - 55% homologous to recognized functions
  - 15% homologous to hypothetical proteins
  - 30% have no known homolog
- 1091 annotated genes

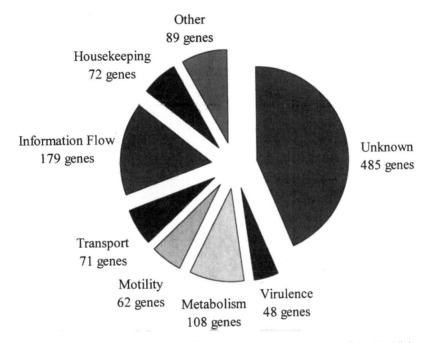

FIGURE 4 The genome of *T. pallidum*. Summary of the analysis of the *T. pallidum* genome sequence. The range in genome size is due to the presence of frameshifts: depending on whether and how these are corrected, one arrives at a slightly different genome length. Note that the frameshifts could be due to several sources, such as sequencing errors, decaying genes, genes regulated by translational frameshifting, or contingency genes (3, 12, 18, 22). Hypothetical proteins are those entries in databases that have no assigned function and have thus not been demonstrated to be functional. The total number of annotated genes include protein-coding genes as well as non-translated RNAs (tRNA, rRNA, e.g.).

proaches. However, not all procedures are equally efficient. Some methods are better than others at identifying virulence factors, but are not easily applied to organisms without good genetic systems. Still, many approaches can be applied without relying on sophisticated methods for genetic manipulation.

ANALYSIS BY HOMOLOGY

As described above, all predicted coding sequences are compared to sequence databases, allowing the initial annotation of the genome to be performed from these homology searches. In Fig. 4 the statistics for such an analysis of the *T. pallidum* genome are shown, indicating that only 55% of the predicted coding sequences matched a database entry with a listed function. The other 45% either matched an entry with no function listed (possibly the product of another genome project) or did not match anything. In fact, this is typical of virtually all published genomes: 20 to 30% of the predicted coding sequences are novel, with no database matches. Therefore, there is an intrinsic limit to how much information can be gained from database homologies alone. Of course, the other caveat for homology-based inferences is that these do not prove function, but only provide hypotheses as to function that can be tested.

The initial identification of virulence factors of *T. pallidum* from database homologies (27) yielded relatively few candidates. As shown in Figs. 4 and 5, the limited number of hits represent only a few classes of functions. The 12-member *tpr* gene family, homologous to a protein from another treponemal pathogen, the Msp protein of *Treponema denticola*, is a major finding. The functions of these genes are not known, but in view of the interactions of Msp with the host (6), it is likely that the *tpr* genes play a role in infection. Five more genes match proteins annotated as hemolysins in the databases, with four of these comprising a family with homology to the *tlyC* gene of *Serpulina* (*Brachyspira*) *hyodysenteriae* (26). Closer inspection of the database matches for these four genes shows that the evidence for hemolysins is weak (Table 1). Thus, care must be taken in relying on the annotation information of databases. Another class of potential virulence-related functions are regulators, comprising about eight genes. Some of these have homology to regulators of virulence in other organisms, while several are two-component regulatory systems. Although the target for these regulators is not known, because they respond to the environment through their sensor component, and *T. pallidum* only lives in its human host, these must be monitoring the host environment. Beyond these classes are a number of isolated genes that do not

necessarily have homology to known virulence factors, but are interesting based on the genes they do match. However, this very limited assortment of virulence gene candidates must only be a fraction of the genes needed for infection by *T. pallidum*.

The situation with *T. pallidum* illustrates the challenge of converting genomic information into applications. This bacterium, a spirochete, has been understudied, since it cannot be cultured continuously in the laboratory. It is evolutionarily removed from the well-studied proteobacteria or gram-positive organisms, so it is likely to have less homology to virulence factors from these groups. This is likely the case with many other pathogens. Moreover, independent of the organism, a fraction of genes in all microbes do not have database matches. Some of these are likely to be (species-specific) important in infection. The *tpr* genes mentioned above are examples. Thus, other methods besides database homologies must be used for mining the sequence.

GUILT BY ASSOCIATION: COMPARATIVE GENOMICS

One approach to obtain additional information without relying on homology is comparative genomics. Intraspecies comparisons measure which regions are conserved or which are variable between different strains of the same organism. An example is the enormous differences between the varieties of pathogenic *Escherichia coli* strains (2, 19, 20). In addition, there are interspecies comparisons, different species of the same genus that may cause different but related diseases. Examples of this for treponemes are *T. pallidum* and *Treponema pertenue* (technically two differently subspecies of *T. pallidum*) that cause syphilis and yaws, respectively. Comparisons can be performed in a number of ways once one reference sequence is available. A whole genome fingerprinting approach has been performed for spirochetes (Fig. 6). This analysis showed that *tpr* genes were highly conserved among syphilis strains, whereas most of the variation between syphilis and yaws strains involved *tpr*

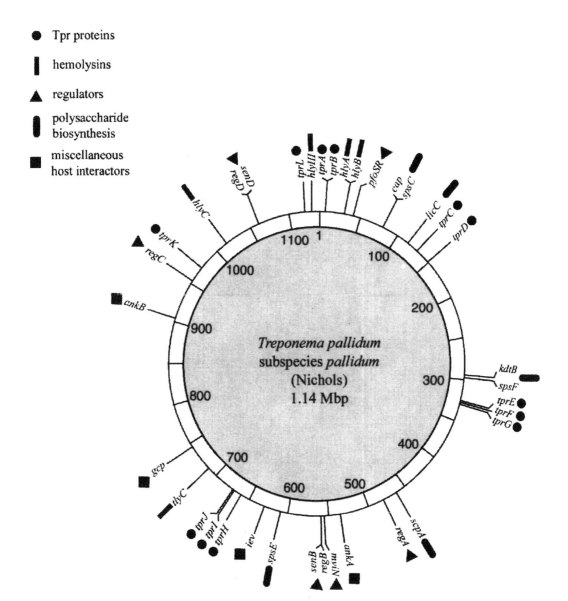

FIGURE 5 Candidate virulence factors of *T. pallidum* based on sequence homology.

genes (unpublished results). Thus, the differences identified by this approach confirm the importance of these genes and identify which genes contribute to the differences between these two infections.

Related to this whole genome search for variability are other inhomogeneities that become apparent on the whole-genome scale.

Genomes are mosaic for a variety of features, ranging from base composition to clustering of virulence genes (pathogenicity islands) (13). These are thought to reflect horizontal transfer of chunks of material from other organisms, with subsequent selection for their stable maintenance. This selection can include changes in host range and disease mechanism

TABLE 1 The TlyC family of putative hemolysins[a]

Organism	No. of genes	Discovered by	Evidence for function
Bacillus subtilis[b]	3	Genome project	Homology
Chlamydia trachomatis	1	Genome project	Homology
Deinococcus radiodurans[b]	2	Genome project	Homology
Escherichia coli	4	Genome project	Homology
Haemophilus influenzae	3	Genome project	Homology
Helicobacter pylori	1	Genome project	Homology
Klebsiella pneumoniae	≥1	Serendipity	Homology
Mycoplasma genitalium	1	Genome project	Homology
Mycoplasma pneumoniae	1	Genome project	Homology
Mycobacterium tuberculosis	≥3	Genome project	Homology
Neisseria gonorrhoeae	2	Genome project	Homology
Neisseria meningiditis	1	Genome project	Homology
Pseudomonas aeruginosa	3	Genome project	Homology
Rickettsia prowazekii	≥1	Genome project	Homology
Rhodobacter sphaeroides[b]	≥1	Genome project	Homology
Serpulina hyodysenteriae	≥1	Hard science	Hemolytic activity
Streptococcus pyogenes	1	Genome project	Homology
Synechocystis sp.[b]	2	Genome project	Homology
Thermatoga maritima[b]	2	Genome project	Homology
Treponema pallidum	4	Genome project	Homology
Vibrio cholerae	1	Genome project	Homology

[a] The organisms show matches to the genes with homology to *tlyC* from *Serpulina* (*Brachyspira*) *hyodysenteriae* (26). In most cases, there are multiple genes in the genome that match (e.g., see Fig. 5 for *T. pallidum*).
[b] Soil bacterium, not considered an animal pathogen.

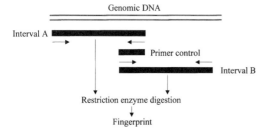

FIGURE 6 Whole genome fingerprinting. The *T. pallidum* sequence was divided into 75 intervals ranging from 5 to 28 kb, with overlaps ranging from 0.2 to 1 kb. Each interval can be amplified by PCR in both *T. pallidum* and *T. pertenue* strains. The amplified fragments are digested with a restriction enzyme to generate the fingerprint, allowing whole genomes to be compared.

when these mosaic segments are involved in virulence. Therefore, identifying such inhomogeneities and careful inspection of the genes in these segments can provide another clue.

HETEROLOGOUS EXPRESSION

The whole genome sequence with predicted coding sequences allows any gene to be cloned. Thus, for any organism, but especially for intractable ones, it becomes possible to clone and introduce coding sequences into expression vectors in a heterologous host such as *E. coli*. The simplest way to accomplish this is by PCR, using genomic DNA as a template.

Heterologous Expression

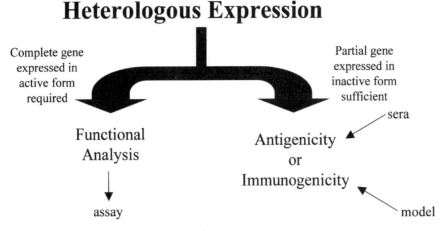

Complete gene expressed in active form required

Functional Analysis

assay

Partial gene expressed in inactive form sufficient

sera

Antigenicity or Immunogenicity

model

FIGURE 7 Applications for heterologous expression of genes.

This is particularly useful for organisms that are hard to grow, since relatively little template DNA is required to clone and express most genes individually from the genome. Expression of whole genes as well as gene fragments is extremely useful (Fig. 7).

This provides many avenues to identify important genes for infection. Expression of an intact, complete polypeptide allows one to assay for activity. Experiments are underway to test the activity of the putative hemolysins of *T. pallidum* using this approach. Although not often discussed, single proteins may not have the full activity but may require accessory proteins. Often, these can be present in an operon together, requiring coexpression in the heterologous host. In addition to this type of difficulty is the unpredictable nature of heterologous expression itself, where a complete polypeptide may aggregate, be degraded, or be toxic to the host. For proteins that are part of hetero-oligomers, this can be accentuated when the other subunits are not present. Despite these difficulties, expression of an active protein provides definite evidence as to function. There are also proteins that are only expressed in infection and would not be present in laboratory grown bacteria. For these, and for organisms that cannot be grown easily in the laboratory, heterologous expression provides a crucial approach to functional analysis.

For many applications it is not necessary to express genes in their entirety; a segment of the gene is sufficient (Fig. 7). This is simpler and less demanding than expression of whole genes. If serum samples are available from infected hosts, this approach allows one to identify antigens, because only an epitope needs to be expressed to allow identification. This also demonstrates that the coding sequence is expressed during infection, demonstrating that gene prediction and annotation are correct. Seroreactivity has been observed for all of the *T. pallidum* hemolysins and some of the *tpr* genes, providing definitive evidence for their expression in infection (unpublished results). It is also worth noting the potential of genome sequences for exhaustive screening for antigens (Fig. 8). Traditional methods, based on making shotgun libraries in plasmid or phage vectors, have a number of biases that make it

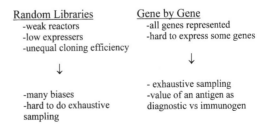

Random Libraries
-weak reactors
-low expressers
-unequal cloning efficiency

↓

-many biases
-hard to do exhaustive sampling

Gene by Gene
-all genes represented
-hard to express some genes

↓

- exhaustive sampling
-value of an antigen as diagnostic vs immunogen

FIGURE 8 Comparison of methods for antigen identification.

difficult to thoroughly sample the genome. In contrast, the approach of expressing each gene in a heterologous host allows unbiased sampling of the whole genome, producing more accurate results. Finally, production of partial polypeptides can be used to test efficacy as a vaccine if a suitable model is available.

CONCLUSIONS

Whole-genome sequencing is having profound effects on the understanding of mechanisms of infection, the identification of virulence factors, and the development of diagnostics and vaccines. Methods to identify virulence factors can rely on classic genetic approaches as well as homology searches and other genome-wide analyses such as comparative genomics. Developing diagnostics and vaccines can be done in general ways that test each protein without bias. Whereas there may be value in limiting this analysis to secreted proteins (as those likely to be useful as diagnostics or vaccine targets), limitations in predicting exported proteins, coupled with the possibility of analyzing proteins wholesale, make the latter alternative attractive. Further advances in these technologies will only make this more feasible. We can anticipate that not only will the whole genome sequences be available in the future, but also clones of each gene, expressed in a host that can be easily manipulated. Taken together, these advances will create a new era in infectious disease research.

REFERENCES

1. **Brown, P. O., and D. Botstein.** 1999. Exploring the new world of the genome with DNA microarrays. *Nat. Genet.* **21:**33–37.
2. **Burland, V., Y. Shao, N. T. Perna, G. Plunkett, H. J. Sofia, and F. R. Blattner.** 1998. The complete DNA sequence and analysis of the large virulence plasmid of Escherichia coli O157: H7. *Nucleic Acids Res.* **26:**4196–4204.
3. **Deitsch, K. W., E. R. Moxon, and T. E. Wellems.** 1997. Shared themes of antigenic variation and virulence in bacterial, protozoal, and fungal infections. *Microbiol. Mol. Biol. Rev.* **61:** 281–293.
4. **Ewing, B., and P. Green.** 1998. Base-calling of automated sequencer traces using phred. II. Error probabilities. *Genome Res.* **8:**186–194.
5. **Ewing, B., L. Hillier, M. C. Wendl, and P. Green.** 1998. Base-calling of automated sequencer traces using phred. I. Accuracy assessment. *Genome Res.* **8:**175–185.
6. **Fenno, J. C., K. H. Muller, and B. C. McBride.** 1996. Sequence analysis, expression, and binding activity of recombinant major outer sheath protein (Msp) of *Treponema denticola. J. Bacteriol.* **178:**2489–2497.
7. **Fleischmann, R. D., M. D. Adams, O. White, R. A. Clayton, E. F. Kirkness, A. R. Kerlavage, C. J. Bult, J. F. Tomb, B. A. Dougherty, and J. M. Merrick.** 1995. Whole-genome random sequencing and assembly of Haemophilus influenzae Rd [see comments]. *Science* **269:**496–512.
8. **Goodman, L.** 1999. A.D. hypothesis-limited research [editorial]. *Genome Res.* **9:**673–674.
9. **Gordon, D., C. Abajian, and P. Green.** 1998. Consed: a graphical tool for sequence finishing. *Genome Res.* **8:**195–202.
10. **Hensel, M.** 1998. Whole genome scan for habitat-specific genes by signature-tagged mutagenesis. *Electrophoresis* **19:**608–612.
11. **Hensel, M., J. E. Shea, C. Gleeson, M. D. Jones, E. Dalton, and D. W. Holden.** 1995. Simultaneous identification of bacterial virulence genes by negative selection. *Science* **269:**400–403.
12. **Hood, D. W., M. E. Deadman, M. P. Jennings, M. Bisercic, R. D. Fleischmann, J. C. Venter, and E. R. Moxon.** 1996. DNA repeats identify novel virulence genes in Haemophilus influenzae. *Proc. Natl. Acad. Sci. USA* **93:**11121–11125.
13. **Lee, C. A.** 1996. Pathogenicity islands and the evolution of bacterial pathogens. *Infect. Agents Dis.* **5:**1–7.
14. **Lukashin, A. V., and M. Borodovsky.** 1998. GeneMark.hmm: new solutions for gene finding. *Nucleic Acids Res.* **26:**1107–1115.
15. **Mahan, M. J., J. M. Slauch, P. C. Hanna, A. Camilli, J. W. Tobias, M. K. Waldor, and J. J. Mekalanos.** 1993. Selection for bacterial genes that are specifically induced in host tissues: the hunt for virulence factors. *Infect. Agents Dis.* **2:**263–268.
16. **Mahan, M. J., J. M. Slauch, and J. J. Mekalanos.** 1993. Selection of bacterial virulence genes that are specifically induced in host tissues [see comments]. *Science* **259:**686–688.
17. **McIninch, J. D., W. S. Hayes, and M. Borodovsky.** 1996. Applications of GeneMark in multispecies environments, p. 165–75. *In Intelligent Systems for Molecular Biology '96.* AAAI Press.

18. Moxon, E. R., P. B. Rainey, M. A. Nowak, and R. E. Lenski. 1994. Adaptive evolution of highly mutable loci in pathogenic bacteria. *Curr. Biol.* **4:**24–33.

19. Perna, N. T., G. F. Mayhew, G. Posfai, S. Elliott, M. S. Donnenberg, J. B. Kaper, and F. R. Blattner. 1998. Molecular evolution of a pathogenicity island from enterohemorrhagic Escherichia coli O157:H7. *Infect. Immun.* **66:** 3810–3817.

20. Plunkett, G., III, D. J. Rose, T. J. Durfee, and F. R. Blattner. 1999. Sequence of Shiga toxin 2 phage 933W from Escherichia coli O157: H7: Shiga toxin as a phage late-gene product. *J. Bacteriol.* **181:**1767–1778.

21. Salzberg, S. L., A. L. Delcher, S. Kasif, and O. White. 1998. Microbial gene identification using interpolated Markov models. *Nucleic Acids Res.* **26:**544–548.

22. Saunders, N. J., J. F. Peden, D. W. Hood, and E. R. Moxon. 1998. Simple sequence repeats in the Helicobacter pylori genome. *Mol. Microbiol.* **27:**1091–1098.

23. Slauch, J. M., M. J. Mahan, and J. J. Mekalanos. 1994. In vivo expression technology for selection of bacterial genes specifically induced in host tissues. *Methods Enzymol.* **235:**481–492.

24. Smith, V., D. Botstein, and P. O. Brown. 1995. Genetic footprinting: a genomic strategy for determining a gene's function given its sequence. *Proc. Natl. Acad. Sci. USA* **92:**6479–6483.

25. Smith, V., K. N. Chou, D. Lashkari, D. Botstein, and P. O. Brown. 1996. Functional analysis of the genes of yeast chromosome V by genetic footprinting. *Science* **274:**2069–2074.

26. ter Huurne, A. A., S. Muir, M. van Houten, B. A. van der Zeijst, W. Gaastra, and J. G. Kusters. 1994. Characterization of three putative Serpulina hyodysenteriae hemolysins. *Microb. Pathog.* **16:**269–282.

27. Weinstock, G. M., J. M. Hardham, M. P. McLeod, E. J. Sodergren, and S. J. Norris. 1998. The genome of Treponema pallidum: new light on the agent of syphilis. *FEMS Microbiol. Lett.* **22:**323–332.

CONCLUDING
PERSPECTIVE

STATE AND FUTURE OF STUDIES ON BACTERIAL PATHOGENICITY: IMPACT OF NEW METHODS OF STUDYING BACTERIAL BEHAVIOR IN VIVO

H. Smith

18

This is the third time that I have assessed the state and future of studies on bacterial pathogenicity at the end of symposia held in Ames, Iowa, on "Virulence Mechanisms of Bacterial Pathogens." On previous occasions (60, 63), I have taken a broad approach to the task, defining the logical steps necessary to achieve the objectives of research on bacterial pathogenicity, which are to recognize virulence determinants, identify them, and relate their structure to function. Then, from the papers of the symposia I assessed how far up this ladder was research on the five essential requirements for pathogenicity, namely infection of mucous surfaces, entry into the host, multiplication in vivo, interference with host defense, and damage caused to the host. This time, I am taking a different tack; I will concentrate on one topic that affects all areas of pathogenicity.

Perhaps the major development in studies of bacterial pathogenicity since this series of symposia began has been a rising interest in the activities of bacterial pathogens in vivo and the design of new methods for studying them.

A spur for this development has been the increased attention given to regulation of virulence-determinant production and the effect of environmental factors on it. This was evident in the last symposium (44, 63) and certainly in this one (Chapters 5, 16, and 17). I will describe the new methods and discuss their impact. First, I will summarize aspects of behavior of bacterial pathogens in vivo to be investigated by both conventional and new methods.

ASPECTS OF BACTERIAL BEHAVIOR IN VIVO THAT SHOULD BE INVESTIGATED

Environmental conditions in vivo (osmolarity, pH, E_h, and nutrient and substrate availability) differ from those in laboratory cultures. The bacteria may be on mucous surfaces, extracellular in the body fluids, or intracellular in either the cytoplasm or within vacuoles. Not only are environmental conditions more complex, they alter as an infection proceeds, due to inflammation and tissue breakdown, and spread from one anatomical site to another; and from host cells changing their gene expression when infected by bacteria. Changes in environment can affect bacterial behavior by varying the availability of nutrients and substrates and by influencing regulation of

H. Smith The Medical School, Edgbaston, The University of Birmingham, Birmingham B15 2TT, United Kingdom.

Virulence Mechanisms of Bacterial Pathogens, 3rd ed., Edited by K. A. Brogden et al.
©2000 ASM Press, Washington, D.C.

gene expression (4, 16). Bacterial pathogens in vivo will therefore be different from those grown in vitro, and this has been confirmed for many species (61, 64). This means that behavior in vivo must be explored to increase knowledge of bacterial pathogenicity.

If we consider the five requirements for pathogenicity, aspects of bacterial growth in vivo that need investigation are (i) the pattern of a developing infection, including growth rates and population sizes at different stages; (ii) identification of key nutrients that become available or depleted; and (iii) how the nutrients are metabolized and by what bacterial determinants. Turning to mucosal colonization, penetration, interference with host defense, and damage to the host, the following points need attention. Some putative determinants of these requirements recognized by experiments in vitro may not be formed in vivo. Some important determinants formed in vivo may not be produced in vitro. Finally, the complement of virulence determinants may change as infection proceeds due to alterations in environmental conditions. The latter raises the question of what regulatory systems operate in vivo. Are they the same as those that function in vitro, e.g., PhoP/PhoQ of *Salmonella enterica* serovar Typhimurium (20) and ToxR/ToxS of *Vibrio cholerae* (59)? Production in vivo of a virulence determinant whose formation in vitro is controlled by a certain regulon does not necessarily mean that the same regulon controls production in vivo. To confirm that the regulon is involved, expression of its genes in vivo should be detected, mutation of these genes should reduce virulence, and relevant environmental parameters should be present at levels at which they are effective in vitro.

Turning to a recently discovered regulatory system, quorum sensing (22, 75), we find that it cannot occur early in infection because the numbers of pathogens are too small. To be certain that it operates later, evidence is needed for expression of relevant genes, mutations in these genes reducing virulence, attainment of requisite population densities, and detection in vivo of signaling molecules. For type III secretion systems (30), the evidence required for proving that they operate in vivo is contact with relevant cells, expression of regulatory genes, and reduction of virulence by mutation of these genes.

NEW METHODS FOR STUDYING BACTERIAL BEHAVIOR IN VIVO

These methods are listed in Table 1 and are described here. Results from their use are discussed in the next section.

Following Pathogenesis of Infection in Animals

The classical method of following the pattern of a developing infection is to take samples of body fluids and tissues during the course of infection and examine them outside the host by in vitro methods. The latter include culture, total and viable counts, and light and electron microscopy of histopathological specimens. Recently, there have been four advances. Three add to the dimensions of the classical method. The fourth is completely new.

TABLE 1 New methods for studying bacterial behavior in vivo

Following infection in animals
- Confocal laser scanning microscopy
- Fluorescence-activated cell sorting
- Laser microprobe mass spectrometry
- Photonic and radio detection of pathogens in vivo

Measuring environmental parameters in vivo
- Quantitative fluorescence microscopy
- Use of reporter genes
- X-ray microanalytical electron microscopy

Detection of genes expressed in vivo
- In vivo expression technology
- Differential fluorescence induction
- Differential display of cDNAs
- Reaction of products with antibodies evoked by infection

Direct identification of virulence genes
- Signature tagged mutagenesis
- Virulence complementation

Global analysis of potential gene expression
- Complete genome sequencing
- Chip technologies

Confocal laser scanning microscopy (CLSM) (45) and computerized image analysis techniques allow the examination of immunostained tissue sections that are 30 μm thick. This is 10 to 30 times thicker than sections used in conventional microscopy and more than 300 times thicker than those used in electron microscopy. Large areas of tissue can be scanned for relatively few bacteria. CLSM allows examination of early stages of infection by small inocula. Formerly, large, unnatural inocula were needed for bacteria to be visualized by the methods available. CLSM was used to follow infection in the livers of mice after intravenous injection of less than 100 serovar Typhimurium. This procedure mimicked infection by the few bacteria that would cross the intestinal epithelium and enter the bloodstream after oral ingestion of the pathogen. After 3 days, salmonellae were seen within liver macrophages, not hepatocytes (53). CLSM can be used for any pathogen and for infection of any tissue.

Fluorescence-activated cell sorting (FACS) coupled with bacteria labeled by fluorescent markers such as green fluorescent protein (GFP) can be used to identify host cells containing pathogens (69). After inoculation of animals, infected tissues are removed, homogenized, and subjected to FACS. Hundreds of cells are sorted in seconds, and subsets targeted by the pathogen are identified. Three days after intraperitoneal injection of GFP-marked serovar Typhimurium into mice, FACS showed that up to 5% of all spleenocytes contained one or more fluorescent organisms (69). After intraperitoneal, intravenous, and orogastric infection of mice with *Yersinia enterocolitica* whose *yopE* genes were fused with the GFP gene, fluorescent bacteria, i.e., those expressing YopE, could be seen in the peritoneum, spleen, liver, and Peyer's patches but less in the intestine. The results were quantified by FACS (31).

Laser microprobe mass spectrometry (LAMMS) of individual bacteria has been used to measure the viability of *Mycobacterium leprae* in skin biopsies from patients and infected tissue from armadillos and mice by determining the ratio of Na^+/K^+ (23, 57). This method can distinguish between dead and live bacteria in tissues and can be extended to other pathogens.

The major advance is to use noninvasive methods to detect pathogens within the living host and to evaluate population changes in different tissues. In photonic imaging (11), bacteria are made luminescent by transformation with a plasmid conferring constitutive expression of luciferase. The wavelength of light emitted by these bacteria (between 500 and 1,500 nm) is relatively nonabsorbed by animal tissues. After injecting the bacteria, photons transmitted through the tissue can be detected by a light-sensitive charge-coupled device (CCD) camera connected to an image processor. Thus, the presence of bacteria in various tissues is demonstrated. This method was used to follow localization of the pathogen during progressive and abortive infections of mice by three strains of serovar Typhimurium of differing virulence, after intraperitoneal or oral inoculation (11). It is a major step forward, which can be extended to infections with other pathogens. There is one cautionary note: oxygen is required for bacteria to exhibit bioluminescence. Although oxygen concentration within animal cells is estimated to be above the predicted minimal level needed for bacterial luminescence (11), the strength of photon emission may reflect the degree of tissue oxygenation as well as the number of bacteria.

A radiolabeling technique for in vivo imaging has also been reported (46). *Salmonella enterica* serovar Abortusovis was labeled with radioactive technetium-99m. The viability of the bacteria was satisfactory and the technetium-99m appeared to be intracellular. After subcutaneous inoculation of the labeled bacteria into sheep, their dissemination in the lymphatic system was detected by radioimaging.

Measurement of Environmental Parameters In Vivo

Much information on environmental parameters is in the literature (33, 34), and more can be obtained by conventional biochemical analysis of body fluids and tissue extracts. Only new methods that have been used to measure the parameters in infected cells are summarized here.

Quantitative fluorescent microscopy (52) of carriers of dyes that react to environmental signals is one method. It was used to measure intraphagosomal pH within macrophages infected with serovar Typhimurium (3). The excitation spectrum of fluorescent-isothiocyanate-conjugated dextran varies with pH. Macrophages were infected with serovar Typhimurium in a medium containing this fluorescent conjugate and entered the phagosomes together with the ingested bacteria. Quantitative fluorescence microscopy visualized fluorescent phagosomes containing bacteria, and the strength of the fluorescence was measured by an image processing system. It was related to pH using a standard curve constructed by incubating macrophages containing fluorescent dextran in their phagosomes in buffers of graded pH. This method could measure other environmental parameters, e.g., oxygen tension, by using different dyes linked to suitable carriers. In addition to cultured macrophages, it might be applied to biopsies from infected animals.

LacZ fusions to genes whose expression depends on certain concentrations of specific factors can indicate their levels in phagosomes. To measure Ca^{2+} levels in monocytes, a gene of *Yersinia pestis* which responded to low levels of Ca^{2+} was coupled to the LacZ operon of *Escherichia coli* (49). Experiments with culture media showed that β-galactosidase production by this fusion derivative was substantial in 1 to 100 μM Ca^{2+} and low in 400 to 2,000 μM. Then, human monocytes were infected with the strain, and at intervals, the intracellular organisms were counted and lysed for β-galactosidase assay. The levels were high, indicating an intraphagosomal Ca^{2+} concentration of 1 to 100 μM.

In studies with serovar Typhimurium and MDCK cells (19), *lacZ* fusions were made in the *iroA* and *mgtB* genes (induced by Fe^{2+} and

Mg^{2+} limitations, respectively). Cells were infected with these strains; at time intervals, samples were taken for viable counts of intracellular bacteria and β-galactosidase assay. The latter was done on lysed bacteria using a sensitive substrate fluorescein-β-galactoside, and the results were related to standard numbers of bacteria. In culture media, the *iroA* gene was suppressed by 33 μM $FeSO_4$, whereas in intracellular bacteria, expression was high, even when 33 μM $FeSO_4$ was added to the cell culture medium. Thus, intravacuolar concentrations of Fe^{2+} are less than 33 μM and not affected by high extracellular concentrations. The *mtgB* gene is suppressed by glucose as well as Mg^{2+}. In media with glucose present, the gene would be suppressed by 10 to 50 μM Mg^{2+}. Since expression was high in intracellular bacteria, an intravacuolar concentration of less than 50 μM Mg^{2+} was indicated.

These two examples show that *lacZ* fusions can give some idea of the levels of environmental factors in intracellular vacuoles but not accurate measurements. The method could be used for other pathogens and environmental factors in cultured cells and possibly biopsies. However, it must be stressed that inside the cells the *lacZ* fusions may respond to factors different from those in vitro.

X-ray microanalysis during electron microscopy (42) can be used to measure elements in situ in cells and subcellular particles. In experiments with neonatal mice and rotavirus, cytoplasmic concentrations of sodium, magnesium, phosphorus, sulphur, chloride, potassium, and calcium in infected and control intestinal cells were measured in specimens taken daily for 6 days after inoculation (65). Sodium and chloride concentrations in villus base cells were significantly higher for infected animals, and phosphorus, sulphur, and potassium concentrations were lower. There were no differences in magnesium concentrations. It was suggested that loss of excess NaCl from the villus cells into the lumen was the driving force for fluid loss. This technique could be used for bacterial infections of the intestine and elsewhere.

Detection of Genes Expressed In Vivo

For some time now, expression of known bacterial genes has been detected in animal cells either by monitoring cDNAs derived from mRNAs or by coupling the genes to *lacZ, lux* or to genes that code for epitopes recognized by specific antibodies (18, 70). The new advance has been detection of expression in vivo of unknown as well as known genes. The methods described below have been used for infections in animals and for growth in macrophage cell lines. Although the latter may not reflect all the nuances of infection in animals, much useful information has accrued. Indeed, bacterial gene expression detected in macrophages has often been demonstrated in animals (26, 70).

IN VIVO EXPRESSION
TECHNOLOGY (IVET)
This technology has three variations. The first uses auxotrophic mutants for selection (37, 38). It was established using DNA fragments from a wild-type serovar Typhimurium, which were attached to a synthetic operon containing a promoterless *purA* gene and a promoterless *lacZ* operon. The resulting constructs were integrated into the chromosome of a *purA* auxotrophic mutant of serovar Typhimurium which, unless complemented, did not grow in vivo. In mice, some bacteria grew in the spleen, indicating *purA* expression by the synthetic gene due to promoters provided by wild-type genes that were also being expressed in vivo. Bacteria with promoters not expressed in vivo were eliminated. Plating bacteria recovered from the spleen on McConkey lactose agar distinguished bacteria with promoters expressed in vitro as well as in vivo (red colonies, about 95%) and those having promoters expressed only in vivo (white colonies, about 5%). The genes of the latter were termed *ivi* (in vivo-induced) genes. Fifteen were obtained from serovar Typhimurium (37, 38). The method was extended to

V. cholerae using thymine auxotrophs. After administering a pool of *thyA-lacZ* fusions orally to infant mice and 1 day later recovering the bacteria from the small intestine, many *ivi* genes were revealed (37). The auxotrophic method also selected *ivi* genes of *Pseudomonas aeruginosa* induced during mouse infection (73) and by growth in respiratory mucus of cystic fibrosis patients (72).

In the second IVET system (37), a promoterless chloramphenicol acetyl transferase (*cat*) gene replaced the *purA* gene, and mice were dosed with chloramphenicol. The promoters provided by serovar Typhimurium genes induced in vivo rendered the bacteria chloramphenicol-resistant. Plating the recovered bacteria on MacConkey lactose agar again distinguished between organisms containing genes expressed in vivo and in vitro (about 95%) and those with *ivi* genes (about 5%). The method was extended to serovar Typhimurium within cultured macrophages and depended on the fact that chloramphenicol, unlike some antibiotics, penetrates into macrophages (39). Later (24), a combination of auxotroph and antibiotic selection methods detected more than 100 serovar Typhimurium *ivi* genes expressed during infection of mice and/or macrophages. The antibiotic method has been used to investigate the contribution of chromosomal genes of *Y. enterocolitica* to virulence (76). A library of promoter fusions with the *cat* gene was constructed using DNA from a plasmid-cured strain of *Y. enterocolitica* and mated into a strain that had the virulence plasmid. After intragastric inoculation into chloramphenicol treated mice and harvesting bacteria from Peyer's patches, *ivi* genes were detected. Forty-eight were cloned and characterized by sequence analysis.

The third variation of IVET overcame a disadvantage of the auxotroph and antibiotic methods. They depend on a gene being expressed for sufficient time and at an adequate level to complement the parental deficiency. Hence, genes transiently induced and/or expressed in vivo at relatively low levels would not be detected. To recognize such genes, genetic recombination is used to report expression (5). A resolvase is produced from a promoterless copy of the *tnpR* gene of the transposable element γδ under the influence of promoters provided by genes expressed in vivo by the pathogen. The resolvase excises a tetracycline resistance reporter gene, making the organisms antibiotic sensitive. Tissue homogenates are replica plated on medium alone and medium containing tetracycline. Colonies sensitive to tetracycline contain bacteria whose genes are expressed in vivo. This system can be used for relatively few bacteria, in different tissues and at different stages of infection. However, although it detects genes expressed in vivo, unlike the first two methods, it does not exclude any of these genes that can also be expressed in vitro. It identified 13 induced genes in *V. cholerae* obtained from intestinal infections of infant mice (5, 6). Recently (36), 45 induced staphylococcal genes were detected during infection in a murine renal abscess model; of these, only 6 were known previously.

IVET is a great boon but it has disadvantages. A high basic level of transcription of reporter genes in vitro may prevent recognition of expression in vivo of some genes because increases in transcription are too small (6). Also, regarding broad application, auxotrophic mutants are not always available, nor are the necessary gene transfer systems. The technology demands recovery of large numbers of IVET-generated plasmid-gene fusions from the chromosomes of the host bacteria. This recovery is difficult for species that do not easily undergo transduction. To overcome the difficulty, a single-step triparental conjugated method for recovering chromosomally integrated fusion plasmids has been devised (50). Used for *Pseudomonas fluorescens* and serovar Typhimurium, it appears to be applicable to other pathogens.

DIFFERENTIAL FLUORESCENCE INDUCTION (DFI)

This is another promoter trap that detects genes induced in vivo. The *ivi* promoters drive expression of GFP, a highly sensitive reporter system, and the fluorescent bacteria are separated by FACS (69, 70). Unlike other reporting systems such as those using *lacZ* or *lux*, exogenous substrates and cofactors are not required. The method was developed for bacteria from the use of GFP of *Aequorea victoria* in eukaryotic systems to study gene expression and protein localization (14). Precipitation of GFP in nonfluorescent inclusion bodies and slow chromophore formation made adaptation to bacteria difficult. However, GFP mutants of *E. coli* (*gfp mut*) were isolated that had enhanced fluorescence emission, high solubility of GFP, and rapid chromophore formation (12).

Random bacterial DNA fragments from serovar Typhimurium were fused to the *gfp mut* of *E. coli* and inserted into serovar Typhimurium. Macrophages were infected and subjected to FACS (69). Macrophages containing fluorescent bacteria were lysed, and the bacteria were recovered on plates, grown in a liquid medium, and subjected to FACS. Bacteria with the lowest fluorescence in the absence of cells (about 15%) were used to infect fresh macrophages. FACS showed an enrichment for fluorescent infected cells, compared with the original library. The fluorescent cells were lysed and the bacteria were plated. Independent clones exhibited varied levels of intracellular-dependent gene induction. Fourteen of the macrophage-induced genes were characterized by sequencing; some were shown to be active in the spleenocytes and hepatocytes of infected mice (69).

DTI should be able to identify gene expression in animal infection and be applicable to different bacterial pathogens. The highly sensitive reporter system and the ability of FACS to deal rapidly with many host cells or bacteria should allow measurement of the genetic response of individual bacteria in large samples; monitoring of gene expression at different times during infection; and, subject to spectrally distinct fluorescent genes becoming available (14), the study of multiple gene expression and interaction within cells (70).

DIFFERENTIAL DISPLAY OF cDNAs

This is another method that has been used to detect gene expression in macrophages and could be adapted to infections in animals. RNA is isolated from intracellular bacteria and cDNA obtained by reverse transcription. This cDNA has been treated in two ways. In identifying a previously unknown gene of *Mycobacterium avium* induced in human macrophages (48), the cDNA underwent subtractive hybridization with cDNA of laboratory-cultured bacteria. The cDNA that was not subtracted was amplified by PCR and used as a probe to recover the required gene. To identify a gene required for survival of *Legionella pneumophila* in a macrophage cell line (1), the cDNA from the organisms grown in macrophages was amplified by PCR and compared directly with similarly treated cDNA from organisms grown in a laboratory medium.

REACTION WITH ANTIBODIES PRODUCED BY INFECTION

Proteins coded by gene expression libraries in *E. coli* are separated by sodium dodecyl sulfate-polyacrylamide gel electrophoresis (SDS-PAGE) and immunoblotted with sera obtained from live, infected animals and with antisera raised against heat-killed bacteria. Comparison of the immunoblots indicated products formed only in vivo, i.e., those reacting with the former but not the latter sera. This method revealed hitherto unknown genes of *Borrelia burgdorferi*, which were induced in infected mice and patients (2, 67, 71). It is surprising that this method, which depends only on the gene product being antigenic, has not been used more extensively, especially as it can be applied to human infections.

It should be emphasized that the methods described merely detect genes that are induced in vivo. They do not distinguish virulence genes from others. To distinguish which in vivo-expressed genes are virulence genes, the bacteria carrying them must be compared with deletion mutants in animal tests. This has been done for serovar Typhimurium, *V. cholerae, P. aeruginosa, Staphylococcus aureus,* and *Y. enterocolitica* (6, 36, 37, 38, 39, 69, 73, 76), and in all cases, some genes were proved to be virulence genes by the reduced virulence of the mutants, but others were not.

Direct Identification of Virulence Genes

Signature-tagged mutagenesis (STM) identifies genes induced in vivo by using the fact that they are needed for virulence (28). Transposons carrying unique DNA sequence tags allow individual insertion mutants of bacterial pathogens to be differentiated from one another. The tags have 40-bp variable central sequences between two invariable arms of 20 bp, which enable the central portion to be coamplified by PCR. Individual tagged mutant strains are assembled in microtitre dishes and then combined into the inoculum pool. This is injected into an animal and after an appropriate interval, bacteria are recovered from the animal by plating spleen homogenates. The bacterial colonies are combined to form the recovered pool. The tags in both inoculum and recovered pools are separately amplified and radiolabeled by PCR. The arms are released by digestion with *Hin*dIII, and the central tags are used to probe replica colony blots from all the original individual microtitre dishes. Mutants with attenuated virulence (i.e., those whose wild type contain the virulence gene that had been inactivated by the transposon) are those with tags that hybridize with tags from the inoculum pool and not with those from the recovered pool. The genes are characterized by sequencing 300 to 600 bp flanking the transposon and making comparisons with databases. First applied to infections of serovar Typhimurium in mice, STM identified 13 known virulence genes, 5 genes with sequences similar to known genes, and 9 new virulence genes (28). Further mapping of the transposon insertion points of 16 mutants revealed a second pathogenicity island of serovar Typhimurium (58). Used for *S. aureus* in a murine model of bacteremia, about

50 virulence genes were indicated, many of unknown function (40). Another STM study of *S. aureus* in murine abscess, bacteremia, and wound infection models focused on one virulence gene that coded for proline permease (56). In a study of *V. cholerae* colonization in suckling mice, approximately 1,100 mutants were screened and about 20 had significant colonization defects. They included five that had insertions in the toxin-coregulated pilus gene and several in novel virulence genes (8). A total of 16 virulence genes of *L. pneumophila* were detected by STM in a guinea pig pneumonia model; all 16 failed to multiply in the lung and spleen, but 4 could multiply in macrophages (17).

Obviously, STM is a powerful tool. However, it may not demonstrate all virulence genes. It only detects genes whose mutants are viable in vitro and those that are required for growth and survival in vivo. Insertion of transposons may not be random (i.e., some genes would never be mutated), and the number of mutants screened may be too low to provide a comprehensive survey. Some virulence genes may be manifested in certain animal models and not in others. In screening 1,520 mutants of *S. aureus* in three mouse models, abscess, bacteremia, and wound, 237 in vivo-attenuated mutants were identified but less than 10% were attenuated in all three models (13). Also, a mutation in a virulence gene in one organism may be complemented by intact genes present in other members of the population that have mutations in different genes. Finally, because of polar effects, the gene into which the transposon is inserted may not always be the virulence gene which is affected and leads to attenuation. Complementation is essential to confirm virulence genes. This has been done in some studies, e.g., for the genes encoding the potentially effector proteins of the type III secretion system of SPI2 (29).

A method for recognizing virulence genes, much less used than STM, is complementation of an avirulent strain by a gene library from a virulent strain. Used for *Mycobacterium tuberculosis*, avirulent strain H37Ra was treated with a gene library from virulent H37Rv using an integrating cosmid vector. Repeated infection of mice and recovery from spleens and lungs selected individual faster growing recombinants, from which H37Rv DNA inserts were retrieved (43). A 25-kb growth promoting DNA fragment was recovered, but the virulence gene in it was not identified. However, a similar study of infection of guinea pigs with *Mycobacterium bovis* identified a virulence gene, *spoV* (9), which is also present in *M. tuberculosis*.

Global Analysis of Potential Gene Expression

Already the new methods have shown that many genes are induced by conditions in vivo and that some have sequences that are not in present databases. However, databases will soon be increased. Technologies to deal with expression of many genes are afoot.

Starting with *Haemophilus influenzae*, complete genome sequencing of many pathogens is proceeding rapidly. Already the sequences are known for *H. influenzae*, *B. burgdorferi*, *E. coli*, *Helicobacter pylori*, *M. tuberculosis*, *S. aureus*, and *Streptococcus pneumoniae* (68). Databases will be transformed within 10 years (66). At present, the functions of 30 to 60% of genes in the above-sequenced pathogens are unknown (68). However, coupling complete genome information on a particular pathogen with results from IVET, DFI, STM, and other methods of detecting gene expression should allow precise definition of which genes are expressed at different stages in infection and under changing environmental conditions. Once a promoter- or transposon-disrupted gene has been identified, the corresponding gene should be easily recognized and isolated. Furthermore, homologies of these genes with known virulence genes in databases of other pathogens can indicate which bacterial strains should be tested for virulence in animals or biological tests. However, there is one major constraint. Complete genome sequencing is at present confined to one strain of a pathogen. Practical considerations prevent a significant increase. Hence, the strain selected to provide the database information may not contain all the virulence genes possessed by those strains that cause infection in animals. The most quoted example is *E. coli* K-12. Its complete genome sequence does not contain most of the virulence genes present in the numerous pathogenic varieties of *E. coli*. Also, it is estimated that closely related species such as serovar Typhimurium and *E. coli* may have up to 30% differences in their genome sequences (32). At present, the only recourse is to compare the sequence of the particular gene induced in vivo with database information on the standard strain, coupled with data on virulence genes manifested by other strains of the same pathogen.

More genes induced in vivo would become apparent if strains and mutants derived by IVET, DFI, and STM could be analyzed in greater numbers. It might be possible to adapt chip technologies that have been devised in related fields to assay expression of hundreds of genes. In work on the human genome (55), microarrays containing over 1,000 individual cDNA clones of unknown sequence (prepared from RNA of human peripheral leukocytes) were printed on sialylated glass slides by high-speed robotics. These small DNA chips were used to monitor quantitatively differential expression of their cognate human genes using cDNA from tissue RNA and a highly sensitive two-color hybridization assay. The array elements on the chips displayed differential expression patterns when viewed by a fluorescence laser scanning device. In another chip system (35), human RNA is converted to cDNA, then transcribed back to RNA using ribonucleotide triphosphates. The RNA is fragmented by heat and the pieces hybridized against small, high-density arrays of tens of thousands of synthetic oligonucleotides, which are based on sequence information and made by combinatorial oligonucleotide chemistry. The arrays are assessed by scanning confocal microscopy or dedicated chip readers.

These methods are being extended to detect simultaneous expression of many genes in plants, yeasts, and bacteria as well as the human genome (51). DNA arrays are made on glass or nylon substrates by high-speed robotics, and labeled probes are used to determine complementary binding. Oligonucleotide microarrays are made either by in situ light-directed combinatorial synthesis or by conventional synthesis and immobilized on glass substrates; sample DNA is amplified by PCR and a fluorescent label inserted before hybridization to the array (51). To detect 100 genes of *S. pneumoniae*, an array of 64,000 oligonucleotide probes was fabricated (15) and total, biotin-labeled RNA was the hybridization probe. There was no need to purify mRNA from total RNA. Duplicates agreed within 25%, and the quantitative results agreed with conventional northern blot analysis of selected genes. Expression of various genes during different phases of growth was followed. During growth, competency genes *cinA*, *recA*, and *lytA* were induced at 30-, 18-, and 10-fold levels, respectively. In the stationary phase, expression of genes coding for enzymes involved in synthesis of capsular polysaccharide and long-chain fatty acids (and those concerned with cell division) was reduced.

THE IMPACT OF THE NEW METHODS

This section describes results from the new methods and their impact on knowledge of bacterial behavior in vivo. First, it should be noted that some of the methods require pathogens to have robust genetics that are easily manipulated, which is not always the case, e.g., *Campylobacter jejuni*.

Pathogenesis

Already there are some interesting results. Photonic imaging provided a surprise about the pattern of infection in salmonellosis in mice. The levels of bioluminescence and its persistence indicated a heavy infection of the caecum, rather than the Peyer's patches of the ileum, which is the conventional view (11). CLSM of immunostained sections of livers of mice infected with realistically small doses of serovar Typhimurium showed that the pathogen resides intracellularly in macrophages and is cytoxic to them (53), as occurs with cultured macrophages (7). Hence, efforts to learn the molecular determinants of cell culture cytotoxicity are now relevant to behavior in vivo.

There could be further benefits. Photonic imaging might show whether populations of pathogens do reach the levels in vivo at which quorum sensing could operate. With regard to detecting type III secretion induced by bacteria/host cell contact in vivo, the use of *lux* reporter genes in *Yersinia* infections and examination of tissue samples by microscopy with a light-sensitive CCD camera might reveal the products of Yop gene expression when bacteria are in contact with phagocytes, as for HeLa cells in vitro (47). This method and possibly CLSM, coupled with immunostaining for products of gene expression, might be extended to examination of type III secretion in vivo of other pathogens. The use of CLSM to detect cytotoxicity of serovar Typhimurium for liver macrophages in vivo should encourage its use to reveal possible parallelisms between cell damage wrought by bacteria and their toxins in cell cultures and that occurring during infections. We know that there is an enormous amount of literature on the mechanisms of damage caused by toxins to tissue culture cells. This knowledge needs to be translated to the situation in vivo.

Preponderance of Genes Involved in Growth and Metabolism

One major impact of the new methods has been to emphasize the importance of bacterial growth and metabolism in the multifactorial nature of virulence. Always, some of the in vivo-induced genes revealed by these methods are involved with growth and metabolism. Often, they outnumber those concerned with other virulence determinants and regulation. In mice and macrophages infected with serovar Typhimurium, IVET detected more than 100 *ivi* genes (24, 26). The most preponderant of the known genes were those concerned with acquisition of metals (Fe^{2+}, Mg^{2+}, and Cu^{2+}), synthesis and acquisition of nucleotides, and cofactors such as heme and vitamin B, DNA repair, RNA processing, membrane modification, protein targeting, and thermo-, osmotic-, and acid tolerance—all aspects of growth and metabolism. Some of the virulence genes of serovar Typhimurium identified by STM in mice were concerned with purine synthesis and osmoregulation (28). Four of 13 genes detected by IVET in infant mouse cholera were predicted to encode products involved with biosynthesis of L-cysteine and L-arginine, the tricarboxylic acid (TCA) cycle, and other aspects of metabolism (6). Use of STM in the same cholera model showed that mutation in genes concerned with biotin and purine biosynthesis and with phosphate transfer caused colonization defects (8). STM identified 50 or so virulence-associated genes in one model of staphylococcal infection in mice; about half of them were involved in nutrient biosynthesis (tryptophan, lysine, threonine, purine, and diaminopimelic acid), peptido-glycan cross-linking, DNA repair, the TCA cycle, membrane transport, and cell surface modification (40). In an STM study of staphylococci in another model of murine infection (56), the prominent identified virulence gene coded for a proline permease. Also, in yet another model of staphylococcal infection in mice, IVET detected *ivi* genes that were predicted from homologies with nonstaphylococcal genes to code for products involved in sugar transport and metabolism; Na^+, K^+, and Cu^{2+} transport; energy metabolism; and ribosome recycling (36). When STM was used in four complementary models of staphylococcal infection, three in mice (abscess, bacteremia, and wound) and one in rabbits (endocarditis), the virulence genes identified dealt with peptide and amino acid transport, energy metabolism, amino acid biosynthesis, purine and pyrimidine synthesis, translation, replication, and cell envelope formation (13). In murine infection with *P. aeruginosa*, IVET identified *ivi* genes concerned with iron uptake, amino acid biosynthesis, carbon source utilization, and energy metabolism (73). In studies of the contribution of chromosomal genes to the pathogenicity of *Y. enterocolitica* in mice, IVET detected many genes in bacteria from Peyer's patches that were necessary for iron acquisition, heme synthesis, protection from environmental stresses, DNA repair, synthesis of cell envelope components including lipopolysaccharide, and other diverse metabolic activities (76).

In my previous articles (60, 63), I emphasized the importance of bacterial growth and metabolism in pathogenicity and deplored the neglect of the subject. The above results should encourage more work in this area. There is much to do. Conventional methods, which have been applied to the acquisition of iron in vivo (74), could be extended to other important nutrients with equal success. The new methods are already providing information. The recognition by STM that proline permease is a virulence determinant of staphylococci (56) indicates that scavenging for proline is essential for virulence. A biotin auxotroph was identified by STM in the intestine of mice infected with *V. cholerae* (8). This suggests that biotin synthesis is a virulence attribute of *V. cholerae* and that there is little in the mouse intestine, which was supported by enhanced colonization when biotin was added to the inoculum (8). The new methods could also supplement conventional approaches for measuring the concentrations of ions, nutrients, and metabolites in various tissues, especially those within cells. It would be a great advance if photonic imaging or other noninvasive methods could be adapted to measure growth rates in vivo as well as population changes. Also, it would be advantageous if, after injecting radiolabeled metabolites, microscopic or electronmicroscopic methods could be designed to identify labeled components on bacteria in situ in infected tissues, without the need to separate them for in vitro analysis.

Here there may be a role for mass spectrometry. LAMMS might be able to identify products formed from metabolites that had been labeled with either deuterium or tritium.

Genes Coding for Other Virulence Determinants

The new methods have revealed many virulence genes. First, there are those that code for known determinants of the particular pathogen. Then there are genes identified as potential virulence genes because of sequence homology with proven virulence genes of other pathogens. Finally, there are the new ones, those that do not relate to known genes but when mutated cause loss of virulence in animal tests. Examples of known virulence genes are *tcp* detected by STM for *V. cholerae* (8); *spvB* detected by IVET (26); *ssaH* (a member of a type III secretion system) by DFI (69); *spvA*, *spvD*, and *spvR* by STM (28) for serovar Typhimurium; and a gene coding for glycerol ester hydrolase by IVET for *S. aureus* (36). Two potential virulence genes of serovar Typhimurium detected by IVET are *iviVI-A* and *iviVI-B*; they have homologies with known adhesins of *E. coli* and *Plasmodium falciparum*, respectively (24, 26). Many hitherto unknown virulence genes induced in vivo have been revealed by IVET and STM for *V. cholerae* (6, 8), serovar Typhimurium (28, 38), *Y. enterocolitica* (75), *S. aureus* (36, 40), and *L. pneumophila* (17).

As the use of IVET, DFI, and STM increases, an interesting fact is emerging. They are not detecting expression of some well-known virulence genes, many of which have been proved to be virulence determinants by tests in vivo using deficient mutants. For example, the recombinant IVET method did not detect expression of the cholera toxin gene nor of other *toxR*/*toxS* regulated genes in intestines of infected mice (6), but STM in the same model detected 5 mutants that had insertions in *tcp* of 1,100 screened and 1 mutant with an insertion in *toxT* (8). In the case of serovar Typhimurium, some of the important invasion genes such as *invA* and *invG* were not included in the genes whose expression in mice and macrophages has been revealed by the auxotroph and *cat* IVET methods (26), STM (28), and GFP (69).

The chromosome of *Y. enterocolitica* contains genes that code for production of invasin, Myf fibrillar structure, and an enterotoxin, but they were not among the 61 *ivi* genes detected by the *cat* IVET method during invasion of Peyer's patches of mice by *Y. enterocolitica* carrying fusions derived from a plasmid-deficient strain (76). The genes coding for type IV pili, elastase, exotoxin A, and exoenzyme S of *P. aeruginosa* were not detected by the auxotroph IVET method in infection of mice (73). Expression of genes coding for the fibrinectin-binding proteins A and B, protein A, leucocidin, α-toxin, and the numerous other toxins of *S. aureus* has not been recognized by either the recombinant IVET method (36) or STM (13, 40, 56) in three different mouse models and one rabbit model of staphylococcal infection.

Why are these genes not being detected? The auxotroph and *cat* methods of IVET detect only those genes that are expressed in vivo and not in vitro, so the lack of detection of genes of well known virulence determinants may be because they are expressed in vitro. However, this does not apply to some genes detected by the recombinant method of IVET, which may be expressed in vitro as well as in vivo like those identified by STM and GFP. Since animals in the various models do not die before bacteria are harvested, the expressed genes will be those involved in invasion of tissues, growth, and survival against host defense, rather than toxins responsible for death. However, some of the well-known virulence determinants that have not been detected are involved in invasion of tissues and inhibition of host defense. Expression of toxin genes could occur early in infection, especially as some toxins can inhibit host defense (62). It is possible that the invasion genes of serovar Typhimurium may not have been detected

because they are involved in mucosal invasion and might only be induced if mice are inoculated orally. Reasons why STM may not detect all expressed genes have been given before. Probably the most important factor, which applies to the other methods, is that insufficient mutants are screened. This is where the new chip technologies could make an impact. We await further developments, but present results raise the possibility that some well-known virulence determinants may not be as important in infectious disease as was first thought.

Analysis of some in vivo-induced virulence determinant genes has provided evidence of evolutionary trends involving horizontal transfer of transposable elements. For example, STM revealed a second acquired pathogenicity island in serovar Typhimurium (58). Also, *ivi* genes of various examined *Salmonella* spp., which encode for predicted adhesin and invasion-like functions and are required for full virulence, are located in genome sequences of atypical base composition that are associated with mobile elements (10).

Regulation

In examining the relevance in vivo of regulatory systems identified in vitro, results from the new methods show two trends. First, some important regulatory genes have been shown to be expressed in vivo. Examples are *phoP/phoQ* and some of the regulatory genes of the type III secretion system in SPI2 of serovar Typhimurium (24, 26, 58); the *agrA* gene of *S. aureus*; part of the *agr* locus involved in quorum sensing (36); and *np20* of *P. aeruginosa*, which is concerned with iron transport and has sequence homology with *fur* (72). Second, there have been anomalies that may indicate that either the known regulon is influenced by unknown environmental parameters or different regulatory systems exist in vivo. IVET detected expression of the iron-regulated *irgA* gene of *V. cholerae* (21) under iron-limiting conditions in vitro and during intraperitoneal infection of mice, but not during intestinal infection, the disease-relevant site (5). Also, neither conditions that modulate *toxR/toxS*-regulated genes nor iron limitation induced in vitro any of the 14 genes identified by IVET in *V. cholerae* from mouse intestines (6). Furthermore, DFI and IVET identified 2 and 1 virulence genes, respectively, in serovar Typhimurium infection of macrophages that were not controlled by *phoP/phoQ* (25, 69), thus indicating that the PhoP/PhoQ system is not the only regulon operating in macrophages.

Other aspects of regulation in vivo have received attention. For example, switch-on of genes at different times during infection, which probably occurs in animals, has been demonstrated for serovar Typhimurium in macrophages. DFI distinguished two classes of macrophage-induced genes, one within an hour of bacteria entering and the other after 4 h (70). Also, interesting effects of the environment on gene expression have been noticed by comparing bacterial behavior in vivo with that under stimulant conditions in vitro. Acid-inducible GFP fusions of serovar Typhimurium were expressed in macrophages, but the levels were much different from those seen in vitro (70), indicating that host cell environment influences induction of pH dependent genes.

Many Genes of Unknown Function In Vivo

The final important point to make about the new technologies is that they have revealed many *ivi* genes that are not known to be involved in nutrition, metabolism, stress response, regulation, or virulence-determinant production by the pathogen under consideration, *nor of any other bacterial pathogen*. In the use of IVET, about 25% of the 100 or so *ivi* genes detected in serovar Typhimurium infections had not previously been identified (24), 3 of 13 for *V. cholerae* (6), 20 of 61 for *Y. enterocolitica* (76), 26 of 45 for *S. aureus* (36), and 6 of 22 for *P. aeruginosa* (73). Six of 14 genes of serovar Typhimurium induced in

macrophages and detected by DFI had not been recognized (69). In using STM, 9 of the 28 virulence genes detected for serovar Typhimurium had not been seen before (28); one of 21 for *V. cholerae* (8), and 29 of 50 for *S. aureus* (40). In addition, the new methods may not be sufficiently sensitive to detect all the genes that are induced in vivo and are relevant to virulence. Thus, vast areas of the behavior of bacterial pathogens in vivo remain to be explored. This process will be helped by use of complete genome sequences of pathogens and chip technologies as described in the previous section.

Design of New Drugs, Vaccines, and Diagnostic Agents

Most antibacterial drugs are based on interference with some aspect of metabolism, which leads to either death or severe restriction in growth. On the whole, vaccines rely on immunological neutralization of virulence determinants such as adhesins (e.g., K88 antigen of *E. coli*), inhibitors of host defenses (e.g., meningococcal capsular polysaccharide), or toxins (e.g., diphtheria toxin). Diagnostic agents depend on detecting genes and products of bacteria or antibodies in patients that are specific for a particular pathogen. Knowledge of bacterial behavior in vivo can make a major contribution to future developments. Drugs, vaccines, and diagnostic agents can be designed in the certain knowledge that the facet of bacterial behavior being targeted does actually occur in vivo.

The new methods have underlined the importance of nutrition and metabolism in virulence and identified many relevant genes expressed in vivo. These genes could be the target for drug design, especially those dealing with reactions unlikely to occur in host tissues such as cell wall synthesis, because antagonists may have a selective effect. Many of the genes code for enzymes whose chemical reactions could provide rapid screening of products from combinatorial chemistry (41). Examples of potential targets identified by STM are *femA* and *femB* (involved in formation of cell wall pentaglycan, pentaglycine cross bridges) and proline permease of *S. aureus* (40, 56) and enzymes needed for biotin synthesis by *V. cholerae* (8).

Exploitation of genes concerned with virulence-determinant production and regulation is another potential area. Type III secretion systems appear to operate in vivo, so inhibitors of components of these systems might be useful. They are often located in pathogenicity islands. Pathogen-specific targets may be preferable to those found in commensals, because risk of drug resistance would be reduced (68). If quorum sensing occurs in vivo, signaling molecules could be targeted. If, for example, N-acyl-L-homoserine lactones are detected in *P. aeruginosa* infections, antagonists to them may control infection (54).

In designing effective vaccines, the new methods can help us recognize antigenic components that contribute to pathogenicity. Newly recognized determinants of growth in vivo should be investigated as well as the more traditional virulence determinants. Also, regulatory elements and genes of pathogenicity islands involved in type III secretion should be considered. Turning to live attenuated vaccines, knowledge of nutrition in vivo could allow derivation of auxotrophs of attenuated virulence for use either as vaccines or as carriers of additional antigens (68). Deletion of regulatory genes known to operate in vivo might also produce live attenuated vaccines, as has happened for deletion of *phoP/phoQ* from *Salmonella enterica* serovar Typhi (68) and DNA adenine methylase (Dam) from serovar Typhimurium (27).

For identifying bacteria in specimens from patients, traditional microbiological methods requiring culture are being replaced by methods such as PCR, which do not require prior isolation of the pathogen. The explosion in gene discovery and sequencing and the use of techniques such as restriction fragment length polymorphism and activated fragment length polymorphism have increased the capability to distinguish different strains and to make correlations for epidemiological purposes. Each technique can benefit from knowledge of the genes expressed in vivo by particular pathogens. Also, the new methods can contribute to the simplest and most rapid method for diagnosing disease, i.e., recognition of specific antibodies. Detecting gene expression in vivo by reacting the sera of patients against the products of DNA libraries reveals pathogen-specific antibodies. Some may be evoked by genes expressed only in vivo and others by genes expressed in vivo and in vitro, but both types could be useful for diagnosis. Work on Lyme's disease (2, 67, 71) is a paradigm of what could happen for other diseases.

CONCLUSIONS

The new methods have great potential for providing more knowledge on bacterial behavior in vivo. Significant contributions are already evident and will increase in the future.

A major advance in studying pathogenesis is the design of noninvasive ways of following infection. Also, CLSM allows (for the first time) examination of tissue sections infected with realistic numbers of pathogens, and FACS allows rapid identification of cells containing them. Conventional methods of measuring environmental parameters in vivo have been supplemented by new methods for measuring these parameters in infected cells.

The new methods of recognizing genes expressed in vivo have confirmed the production in vivo of some putative virulence determinants recognized in vitro and identified the genes of hitherto unknown virulence determinants. Also, the genes of some regulatory systems involved in virulence-determinant production in vitro have been shown to be expressed in vivo. However, there are anomalies that indicate that either the known regulons are influenced by unknown environmental parameters or different regulatory systems exist in vivo.

Three striking facts have emerged. First, genes coding for the determinants of growth and metabolism are more preponderant than those coding for other aspects of virulence.

We hope that this highlighting of the importance of growth and metabolism in pathogenicity will end the present neglect of the subject. Second, expression of some well-known virulence genes such as those coding for determinants of invasion, inhibitors of host defense, and toxins has not been detected, but there may be valid reasons for this anomaly. Third, many of the genes expressed in vivo have sequences that had not been recognized before and whose functions are unknown. This means that large areas of pathogenicity have yet to be explored.

The increasing knowledge of bacterial behavior in vivo will help the design of new drugs, vaccines, and diagnostic methods.

REFERENCES

1. **Abu-Kwaik, Y., and L. L. Pedersen.** 1996. The use of differential display- PCR to isolate and characterize a *Legionella pneumophila* locus induced during intracellular infection of macrophages. *Mol. Microbiol.* 21:543–556.
2. **Akins, D. R., S. F. Porcella, T. G. Popova, D. Shevchenko, S. I. Buker, M. Li, M. V. Norgard, and J. D. Radolf.** 1995. Evidence for *in vivo* but not in vitro expression of *Borrelia burgdorferi* outer surface protein F(OspF) homologue. *Mol. Microbiol.* 18:507–520.
3. **Aranda, C. M. A., J. A. Swanson, W. P. Loomis, and S. I. Miller.** 1992. *Salmonella typhimurium* activates virulence gene transcription within acidified macrophage phagolysomes. *Proc. Natl. Acad. Sci. USA* 89:10079–10083.
4. **Busby, S. J. W., M. C. Thomas, and N. L. Brown.** 1998. *Molecular Biology NATO ASI Series. Series H Cell Biology*, vol. 103. Springer Verlag, Heidelberg, Germany.
5. **Camilli, A., D. T. Beattie, and J. J. Mekalanos.** 1994. Use of genetic recombination as a reporter of gene expression. *Proc. Natl. Acad. Sci. USA* 91:2634–2638.
6. **Camilli, A., and J. J. Mekalanos.** 1995. Use of recombinase gene fusions to identify *Vibrio cholerae* genes induced during infection. *Mol. Microbiol.* 18:671–683.
7. **Chen L. M., K. Kaniga, and J. E. Galan.** 1996. *Salmonella* spp. are cytotoxic for cultured macrophages. *Mol. Microbiol.* 21:1101–1115.
8. **Chiang, S. L., and J. J. Mekalanos.** 1998. Use of signature-tagged transposon mutagenesis to identify *Vibrio cholerae* genes critical for colonization. *Mol. Microbiol.* 27:797–805.
9. **Collins, D. M.** 1996. In search of tuberculosis virulence genes. *Trends Microbiol.* 4:426–430.
10. **Conner, C. P., D. M. Heithoff, S. M. Julio, R. L. Sinsheimer, and M. J. Mahan.** 1998. Different patterns of acquired virulence genes distinguish *Salmonella* strains. *Proc. Natl. Acad. Sci. USA* 95:4641–4645.
11. **Contag, C. H., P. R. Contag, J. I. Mullins, S. D. Spilman, D. K. Stevenson, and D. A. Benaron.** 1995. Photonic detection of bacterial pathogens in living hosts. *Mol. Microbiol.* 18:593–603.
12. **Cormack, B., R. H. Valdivia, and S. Falkow.** 1996. FACS optimized mutants of the green fluorescent protein (GFP). *Gene* 173:33–38.
13. **Coulter, S. N., W. R. Schwan, E. Y. W. Ng, M. H. Langhorne, H. D. Ritchie, S. Westbrook-Wadman, W. O. Hufnagle, K. R. Folger, A. S. Bayer, and C. K. Stover.** 1998. *Staphylococcus aureus* genetic loci impacting growth and survival in multiple infection environments. *Mol. Microbiol.* 30:393–404.
14. **Cubitt, A. B., R. Heim, S. R. Adams, A. E. Boyd, L. A. Gross, and R. Y. Tsien.** 1995. Understanding, improving and using green fluorescent proteins. *Trends Biochem. Sci.* 20:448–455.
15. **De Saizieu, A., U. Certa, J. Warrington, C. Gray, W. Keck, and J. Mous.** 1998. Bacterial transcript imagery by hybridization of total RNA to oligonucleotide arrays. *Nat. Biotech.* 16:45–48.
16. **Dorman, C. J.** 1994. *Genetics of Bacterial Virulence.* Blackwell Scientific Publications, Oxford, England.
17. **Edelstein, P. H., M. A. C. Edelstein, F. Higa, and S. Falkow.** 1999. Discovery of virulence genes of *Legionella pneumophila* by using signature tagged mutagenesis in a guinea pig pneumonia model. *Proc. Natl. Acad. Sci. USA* 96:8190–8195.
18. **Finlay, B. B., and S. Falkow.** 1997. Common themes in microbial pathogenicity revisited. *Microbiol. Mol. Biol. Rev.* 61:136–169.
19. **Garcia-del Portillo, F., J. W. Foster, M. E. Maguire, and B. B. Finlay.** 1992. Characterization of the microenvironment of *Salmonella typhimurium* containing vacuoles within MDCK epithelial cells. *Mol. Microbiol.* 6:3289–3297.
20. **Garcia Vescovi, E., F. C. Soncini, and E. A. Groisman.** 1994. The role of the PhoP/PhoQ regulon in *Salmonella* virulence. *Res. Microbiol.* 145:473–480.
21. **Goldberg, M. B., S. A. Boyko, and S. B. Calderwood.** 1991. Positive transcriptional regulation of an iron-regulated virulence gene in *Vibrio cholerae. Proc. Natl. Acad. Sci. USA* 88:1125–1129.

22. **Guangyong, J., R. C. Beavis, and R. P. Novick.** 1995. Cell density control of staphylococcal virulence mediated by an octapeptide pheromone. *Proc. Natl. Acad. Sci. USA* **92:** 12055–12059.

23. **Haas, M., B. Lindner, U. Seydel, and L. Levy.** 1993. Comparison of the intrabacterial Na^+K^+ ratio and multiplication in the mouse foot pad as measures of the proportion of viable *Mycobacterium leprae. Int. J. Antimicrobiol. Agents* **2:** 117–128.

24. **Heithoff, D. M., C. P. Conner, P. C. Hanna, S. M. Julio, U. Henschel, and M. J. Mahan.** 1997. Bacterial infection assessed by *in vivo* gene expression. *Proc. Natl. Acad. Sci. USA* **94:**934–939.

25. **Heithoff, D. M., C. P. Conner, U. Hentschel, F. Govantes, P. C. Hanna, and M. J. Mahan.** 1999. Coordinate intracellular expression of *Salmonella* genes induced during infection in vivo. *J. Bacteriol.* **181:**799–807.

26. **Heithoff, D. M., C. P. Conner, and M. J. Mahan.** 1997. Dissecting the pathology of a pathogen during infection. *Trends Microbiol.* **5:** 509–513.

27. **Heithoff, D. M., R. L. Sinsheimer, D. A. Low, and M. J. Mahan.** 1999. An essential role for DNA adenine methylation in bacterial virulence. *Science* **284:**967–970.

28. **Hensel, M., J. E. Shea, C. Gleeson, M. D. Jones, E. Dalton, and D. W. Holden.** 1995. Simultaneous identification of bacterial virulence genes by negative selection. *Science* **269:**400–403.

29. **Hensel, M., J. E. Shea, S. R. Waterman, R. Mundy, T. Nikolaus, G. Banks, A. Vazquez-Torres, C. Gleeson, F. C. Fang, and D. W. Holden.** 1998. Genes encoding putative effector proteins of the type III secretion system of Salmonella pathogenicity island 2 are required for bacterial virulence and proliferation in macrophages. *Mol. Microbiol.* **30:**163–174.

30. **Hueck, C. J.** 1998. Type III protein secretion systems in bacterial pathogens of animals and plants. *Microbiol. Mol. Biol. Rev.* **62:**379–433.

31. **Jacobi, C. A., A. Roggenkamp, A. Rakin, R. Zumbihi, L. Leitritz, and J. Heesemann.** 1998. *In vitro* and *in vivo* expression studies of *yopE* from *Yersinia enterocolitica* using the *gfp* reporter gene. *Mol. Microbiol.* **30:**865–882.

32. **Lawrence, J. G., and J. A. Roth.** 1996. Selfish operons: horizontal transfer may drive the evolution of gene clusters. *Genetics* **143:**1843–1860.

33. **Lentner C.** 1981. *Geigy Scientific Tables*, vol. 1. *Units of Measurement, Body Fluids, Composition of the Body, Nutrition*. Ciba Geigy, Basel, Switzerland.

34. **Lentner, C.** 1984. *Geigy Scientific Tables*, vol. 3. *Physical Chemistry, Composition of the Blood, Haematology, Sonatometric Data*. Ciba Geigy, Basel, Switzerland.

35. **Lockhardt, D. J., H. Dong, M. C. Byrn, M. T. Follettie, M. W. Gallo, M. S. Chie, M. Mittmann, C. Wang, M. Kobayashi, H. Horton, and E. L. Brown.** 1996. Expression monitoring by hybridisation to high density oligonucleotide arrays. *Nature Biotech.* **14:**1675–1680.

36. **Lowe, A. M., D. T. Beattie, and R. C. Deresiewicz.** 1998. Identification of novel staphylococcal virulence genes by *in vivo* expression technology. *Mol. Microbiol.* **27:**967–976.

37. **Mahan, M. J., J. M. Slauch, P. C. Hanna, A. Camilli, J. W. Tobias, M. K. Waldor, and J. J. Mekalanos.** 1994. Selection for bacterial genes that are specifically induced in host tissues: the hunt for virulence factors. *Infect. Agents Dis.* **2:**263–268.

38. **Mahan, M. J., J. M. Slauch, and J. J. Mekalanos.** 1993. Selection of bacterial virulence genes that are specifically induced in the host tissues. *Science* **259:**686–688.

39. **Mahan, M. J., J. W. Tobias, J. M. Slauch, P. C. Hanna, R. J. Collier, and J. J. Mekalanos.** 1995. Antibiotic-based selection for bacterial genes that are specifically induced during infection of a host. *Proc. Natl. Acad. Sci. USA* **92:** 669–673.

40. **Mei, J. M., F. Nourbakhsh, C. W. Ford, and D. W. Holden.** 1997. Identification of *Staphylococcus aureus* genes in a murine model of bacteraemia using signature-tagged mutagenesis. *Mol. Microbiol.* **26:**399–407.

41. **Michels, P. C., Y. L. Khmelnitsky, J. S. Dordick, and D. S. Clark.** 1998. Combinatorial biocatalysis: a natural approach to drug discovery. *TIBTECH* **16:**210–215.

42. **Morgan, A. J.** 1985. *X-ray Microanalysis: Electron Microscopy for Biologists*. Oxford University Press, Oxford, England.

43. **Pascopella, L., F. M. Collins, J. M. Martin, M. H. Lee, G. F. Hatfull, C. K. Stover, B. R. Bloom, and W. R. Jacobs.** 1994. Use of in vivo complementation in *Mycobacterium tuberculosis* to identify a genome fragment associated with virulence. *Infect. Immun.* **62:**1313–1319.

44. **Passador, L., and B. H. Iglewski.** 1995. Quorum sensing and virulence gene regulation in *Pseudomonas aeruginosa*, p. 65–78. *In* J. A. Roth, C. A. Bolin, K. A. Brogden, F. C. Minion, and M. J. Wannemuehler (ed.), *Virulence Mechanisms of Bacterial Pathogens*, 2nd ed. ASM Press, Washington, D. C.

45. **Pawley, J. B.** 1995. *Handbook of Biological Confocal Microscopy.* Plenum Press. New York, N.Y.

46. **Perin, F., D. Laurence, I. Savary, S. Bernard, and A. Le Pape.** 1997. Radioactive technetium-99m labelling of *Salmonella abortusovis* for assessment of bacterial dissemination in sheep by *in vivo* imaging. *Vet. Microbiol.* **51:**171–180.

47. **Petterson, J., R. Nordfelth, E. Dubinina, T. Bergman, M. Gustafsson, K. E. Magnusson, and H. Wolf-Watz.** 1996. Modulation of virulence factor expression by pathogen target cell contact. *Science* **273:**1231–1233.

48. **Plum, G., and J. E. Clark-Curtiss.** 1994. Induction of *Mycobacterium avium* gene expression following phagocytosis by human macrophages. *Infect. Immun.* **62:**476–483.

49. **Pollack, C., S. C. Straley, and M. S. Klempner.** 1986. Probing the phagolysosome environment of human phagocytes with a Ca^{2+} responsive operon fusion in *Yersinia pestis. Nature* (London) **332:**834–836.

50. **Rainey, P. B., D. M. Heithoff, and M. J. Mahan.** 1997. Single step conjugation cloning of bacterial gene fusions involved in microbe-host interactions. *Mol. Gen. Genet.* **256:**84–87.

51. **Ramsay, G.** 1998. DNA chips: State-of-the-art. *Nat. Biotech.* **16:**40–44.

52. **Rest, F. W. D.** *Quantitative Fluorescence Microscopy.* Cambridge University Press, Cambridge, England.

53. **Richter-Dahlfors, A., A. M. J. Buchan, and B. B. Finlay.** 1997. Murine salmonellosis studied by confocal microscopy: *Salmonella typhimurium* resides intracellularly inside macrophages and exerts a cytotoxic effect on phagocytes *in vivo. J. Exp. Med.* **186:**569–580.

54. **Robson, N. D., A. R. J. Cox, S. J. M. McGowan, B. W. Bycroft, and G. P. C. Salmond.** 1997. Bacterial N-acyl-homoserine-lactone-dependent signalling and its potential biotechnological applications. *TIBTECH* **15:**458–464.

55. **Schena, M., D. Shalon, R. Heller, A. Chai, P. O. Brown, and R. W. Davis.** 1996. Parallel human genome analysis: microarray-based expression monitoring of 1000 genes. *Proc. Natl. Acad. Sci. USA* **93:**10614–10619.

56. **Schwan, W. R., S. N. Coulter, E. Y. W. Ng, M. H. Langhorne, H. D. Ritchie, L. L. Brody, S. Westbrock-Wadman, A. S. Bayer, K. R. Folger, and C. K. Stover.** 1998. Identification and characterization of the PutP proline permease that contributes to in vivo survival of *Staphylococcus aureus* in animal models. *Infect. Immun.* **66:**567–572.

57. **Seydel, U., M. Haas, E. T. Rietschel, and B. Lindner.** 1992. Laser probe mass spectrometry of individual bacteria organisms and of isolated bacterial compounds; a tool in microbiology. *J. Microbiol Methods* **15:**167–181.

58. **Shea, J. E., M. Hensel, C. Gleeson, and D. W. Holden.** 1996. Identification of a virulence locus encoding a second type III secretion system in *Salmonella typhimurium. Proc. Natl. Acad. Sci. USA* **93:**2593–2597.

59. **Skorupski, K., and R. K. Taylor.** 1997. Control of the ToxR virulence regulon in *Vibrio cholerae* by environmental stimuli. *Mol. Microbiol.* **25:**1003–1009.

60. **Smith, H.** 1988. The state and future of studies on bacterial pathogenicity, p. 365–382. *In* J. A. Roth (ed.), *Virulence Mechanisms of Bacterial Pathogens.* American Society for Microbiology, Washington, D.C.

61. **Smith, H.** 1990. Pathogenicity and the microbe *in vivo. J. Gen. Microbiol.* **136:**377–393.

62. **Smith, H.** 1995. The revival of interest in mechanisms of bacterial pathogenicity. *Biol. Rev.* **70:**277–316.

63. **Smith, H.** 1995. The state and future of studies on bacterial pathogenicity, p. 335–359. *In* J. A. Roth, C. A. Bolin, K. A. Brogden, F. C. Minion, and M. J. Wannemuehler (ed.), *Virulence Mechanisms of Bacterial Pathogens*, 2nd ed. ASM Press, Washington, D.C.

64. **Smith, H.** 1996. What happens *in vivo* to bacterial pathogens? *Ann. N. Y. Acad. Sci.* **797:**77–92.

65. **Spencer, A. J., M. P. Osbourne, S. J. Haddon, J. Collins, W. G. Starkey, D. C. A. Candy, and J. Stephen.** 1990. Xray microanalysis of rotavirus-infected mouse intestine: a new concept of diarrhoeal secretion. *J. Pediatr. Gastroenterol. Nutr.* **10:**516–529.

66. **Strauss, E. J., and S. Falkow.** 1997. Microbial pathogenesis, genomes and beyond. *Science* **276:**701–712.

67. **Suk, K., S. Das, W. Sun, B. Jwang, S. W. Barthold, R. A. Flavell, and E. Fikrig.** 1995. *Borrelia burgdorferi* genes selectively expressing in the infected host. *Proc. Natl. Acad. Sci. USA* **92:**4269–4273.

68. **Tang, C., and D. W. Holden.** 1999. Pathogen virulence genes-implications for vaccines and drug therapy. *Brit. Med. Bull.* **55:**387–400.

69. **Valdivia, R. H., and S. Falkow.** 1997. Fluorescence-based isolation of bacterial genes expressed within host cells. *Science* **277:** 2007–2011.

70. **Valdivia, R. H., and S. Falkow.** 1997. Probing bacterial gene expression within host cells. *Trends Microbiol.* **5:**360–363.

71. **Wallich, R., C. Brenner, M. D. Kramer, and M. M. Simon.** 1995. Molecular cloning and immunological characterization of a novel linear-

plasmid-encoded gene *pG* of *Borrelia burgdorferi* expressed only in vivo. *Infect. Immun.* **63:**3327-3335.

72. **Wang, J., S. Lory, R. Ramphal, and S. Jin.** 1996. Isolation and characterization of *Pseudomonas aeruginosa* genes inducible by respiratory mucus derived from cystic fibrosis patients. *Mol. Microbiol.* **22:**1005–1012.

73. **Wang, J., A. Mushegian, S. Lory, and S. Jin.** 1996. Large scale isolation of candidate virulence genes of *Pseudomonas aeruginosa* by *in vivo* selection. *Proc. Natl. Acad. Sci. USA* **93:**10434–10439.

74. **Weinberg, E. D.** 1995. Acquisition of iron and other nutrients in vivo, p. 79–93. *In* J. A. Roth, C. A. Bolin, R. A. Brogden, F. C. Minion, and M. J. Wannemuehler (ed.), *Virulence Mechanisms of Bacterial Pathogens*, 2nd ed. ASM Press, Washington, D.C.

75. **Winson, M. K., M. Cormara, A. Latifi, M. Foglino, S. R. Chhabra, M. Daykin, M. Bally, V. Chapon, G. P. C. Salmond, B. W. Bycroft, A. Lazdunski, G. S. A. B. Stewart, and P. Williams.** 1995. Multiple N–acyl–L–homoserinelactone signal molecules regulate production of virulence determinants and secondary metabolites in *Pseudomonas aeruginosa*. *Proc. Natl. Acad. Sci. USA* **92:**9427–9431.

76. **Young, G. M., and V. L. Miller.** 1997. Identification of novel chromosome loci affecting *Yersinia entercolitica* pathogenesis. *Mol. Microbiol.* **25:**319–328.

INDEX

Abaecin, 23
Acid resistance, 61, 68
Acidified nitrite, 137
acr operon, 95
Actin, 5, 11, 175
   actin-based propulsion, 5
   binding protein, 211
   filamentous, 182
   G-actin, 182
   polymerization, 52
   rearrangement, 241
Actinin, 9, 177
Actinobacillus actinomycetemcomitans, 62
Actinobacteria, 136
Acyl-homoserine lactones (acyl-HSLs), 81, 85, 97
   quorum sensing, 77
   synthase,82
   synthesis, 84
Adaptive immune response, 42
Adenine methylase, 147
Adenylate cyclase cytotoxins, 164–165, 171–172, 211
Adherence, 44, 119, 176
Adhesins, 277
   invasion-like functions, 276
Adhesion, 4
ADP-ribosylation, 164, 168
ADP-ribosyltransferase, 169, 171
A/E lesion, 4, 6–7
   effector, 210
   formation, 211
Aequorea victoria, 270
Aerolysin, 171

Agglutination, 119
Agr, 276
agr, 191
*agr*A, 276
Agrobacterium tumefaciens, 84, 204
ahpC, 134–136
AhpF, 136
AIDA-I, 235
Albumin, 121
AlgR, 98
AlgU, 98
Alkaline protease, 79, 81
Alkyl hydroperoxide reductase, 134
Anaphylatoxin
   C3a, 125
   C5a, 125
Anaplasma marginale, 125
Anionic peptide, 21
Anthrax, 120
   toxin, 163, 165
Antibiotic, 245
   resistance, 95
Antibody, 119
Antigen presenting cells (APC), 45
Antigenic variation, 123
Antimicrobial activity, 131
Antimicrobial proteins and peptides, 19, 189
Antisecretion complex, 210
AP-1 a transcriptional activator, 194
AphC, 113
Apidaecin, 23
Apoptosis, 6, 12, 43, 111, 176, 189
*apr*A, 79

Aquaporins, 66
AraC-like protein InvF, 241
ARF (ADP-ribosylation factor), 166
aroA, 232
Arp2/3 complex, 181–182
AsialoGM1, 192, 198
    receptors, 193, 200
Asialylated glycolipids, 192
Aspartokinase II-homoserine dehydrogenase II,
    136
ATPase, 204
ATP-binding cassette (ABS) transporter, 31
Autoinducer, 81, 97, 101
    3-oxododecanoyl-HSL, 87
Autooxidation, 131
Autotransporters, 181
Auxotroph, 269, 276, 279
    biotin auxotroph, 275
    IVET auxotroph, 276
    mutants, 269
Azurocidin, 19

Bac5, 23
Bac7, 23
Bacillary dysentery, 175
Bacillus spp.
    anthracis, 120, 163, 172
    subtilis, 65
Bacterial adhesin, 211
Bactericidal/permeability-increasing protein, 19
Bacteriophage, 164
Bacteroides forsythus, 62
Bafilomycin A, 110
BAPTA/AM, 196
BarA, 241
barA, 241
Basolateral membranes, 176
Bax, 48
Bcl-2, 48
Beta-galactosidase, 268
Betaine, 67
Beta-lactamase, 94
Biofilm, 78, 86, 97
Bioluminescence, 273
Biphosphate (PI(4,5)P^2), 181
Bladder infection, 45
Blocking antibodies, 122
B-lymphocytes, 46
Bordetella spp. 171
    bronchiseptica, 28
    pertussis, 28, 172, 204
Borrelia spp.
    burgdorferi, 271–272
    hermsii, 123
Botulinum neurotoxin, 163, 168, 170
BRSV, 35

Brucella abortus, 19, 27, 122
    strain 45/20, 28
    strain S19, 28
Bundle-forming pilus (BFP), 7
Burkholderia cepacia, 30, 52
Butyryl
    -ACP, 85
    -HSL, 84–85

C- and MS-rings, 204
C3 cytotoxin, 170
Cadherin, 183
Calcineurin, 195
Calcium, 268
    signals, 194–195
Calmodulin, 172
cAMP, 164
Campylobacter spp.
    fetus, 123
    jejuni, 273
        LPS, 121
    rectus, 62
CAP18, 27
CAP37, 23
Capnocytophaga matruchoti, 64
Capsular polysaccharide, 132, 277
Capsule, 123
Carrier state, 20–21
Casein, 23
Casocidin I, 23
Caspase 1 (ICE), 12, 45, 111, 176
cat, 244–245, 269, 276
    IVET, 276
Catalytic domain, 164
Cathelicidins, 21
Cathepsin
    A, 107
    B, 107
    C, 107
    D, 107, 110
    G, 19
    H, 107
    L, 107
    S, 107
Cationic peptides, 243
Caveoli, 198
CD1, 111, 124
CD14, 191
Cdc42, 11, 177
Cecropin, 21–22
    A, 27
    P1, 27
    D, 27
Cell cycle mutants, 254
Cell lines
    1HAEo, 195

9HTEO, 197
Caco-2 monolayers, 180
CHO (Chinese hamster ovary), 177
MDCK cells, 7, 268
CEMA, 100
Ceramidase, 107
Ceramide conjugate, 198
CFTR, 11, 43–44, 49, 190, 194
cGMP, 172
Chaperone, 79, 204, 213
Charge-coupled device (CCD), 267
Chediak-Higashi syndrome, 34
Chemokines, 190
Chlamydia spp., 212
    trachomatis, 110
Chloramphenicol acetyl transferase, 269
Chloride, 268
Cholera toxin, 4, 164–166, 168, 243, 275
Choline, 26, 191
ChoP⁻, 191
ChoP⁺, 191
Chromosomal product, 209
Chronic granulomatous disease, 34
cinA, 273
Citrobacter rodentium, 6
Clostridium spp., 170
    botulinum, 164
CLSM, 273, 279
CNF1, 168
Coaggregation, 61
Coleoptericin, 23
Collagen type, 209
Colonization, 61
    factors, 78
Columnar absorptive cells, 3
Commensal, 21
Comparative genomics, 143, 256, 260
Complement
    C3, 125
        C3a, 125
        C3b, 109, 122
        iC3b, 109, 122
    C5a, 122, 125
    receptors, 108
Confocal laser scanning microscopy (CLSM), 267
Conjugative plasmids, 208
Copper, 274
Cortactin, 177
    P, 177
Corynebacterium diphtheriae, 163
Coxiella burnetii, 109
CpxA, 183
CpxR, 183
crcB, 30
CREB-binding protein, 112
CRP-like protein, 82

cspE, 30
CyaA, 172
Cyclosporin, 195
Cysteine proteinase, 125
Cystic fibrosis, 20, 34, 43, 77, 85, 189, 269
Cystic fibrosis transmembrane conductance
    regulator (CFTR), 11, 43–44, 49, 190,
    194
Cytokeratine, 122
Cytokine networks, 124

Dam, 147
dam, 147
Deamidation, 168
Defensin, 19, 21, 23, 43, 101
Dendritic cells, 46
Denitrification, 137
Depurination, 168
Desquamation, 45
DFI, 272, 275, 277
Diagnostics, 251
Diapedesis, 48
Diarrhea, 228
Diazepam-binding inhibitor, 23
Differential fluorescence induction (DFI), 270
Diphtheria, 163, 170
    toxin, 163–165, 169–170
dlt, 31
DltA, 31
DltB, 31
DltC, 31
DltD, 31
DNA, 147
    adenine methylase, 279
    gyrase, 150
    polymerase, 145–146
    repair, 274
    transformation, 208
dot, 110
Dysentery, 228

E-cadherin, 121
Edema factor, 165, 172
Efflux pump, 94
    resistance-nodulation-division (RND), 95, 32
        AcrA-AcrB-TolC, 95
        MexA-MexB-OprM, 95
    staphylococcal multidrug resistance (SMR), 95
EGTA, 196
Eh, 265
Eikenella corrodens, 62
Elastase, 19, 79, 81, 276
Elastolytic proteases, 79
Electrochemical gradient, 97
Electron transport system, 96
Elongation factor-2, 163–164

EMBL, 254
Endocytic vacuoles, 49
Endopeptidase, 107
Endoprotease activity, 165
Endoproteolysis, 168
Endothelial cells, 121
Endotoxin, 27
Entamoeba histolytica, 125
Enterotoxin, 165–166, 172
Environmental conditions in vivo, 265
EnvZ/OmpR, 183, 184
Epidermal
    cell internalization of bacteria, 43
    cell layer, 43
    cell ruffling, 11
    cells, 121
    growth factor, 167
Escherichia coli, 27, 30, 33, 45, 61, 81, 86, 96,
        133, 134, 135, 137, 144, 145, 150, 153,
        165, 166, 168, 169, 172, 177, 189, 191,
        204, 227, 235, 243, 246, 256, 270, 272,
        273, 275
    EHEC, 6
    EPEC, 4, 6–7, 177, 210
    K-12, 149
    K88 antigen, 277
    O157:H7, 68
Esp, 7, 9
    EspA, 7, 9, 211–212
    EspB, 7, 9
    EspD, 7, 177
    EspF, 7
Eubacterium spp., 62
Euprymna scolopes, 81
Exoenzyme
    S, 80–81, 171, 193, 276
    T, 193
    U, 193
Exopeptidases 107
Exo
    S, 171
    T, 171
    U, 171
    Y, 171, 172
Exotoxin, 164
    A, 79, 81, 164–165, 168–170, 276
    S, 12
Export, 203
Extracellular polymeric substances (EPS), 78
Ezrin, 8, 177

FACS, 270
Fc fragment, 126
femA, 278
femB, 278
Ferrioxamine, 236

Ferritin, 80
Fibrinectin
    binding protein A, 276
    binding protein B, 276
Fibrinogen, 121
Fibroblasts, 121
Fibronectin, 122, 209
Filipin, 198
Filopodia, 181
Fimbriae (pili), 44, 79, 122, 182, 191, 200, 203
    biosynthetic genes, 246
    type I, 191
    type IV, 78, 81, 191, 204, 218, 276
FK506, 195
Flagella, 203
Flagellin, 44, 78, 191
Flagellum, 78
Flavohemoglobin, 135
Flavohemoprotein (Hmp), 134, 137
FlhB, 213
FliD, 78
FliF, 218
fljA, 234
fljB, 234
fljC, 78, 234
Fluid flow, 61
Fluorescence-activated cell sorting (FACS), 267
Focal adhesion kinase, 211
Francisella tularensis, 132
Freund's complete adjuvant, 111
FucNAcMe, 28
Fur repressor, 80
fur, 276
Fusobacterium nucleatum, 62–63

G proteins, 198
GacA, 82, 101
GacS, 101
GalNAc1-4Gal, 192
GALT, 235
galU, 178, 180
Gangliosides, 121
Gastric inhibitory polypeptide, 23
GenBank, 254
Gene expression, 266
General secretion pathway (GSP), 204
Genetic footprinting with a more genomic view,
        254
Genomic sequencing, 254
GFP, 270, 276
    systems, 245
gfp mut, 270
Gingival crevice, 61
gipA, 247
Glucose-6-phosphate dedydrogenase (G6PD), 135
Glucosylation, 168

Glutathione, 136
glyA, 135
Glycocalyx, 4
Glycolipids, 78
Glycopeptides, 189
Glycosphingolipid asialo-GM1, 44
GM1, 78
gnd, 233
Goblet cells, 3
Granuloma, 112
Green fluorescent protein (GFP), 245, 267
Gs heterotrimeric protein, 164
Gsα, 164, 168
gshA, 136
GSNO (S-nitrosoglutathione), 134, 136, 137
GTP-binding protein, 80, 110
  Cdc42, 181
GTPase
  activating proteins, 171
  rab 5, 109
  Rgr, 177
  Rho, 177
Guanine nucleotide exchange factor, 171
Guillain-Barré syndrome, 121
Gut-associated lymphoid tissue (GALT), 235
gyrA, 96, 150

Haemophilus spp., 27, 125
  influenzae, 20, 24, 26, 34, 125, 153, 191, 195,
      197, 272
  somnus, 19, 123
Hafnia alvei, 6
Helicobacter spp., 69
  felis, 69
  mustelae, 70
  pylori, 48, 61, 69, 122, 189, 204, 272
Heme, 274
Hemolysin, 79, 171, 177, 256
  alpha-hemolysin, 204
Heparin, 121, 122, 196
Hepatocytes, 121
Heterologous expression, 258
Heterotrimeric G proteins, 166
Hfq, 155
HilA, 99, 229, 241
hilA, 99, 229, 241
hilC, 241
hilD, 241
hin, 234
Histidine mutation, 144
Histidine-kinase, 98
Histone-like proteins, 209
Hly, 121
Hmp (flavohemoglobin), 113
hmp, 134–135
  hmp-lacZ, 134

H-NS, 184, 209
  H-NS DNA, 183
Holotricin 2, 23
Homocysteine, 136–137
Homoserine lactone, 81, 84, 191
Horizontal gene transfer, 144, 208
HrcJ, 218
HrpA, 212
hsd, 153
HtrB, 30
Hydrogen peroxide, 108, 137
Hydrolases, 107
Hydroperoxidases, 135–136
Hydroxyquinolone chromophore, 80
Hymenoptaecin, 23

ICAM-1, 47
icm, 110
IcsA, 52, 179, 235
  IcsA-mediated intracellular motility, 178
icsA, 183
IcsB, 183
IcsP, 179
icsP mutant, 178
IgA, 122
  protease, 181
IgA1, 125
  IgA1 proteinase, 125
IgG, 121, 122
  IgG1, 121, 125–126
  IgG2, 121, 125–126
IkBs, 194
IM translocase, 204
Immunoblots, 271
Immunoglobulin binding proteins (IgBPs), 126
In silico, 151
In vivo expression technology 241, 243–245,
      254, 269, 272, 274–277
Indolicidin, 23
Inducible nitric oxide synthase (iNOS), 108
Inert surfaces/viscous matrices, 61
Inflammation, 265
Inhibitors
  D609, 199
  PD98059, 197
  SB202190, 197
InlA, 121
InlB, 121
Innate host defense, 42–43
Innate killing mechanisms, 107
iNOS, 108, 112, 131
Inositol phosphate, 7
  phosphatase, 211
Integrin
  α5β1, 177
Intercellular spread, 178

Interferon-γ, 108, 112
Interleukin
    IL-1, 98
    IL-1β, 12, 45–46, 176, 190
    IL-4, 111, 125
    IL-5, 111
    IL-6, 111–112
    IL-8, 45–46, 190–191
    IL-10 111
    IL-12, 111
Internalin A, 121
Intimin, 7, 211
    Tir, 9
Intracellular trafficking, 13
Intraphagosomal pH, 268
Intravacuolar concentrations of iron, 268
InvA, 213
invA, 67, 276
Invasin, 177, 276
    Ipa, 209
    secretion, 209
Invasion, 176, 241
    associated type III secretion system, 229
    locus, 241
    of adjacent epithelial cells, 182
Invasomes, 212
InvE, 210
InvF, 241
invG, 276
Invertase, 234
ipa operon, 183
Ipa proteins, 177
    secretion, 209
    invasiveness, 218
    IpaA, 177 209
    IpaB, 111, 176–177, 209–210, 220
    ipaB, 210
    IpaC, 110, 177, 209–210
    IpaD, 110, 209–210, 220
    ipaD, 210
    IpaH, 210
irgA, 276
iroA, 268
Iron, 196, 268, 274
    uptake systems, 246
Isogenic mutants, 70
Isoniazid, 113
IVET (in vivo expression technology), 241, 243–
    245, 254, 269, 272, 274–277
ivi, 154, 269, 274, 276
    VI-A, 275
    VI-B, 275

JNK family, 194

Kat, 65
    KatG, 113
kat, 136
    katG, 133, 136
KDEL (Lys-Asp-Glu-Leu), 168
Klebsiella pneumoniae, 30

Lactoferricin B, 27
Lactoferrin, 19, 43
Lactonization, 84
lacY, 244
LacZ, 268
lacZ, 244, 269–270
Lamin A, 43
Laminin, 209
LAMMS, 275
LAMP-1, 13
las, 82–83
LasA, 79
    protease, 79, 81
lasA, 82
LasB, 79
lasB, 79, 81–82
lasD, 79
lasI, 82, 84, 97, 186
LasR, 81–82, 97
lasR, 81, 191
Laser microprobe mass spectrometry (LAMMS),
    267
Latency, 19, 21
LcrD, 213
LcrG, 210
LcrQ, 210
LDL-like receptor, 167
LEE, 7
Legionella pneumophila, 52, 204, 271, 275
Leishmania spp.
    donovana, 108
    major, 108
Lethal factor, 165
Leucocidin, 276
Leukotoxin, 120
Lingual antimicrobial peptide, 23
Lipid A, 28, 132
    2-OH myristate, 28
    aminoarabinose, 28, 30
    glucosamine I, 28
    heptaacylated, 28
    myristate, 28, 30
    palmitate, 28
Lipidoglycan, 109
Lipopolysaccharide, 26–27, 44, 47, 120, 132,
    178, 243
Listeria monocytogenes, 52, 121, 132, 177

Listeriolysin, 110, 121
LL-37, 34
LOS, 121–123
lpf fimbrial operon, 11
LPS, 26–27, 44, 47, 120, 132, 178, 243
   LPS-binding protein, 111
LspA1, 127
LspA2, 127
Luciferase, 81
lux box, 82
lux, 269–270, 274
   luxA, 81
   luxB, 81
   luxC, 81
   luxD, 81
   luxE, 81
   LuxI 84
   luxI, 81
   LuxR, 81
   luxR, 81
Lyme's disease, 279
Lysine-specific protease, 79
Lysogenic, 164
   phage, 164, 243
Lysosome, 23, 107, 109
   glycosidases, 107
   network, 107–108
Lysozyme, 3, 19, 43
LytA autolysin, 98
lytA, 98, 273

M cells, 3–4, 49, 175–176
M proteins, 122
MAC, 123, 127
Macrophage-derived NO, 137
Macrophages, 107, 176
   derived NO, 137
Macropinocytotic events, 211
Macropinosomes, 11
Magainin, 27–28
Magaininase, 33
Magnesium, 268, 274
Major facilitator family, 95
MAPK, 194, 197
   ERK, 197
Melittin, 27
Membrane
   acting toxin, 164
   attack complex, 120
   cytoplasmic, 203
   outer membrane, 203
      proteins, 27
      vesicles (MVs), 123
   protein III (PIII), 122

ruffles, 52, 177, 211–212
Membranous (M) cells, 3–4, 49, 175–176
Meningococci, 154
Meroclone, 151
Metabolic, 245
Methicillin-resistant *Staphylococcus aureus* (MRSA), 95
Methylation, 147
Methyl-directed mismatch repair (MMR), 144–145
metL, 136
MetR, 135
mgtA, 30
mgtB, 268
mgtCB, 30
MHC (major histocompatibility class)
   class I, 111
   class II, 111, 124
Microbial genomes, 251
Microsensor probes, 86
Microvilli, 4
MisL, 235
Mitogen-activated protein kinase, 120
Mn, 113
MRSA, 95
MsbB, 30
Msp, 256
mtgB, 268
mtr, 32
MtrC, 32
MtrD, 32
MtrE, 32
MucD, 34
mucD, 34
Mucin layer, 78, 176
Mucociliary clearance system, 42
Mutation, 143–144
   rate, 144
Mutator phenotype, 145, 150–151
MutH, 148, 154
mutH, 148, 153, 154
MutL, 148, 154
mutL, 144, 148–149, 153
MutS, 148, 154
mutS, 148–149, 151, 153, 154
mutT, 150
MxiA, 213, 220
MxiC, 210
MxiJ, 218
Mxi-Spa, 204, 210, 213
mxi-spa, 183
Mycobacterium spp.
   avium, 109, 271
   bovis, 108, 109, 272

smegmatis, 137
tuberculosis, 108, 109, 111, 112, 113, 125, 134, 272
leprae, 125
Mycoplasma spp., 26
Mycothiols, 136
Myf fibrillar, 276
Myosin
    light chain kinase, 7
    light chain, 9

N-acetyl-D-glucosamine, 28, 122
N-acyl-L-homoserine lactones, 278
NAD, 170
N-butyrylhomoserine, 101
Needle complex, 220
Neisseria spp., 122, 123, 125, 191
    gonorrhoeae, 32, 121, 122, 123, 181
    meningitidis, 19, 26, 27, 122, 123, 153
Neutralization, 119
Neutrophil, 46, 190
    elastase, 190
NF-κB, 194
nfo, 132
Nitrate respiration, 137
Nitric oxide, 131
Nitrite, 134
Nitrogen dioxide, 137
Nocardioforms, 136
Nonlysogenic, 164
NOS2, 131
Nosocomial infections, 77
NoxR1 gene, 113
noxR1, 137
noxR3, 137
*np*20, 276
Nramp1, 108
Nucleolin, 43
Nutrient and substrate availability, 265
N-WASP, 181–182

O4 antigen, 232
O5 antigen, 232
O9 antigen, 232
O12 antigen, 232
O-antigen biosynthesis, 153
OAV-6, 35
OMP p78, 125
ompR, 246
Opsonization, 123, 126
Opsonophagocytosis, 126
ORF, 127
ORF5, 30
Osmoadaptation, 61, 65
Osmolality, 61, 265
Osmoprotectant, 67

OsyR, 132
Oxidative damage, 145
Oxo-hexanoyl-CoA, 84
Oxygen
    intermediates, 190
    redox potential, 61
    tension, 268
OxyR, 113, 133

PIII, 122
p300, 112
PA1244, 191
pag/pqa, 28
pagC, 28
PagP, 30
pagP, 30
PAK, 191
    fliA, 191
    fliC, 191
Paneth cells, 3
PAO, 191
Parsimony, 227
Pasteurella spp.
    haemolytica, 19, 27, 120
    multocida, 19
Pathogenicity island 5, 9, 12–13, 153, 208, 243, 257, 271, 276, 278–279
    chromosomal, 208
    SPI1, 11–12, 99, 211, 229, 241, 243
    SPI-2, 13, 109, 230, 246, 272, 276
    SPI-5, 12
Pathovar, 227, 229
Pedestals, 4, 7, 9
Pentaglycan, 278
Pentaglycine, 278
Peptidoglycan, 203
Peptostreptococcus micros, 62
Periciliary, 43
Perinuclear region, 211
Periodic selection theory, 144
Periodontal disease, 62
Periodontopathogens, 62
Periplasmic space, 203
Peristalsis, 3
Peroxynitrite, 108, 131, 135, 137
Persistent infections, 97
Pertussis toxin, 165–166, 168, 170
Peyer's patches, 4
pH, 61, 265
Phagocytosis, 175
Phagosome, 19, 109
Phase variation, 123
phoB, 80
phoN, 30
PhoP/PhoQ, 28, 98–99, 154, 241, 246, 266, 277
phoP/phoQ, 276–277, 279

Phosphatidylinositol, 181
Phospholipase, 110
  A1, 107
  A2, 107
  C, 7, 80, 107, 199
Phosphorus, 268
Phosphorylethanolamine, 28
Photonic imaging, 267, 273, 275
Phylogenetic tree, 227
Pi protein, 244
Pili, 44, 79, 122, 182, 191, 200, 203
  type I, 191
  type IV, 78, 81, 191, 204, 218, 276
PIV-3, 35
pIVET1, 243
Plasmids, 164
Plasmodium falciparum, 275
plcS gene, 80
PmrA, 30, 34, 99
PmrB, 30, 99
PmrD, 33
pmrE, 30
PmrF, 33
pmrF, 30
PMN, 46, 190
Polyglutamic acid capsule, 163
Polymorphonuclear leukocytes, 46, 190
Polymyxin B, 27, 30, 99
Polysialyltransferase, 154
Porin, 48, 122
  OmpC, 27
  OprD, 93
Porphyromonas gingivalis, 41, 62
Potassium, 66, 268, 274
pp60^{c-src}, 177
pp60-src(18), 52
pqaB, 30
PR-39, 23
Prevotella spp.
  intermedia, 62
  nigrescens, 63
prg/*pqr*, 28
PrgH, 229
prgH, 229
prgHIJK, 30, 229
Proenzymes, 166
Pro-inflammatory
  C3a, 126
  C5a, 126
  cytokines, 190
  IL-8, 176
Proline, 67
proP, 68
Prophenin, 23
Protamine, 34
Proteases, 79

Protective antigen, 165
Protegrins, 19
Protein
  A, 126, 276
  G, 126
  H, 126
  kinase C, 7
  kinases, 7
  phosphatase, 171
proteinases, 125
Proteobacteria, 256
Proteosome protease, 194
Proteus vulgaris, 30
Proton motive force (PMF), 33, 95
Ps-1, 79
Pseudo-*icsP* phenotype, 178
Pseudomonas spp., 171
  aeruginosa, 12, 20, 27, 43, 77, 84, 97, 137, 164,
    165, 168–172, 191, 192, 195, 197, 210,
    211, 269, 271, 276, 279
  fluorescens, 86, 101, 270
  syringae, 212
Pulmonary surfactant, 21
purA, 243, 244, 245, 269
purA-based pIVET1, 246
purA-lacZY, 244
PvdS, 80
Pyochelin, 80
Pyocyanin, 80–81, 137
Pyoverdine, 80–81
PYVe, 27

Quantitative fluorescence microscopy, 268
Quorum sensing, 77, 81, 83, 97, 101, 191, 266,
  274, 276

Rab5, 110
Rab7, 13
Rac, 11, 177
Radio-imaging, 267
Ras 80, 211
RDEC-1, 6
Reactive oxygen intermediates, 108
RecA, 153
recA, 273
RecB, 137
recB, 137
RecC, 137
recC, 137
Receptor-binding domain, 164
Receptor-mediated endocytosis, 166, 177
Recombination-based systems, 245
Recombination, 143–144
Reduction modifiable protein, 122
Regulators, 256
  virulence genes, 241, 246

Regulon, 266
REPEC, 6–7, 9
res, 245
Resolvase, 270
rfb, 153, 233
rfe, 179, 180
rhl, 82
RhlI, 82, 84, 97, 101
RhlR, 82, 97, 101
rhlR, 82
Rhamnolipids, 80, 81
RhoA, 168
  GTPase, 11, 171, 211
Rmp, 122
RNA phages, 145
RNA polymerase, 150
RNA processing, 274
RND efflux pump, 31–32, 95
rpoB, 96, 150
RpoS, 101, 155
RpoS, 101, 153
rpsE, 150
rpsL, 96
RsaL, 82
RTX toxin, 120

sad, 97
S-adenosylhomocysteine, 84
S-adenosylmethionine (SAM), 84–85
Salmonella spp., 153, 171, 189, 227
  bongori, 227
  enterica, 150, 210, 227, 230, 232
    Abortusovis, 231 267
    Choleraesuis, 229–230, 233
    Dublin, 12, 229–230, 233
    Flexneri, 5–6
    Gallinarum, 230, 232
    Heidelberg, 230
    Montevideo, 27–28
    Paratyphi A, 230, 233
    Paratyphi B, 230, 233
    Paratyphi C, 230, 233
    Typhi, 9, 49, 51, 149, 229–230, 233, 279
    Typhi *pqaB*, 30
    Typhimurium, 5–7, 9, 12, 19, 28, 34, 52, 67, 96, 98, 108–109, 132, 134, 136–137, 144, 149, 153, 209, 229, 230–233, 235, 241, 243, 246, 266–279
    Typhimurium pmrF operon, 30
    Typhimurium rough LPS mutants (Ra to Re), 28
    Typhimurium *umuC*, 137
    Typhimurium *aroA* vaccine, 232
    subspecies I, IIIa, IIIb, 229
Salmonella plasmid virulence (*spv*), 230
Salmonella-containing vacuole, 12–13, 96

SAM, 84
SapABCDF, 33
*sap*ABCDF, 31
SapA, 31
SapB, 31
SapC, 31
SapD, 31
SapF, 31
sar, 191
Scanning confocal laser microscopy (SCLM), 85–86
Scavengers, 136
SCV, 12–13, 96
SecA, 204
  ATPase, 213
SecB, 204, 213
Sec-like signal, 181
Secretins, 218
Secretion, 5
Secretion systems
  type I, 170, 203, 204
  type II, 79, 170, 203, 204, 218
  type III, 5, 7, 9, 13, 52, 164, 170, 172, 193, 203–205, 209, 218, 230, 241, 266, 272, 274–275, 276, 278, 279
  type IV, 203, 204
Selenomonas spp., 62
Self-promoted uptake, 94
Sensor kinase, 241
Sensor kinase/effector proteins, 183
Sequencing, 251
  gazing, 254
  microbial genomes, 252
Serpulina (Brachyspira) hyodysenteriae, 256
Serum resistance, 121
shdA, 235
Shiga toxin, 168
Shigella spp., 52, 153, 171, 175
  boydii, 175
  dysenteriae, 175
  flexneri, 5, 52, 175, 180, 204, 209, 235
  sonnei, 175
Shotgun
  DNA sequencing, 251
  libraries, 259
siaD, 154
Sialyltion, 122
Siderophores, 80
Sifs (Salmonella-induced filaments), 13
SigD, 12
Sigma factors, 209
Signaling kinases, 194
Signature-tagged mutagenesis, 271
Sinefungin, 84
SipA, 12
SipB, 12

SirA, 241
sirA, 241
sirC, 241
Sister chromatid exchange (SCE), 152
sitABCD, 243, 247
Sites of infection, 42
SMAP29, 27
S-Nitrosocysteine, 134
sodA, 132, 246
SodC, 135
Sodium nitroprusside, 134
Sodium, 268, 274
SopA, 178
SopB, 12
  SigD, 12
SopE, 11, 171
SoxRS, 132
soxS, 132
Spa, 218
Spa9, 213, 220
Spa24, 213, 220
Spa29, 213, 220
Spa40, 213, 220
Sphingomyelin, 110
Sphingomyelinase, 107
SpiC, 13, 109
sprA, 241
SptP, 11, 12
spvA, 230, 275
spvB, 230, 275
spvC, 230
spvD, 230, 275
spvR, 230, 275
Src, 177
  kinases, 177, 199
ssaH, 275
SSRRASS, 180
Staphylolysin, 79
Staphylococcal enterotoxin, 163
Staphylococcus spp., 122, 172
  aureus, 20, 31, 96, 126, 164, 171, 191, 195,
      197, 271–276
  xylosus, 31
STM, 254, 272, 275–277
Streptococcal M protein, 121
Streptococcus spp., 41, 172
  group C, 126
  group G, 126
  intermedius, 62
  pneumoniae, 26, 98, 100, 272–273
  pyogenes, 122, 126
  sanguis, 63
Stress regulons, 132
Sulfatases, 107
Sulfur, 268
Superantigens, 124, 171–172

Superoxide, 108, 132, 190
  dismutase, 135, 246
  Cu/Zn dismutases, 113
Supramolecular secretion complex, 213
Surface motility, 78
Syphilis, 256

T cell antigen receptor 124
T cells, 111
T domain, 165
TACO, 110
Talin, 9
tcp, 275
TCR, 124
Teichoic acids, 31
Temperature sensitive, 254
TEMPO, 196
Tetanus toxin, 168, 170
TH1 helper cells, 111, 124
TH2 helper cells, 111, 124
Thapsigargin, 195, 197
Therapeutics, 251
Thioredoxin reductase, 135
Thioredoxin, 135
thrA, 136
thyA-lacZ, 269
Tir, 210
Tissue breakdown, 265
TLCK, 194
tlyC, 256
TNFα, 45, 132, 190
tnpR, 270
toxA, 79, 82
Toxin, 79, 170, 276–277
  AB structure-function properties, 164
  alpha toxin, 171
  holotoxin, 166
  pore-forming, 164, 171
Toxoids, 170
ToxR, 266
toxR, 275, 277
ToxS, 266
toxS, 275, 277
toxT, 275
TPCK, 194
T-plastin, 177
tpr, 256
Tracheal antimicrobial peptide, 23
TraI, 84
traI, 84
Transcription factor, 194
Transducing phage, 208
Transferrin, 80
Translocase, 204
Translocated intimin receptor (Tir), 7
Translocation, 5, 164

Transmembrane secretion, 203
Transposon (Tn*5*), 28, 136
Trehalose, 67
Treponema spp.
   denticola, 63, 256
   pallidum, 254, 256
Tricarboxylic acid (TCA) cycle, 274
Trichomonas spp.
   vaginalis, 126
   foetus, 121, 125
Tropomyosin, 122
Tumor necrosis factor-α, 45–46, 98, 108
TUNEL, 43
Turgor pressure, 65
Twitching motility, 79
Two-component regulatory system, VanS$_B$-
   VanR$_B$, 100
TyeA, 210
Type I secretion system, 170, 203, 204
Type II secretion system, 79, 170, 203, 204, 218
Type III secretion system, 5, 7, 9, 13, 52, 164,
   170, 172, 193, 203–205, 209, 218, 230,
   241, 266, 272, 274–275, 276, 278, 279
Type IV secretion system, 203, 204
Type I pili, 191
Type IV pili, 78, 81, 191, 204, 218, 276
Typhoid fever, 51
Tyrosine
   kinase pp60[c-src], 177
   kinases, 7, 198
   phosphatase, 5, 132, 211
   phosphorylation, calcium flux, 52

UDP-glucose pyrophosphorylase mutant, 178
ugd, 30
Urea, 69
Urease activity, 69
ureB, 70
uvrB, 137
UvrD, 148, 154
uvrD, 153–154

Vaccines, 251
VAI, 84
VanS histidine kinase, 100
Veillonella parvula (alcalescens), 63
Verprolin, 181–182
Vfr, 82

Vibrio spp.
   cholerae, 4, 153, 243, 266, 270–276
   fischeri, 81
Villin, 9
Vimentin, 53, 80, 177
VirA, 210
virB, 183
VirF, 183
virF, 183
VirG, 178
Vitamin B, 274
Vitronectin, 122
VncR, 98
VncS, 98, 100
vncS, 98
vsmR/I system, 82

Whole-genome fingerprinting, 256
wlbA, 28
wlbL, 28

xcp genes, 79
   xcpP, 82
   xcpR, 82
Xenopus cytoplasmic extracts, 181
X-ray microanalysis during electron microscopy,
   268

YadA, 27
Yaws, 256
Yersinia spp., 12, 19, 122, 171, 177, 274
   enterocolitica, 4–6, 27, 30, 153, 267, 269, 271,
      275, 276
   pestis, 268
YlpA, 27
Yop, 27, 209, 210, 274
YopE, 12
yopE, 267
YopH, 132, 171, 211
YopJ, 211
YopM, 211
YopN, 210
YopT, 211
YpkA, 211

Zinc, 21, 24, 196
   metalloendopeptidases, 79
zwf, 132, 134–135
Zymogen, 21